Charles Hugh Stevenson

The preservation of fishery products for food

Charles Hugh Stevenson

The preservation of fishery products for food

ISBN/EAN: 9783337201258

Printed in Europe, USA, Canada, Australia, Japan

Cover: Foto ©berggeist007 / pixelio.de

More available books at **www.hansebooks.com**

U. S. COMMISSION OF FISH AND FISHERIES,

GEORGE M. BOWERS, Commissioner

THE

PRESERVATION OF FISHERY PRODUCTS FOR FOOD.

BY

CHARLES H. STEVENSON.

Extracted from U. S. Fish Commission Bulletin for 1898. Pages 335 to 563.

WASHINGTON:

GOVERNMENT PRINTING OFFICE.

1899.

THE

PRESERVATION OF FISHERY PRODUCTS FOR FOOD.

By CHARLES H. STEVENSON.

SYNOPSIS.

336

THE PRESERVATION OF FISHERY PRODUCTS FOR FOOD.

By Charles H. Stevenson.

INTRODUCTION.

Methods of preservation are of constantly increasing importance to the prosperity of the fisheries—more so, perhaps, than to any other food-supplying industry. In agriculture, cereals are cured sufficiently in the open air to keep for indefinite periods; vegetables and fruits with proper care will generally remain in edible condition long enough to reach distant markets, and some will last until the following season; the domestic animals intended for food may be transported alive to the place of marketing and there slaughtered; but, under ordinary conditions, fishery products are subject to rapid putrefaction after removal from the water.

It is now a generally accepted opinion that all putrefaction is caused by the development of living organisms known generally as bacteria or putrefactive germs, this theory being announced first in 1837 by the German physiologist, Theodore Schwann. "Putrefaction," says Cohn, "begins as soon as bacteria, even in the smallest numbers, are introduced, and progresses in direct proportion to their multiplication." In living animals there is a tendency to counteract the development of these germs, and maintaining marine animals alive is the simplest form of preservation, although rarely the most economical. After life is extinct, heat, moisture, and air are all more or less necessary to the development of bacteria, and it is principally by removing one or all of these-factors that preservation is accomplished. This gives us three principal methods of preserving dead fish, viz: Refrigeration, which diminishes the heat; desiccation or drying, which decreases the moisture; and canning, which separates the preserved product from the air. Another method of great importance is the application of antiseptics, such as salt, vinegar, etc., this process being known generally as pickling. Other forms of preservation, the most important of which is smoking, partake of the characteristics of the preceding with the addition of further treatment for the purpose of flavoring. These six processes, viz, preserving alive, refrigeration, desiccation, canning, pickling, and smoking, include practically all the general methods of preserving fishery foods.

The qualities of the original products, however, are so varied and subject to such delicate influences that a process well adapted to the preservation of one article may be impracticable or deleterious when applied to another, even of the same class. Thus it would not do to refrigerate salmon, herring, and oysters in the same manner; nor is the process of salting codfish, halibut, herring, and swordfish the same. The manner of preservation also differs according to the market for which the article is intended. Codfish destined for the New England market would not be suitable for the **Gulf States**, and that for the West Indies and Brazilian trades requires still different treatment. Therefore, in this report the general method of each form of preservation is first noted, and then its particular process of application to each species and for each of the principal markets is described. It should be remembered, however, that the excellence of any particular product does not arise wholly from the special mode of preservation, but from care and attention in the process, guided by experience and close observation. No matter what process is employed, careful treatment during all the various stages is of fundamental importance, and without it no first-class article will be produced. A serious difficulty with which the fishery trade has to contend is the competition with products of careless or indifferent preservation. In too many cases superior quality of the product is sacrificed entirely to cheapness of production, and preparers who desire to maintain a high standard suffer from the resulting competition and frequently are compelled to cheapen their own process or retire from the business.

In few countries has greater attention been given to the preparation of fishery food products than in the United States. In the various international expositions our exhibits of this class have excited favorable comment because of the great variety and excellence of the products and the neat and convenient forms in which they are prepared for sale. The large representation of foreign nationalities in the United States has probably been a factor in increasing the number of our methods of preparing marine foods. People immigrating to America and devoting their time to handling fishery products naturally make use of the ideas and methods in vogue in their native countries. The smoking of haddock and some other species was introduced in this way by Scotchmen; the Chinese on the Pacific coast and in Louisiana prepare fish, shrimp, etc., by methods similar to those practiced in the Orient, and the preparation of sturgeon products was first begun here by natives of Germany and adjacent countries. The congregation of people of foreign birth in our coast cities also tends to increase the list of fishery products; a small local sale for certain articles developing among those people, the trade gradually extends until such articles become of recognized importance in the food markets. There are, however, many additional methods of preserving marine food products that could be employed advantageously to meet the wants of new markets. Numerous products highly valued in Europe and Asia are never utilized here, although abundant in the United States waters; and a large part of our fishery resources are undeveloped through a failure to appreciate and follow the foreign methods of preservation. Herring, for instance, is one of the most abundant species of fish on the United States coast, being very frequently obtainable in much larger quantities than the fishermen make use of, yet the United States imports annually over $2,000,000 worth of herring products.

The purpose of this paper is not to instruct the various fishery preparators in the methods of their particular trade, but rather to present the chief processes employed, and thus enable those who are interested to compare the different methods. The author has carefully consulted the fishery literature and has freely availed himself of the data contained therein, yet he has avoided giving a description without actual knowledge of the present processes or inquiry from persons familiar therewith. But no care or labor can wholly avoid mistakes, and as the plan of this work embraces a great variety of subjects concerning which much difference of opinion and practice exists among fishermen and marketmen it is altogether likely that it will be somewhat open to criticism, but it is hoped that the errors will not be so numerous or so gross as to materially impair its utility.

PRESERVING FISHERY PRODUCTS ALIVE.

—

In some foreign countries, especially in Germany, a large portion of the fresh-water fish and some salt water species are supplied to the markets alive. The live-fish trade in China is very extensive, the fish being peddled about the cities and villages in buckets of water, and those not sold are returned to inclosures of water for future sale. In the United States, however, live fish represent an inconsiderable portion of the trade. A few of the New York market fishermen take their catch of cod, sea bass, and blackfish into port alive by means of well-smacks, and some of the shore fishermen at points along the coast or on the interior waters retain their fish for a few days in live-cars or live-boxes; but the quantity of fish sold alive in this country is indeed very small. However, lobsters, crabs, oysters, clams, terrapin, and turtles are sold alive, and unless in that condition are not generally considered marketable as fresh, except in the case of shucked oysters and clams.

When practicable, this is one of the most satisfactory methods of marketing marine foods, not only because of the superior quality of the product, but also because it avoids costly processes of preservation. There is no general or uniform process employed for keeping the animals, each species receiving such treatment as it particularly requires. Fish and lobsters are kept alive in large inclosures or in well-smacks and live boxes, while oysters, clams, terrapin, and turtles ordinarily require little care, unless they are to be held a considerable length of time.

INCLOSED WATER AREAS.

When whitefish were abundant in Lake St. Clair and Detroit River, a practice prevailed of building inclosures one eighth to one-half acre or more in extent, conforming to the shore, for retaining the fish during October and November for sale during the early winter. These pens were usually built of 2-inch by 6-inch hard-wood piles driven into the bottom and projecting above the surface, with about ¾ inch space between the piles to allow the water to freely pass through the area. A platform with a barred entrance was arranged at one side to facilitate the handling of the seine and the admission of the fish into the pen, or this was accomplished by having a gate hinged to a mudsill at the bottom and with the upper part about a foot above the surface of the water and inclined at an angle of about 45°. The gate was opened by pushing it beneath the surface, when the fish might be easily emptied from the seine into the pond. The ponds were usually emptied before the end of December, the fish being removed from the inclosure as the market demand required. The introduction of freezing and the increasing scarcity of whitefish in Lake St. Clair led to the abandonment of these ponds about 1888. Whitefish are still preserved alive in net inclosures in Lake Erie, but this is principally for the purpose of obtaining eggs for use in artificial propagation.

At Port Huron, in 1881, Messrs. Früchtnicht & Neilson, of Sandusky, Ohio, constructed a large pen for retaining sturgeon alive. This inclosure covered an area of about one-fourth acre, and was made by driving 2-inch by 12-inch hard-wood timbers into the ground about 2 inches apart on three sides of the pen, the beach forming the fourth side. The cost approximated $5,000. At one time the pen contained as many as 6,500 sturgeon. The fish were usually not fed at all, even though retained four or five months, and little depreciation occurred either in weight or quality. Feeding them on corn was attempted, but they did not appear to require it. The sturgeon were caught for removal by means of a short seine having a chain on the bottom. They had a tendency to burrow, and it was sometimes difficult to catch one even when there were a hundred or more in the pen. The business was highly successful until the decreasing supply of these fish caused its abandonment about 1887.

At several other points on the Great Lakes there were inclosures for retaining sturgeon, and at some of them a regular practice prevailed of feeding the fish on corn. In the fisheries of North and South Carolina it was formerly customary to provide pens in which sturgeon were confined until a sufficient number had been accumulated for a "killing." Some fishermen whose operations were less extensive, however, did not resort to building a pen, but would merely pass a rope through the lower jaw of each fish and fasten the other end to some convenient fixture.

The trap fishermen of Rhode Island have large pounds, made of twine, and sometimes 60 feet square and 30 feet deep, in which scup and other fish are held for two or three months. Sometimes 12,000 barrels of fish are there held for a month or two. The trap fishermen of other localities sometimes have a similar contrivance on the back of each trap net, in which a few barrels of fish may be held for several weeks.

Quantities of striped bass and perch were formerly kept alive for a week or more in southern Delaware by inclosing them in pens built of pine logs along the river banks.

On the coast of Maine there are several inclosed coves or ponds for confining lobsters several months, if necessary, the principal ones being at Vinal Haven, Southport, House Island, South Pond, Prospect Harbor, and Friendship. Their form and adaptability depend on the coastal formation. The first one was established at Vinal Haven, in Penobscot Bay, in 1875, by Messrs. Johnson & Young, of Boston, and that is yet the largest and most successful on the coast. It is the small end of a cove covering about 500 acres, communicating with the sea through a 150-foot channel, in which the tidal range is about 10 feet and the depth from 30 to 180 feet, averaging about 90 feet. The inclosure devoted to lobsters covers about 9 acres, and is separated from the large cove by a natural shoal surmounted by a stout wire fence about 200 feet long. Its bottom is of soft grayish mud and the water ranges from 0 to 60 feet in depth. It has a capacity for about 300,000 lobsters, but a smaller quantity usually does better. The capacity of the other ponds or coves ranges from 25,000 to 200,000 lobsters.

The lobsters are deposited in the inclosures when the condition of the market warrants and are held for a higher price. They are fed quite regularly on cheap fresh fish of various kinds, principally split hake, hake heads, small cod, herring, flounders, bream, etc. Fat herring are not desirable for lobster food, as practical experience has shown that they cause the lobsters to decrease in weight. For the same reason, when using hake, it is well to remove the livers, as they are rather too oily. The quantity of food required depends largely on the temperature of the water, since lobsters do not eat as freely in cold water as in that of a higher temperature. The food should be well scattered over the pond, as throwing it in heaps causes the lobsters to congregate in large numbers, resulting in their biting and injuring each other in their contests for food. If not fed regularly it is quite difficult to keep the lobsters in the inclosure, but when properly supplied they seem contented and improve both in appearance and weight; yet it is not generally profitable to feed them for an increase in weight alone, the profit coming from the ability to place them on the market when the prices are the highest. In catching them seines, pots, or beam trawls are employed. The latter are usually 12 feet across, with 18 inch runners. If properly attended, the mortality is small and the lobsters improve in weight and condition. It is estimated that in November, 1898, there were 700,000 lobsters retained in the ponds or inclosures in Maine.

At several of the fishery ports along the Gulf of Mexico there are small inclosures for retaining green turtle and terrapin. These are usually 400 or 500 square feet in area, and are made by driving rough poles into the ground near the shore, where the water is 6 or 8 feet deep at low tide, connecting and bracing them by nailing a strip along the line near the top, the poles being 1 or 2 inches from each other and sufficiently long to project a few feet above the surface of the water. For convenience in handling the turtle these pens are generally constructed adjacent to the landing pier leading to a market house. The turtle are placed in the pen and removed therefrom by means of a block and tackle attached to a swinging arm. They are generally fed on algæ, fish, etc., until it is desirable to market them, when they are placed in boxes, barrels, or otherwise secured, and shipped without further care.

WELL-SMACKS.

Well smacks were introduced in England in 1712, being first used at Harwich, where 12 were in operation as early as 1720, but the idea seems to have originated with the Dutch fishermen many years before. According to Dr. Fuller's "History of Berwick," well-smacks were used in carrying live salmon from Berwick to London prior to 1740, those vessels being of about 40 tons burden each.

Previous to the general use of ice on vessels, which began about 1810, most of the New England market vessels, especially those in the halibut fishery, were constructed with a well in the hold, in which the fish were retained alive until delivered at the fishing port. The use of well-smacks, or welled-smacks, in the halibut fishery began at New London, Conn., and Greenport, N. Y., about 1820, and by 1840 the fishery had extended to Georges Bank. Before the employment of these vessels the halibut fishery was prosecuted only during cold weather, the fish being carried in bulk in the hold.

The first well-smack at Gloucester was built in 1855 and was designed to carry about 12,000 pounds of halibut. The fish were caught by means of hand lines and were handled very carefully, being placed in the well immediately on removal from the water. Those dying before reaching market, through injuries or otherwise, were sold at about one-fourth the price of live halibut. On account of the greater convenience of using ice and the general adoption of trawl lines in the halibut fishery the well-smacks have been entirely superseded by tight-bottomed vessels.

Formerly nearly all fishing vessels running to the New York market during cold weather were constructed with wells. But the dwindling of the market cod fishery from that port, due to competition with Boston and other New England points having the benefit of the trade with drying establishments, has led to a large decrease in the number of vessels engaged, and during recent years there have been only eight or ten vessels which during the winter and spring take their catch of cod, sea bass, and blackfish into Fulton Market alive. Well-smacks have been employed also in the red-snapper fisheries of Key West and Pensacola, but they are being discarded, it being found more satisfactory to ice the fish than to keep them alive. The lobster trade along the New England coast still uses a number of well smacks, and in some of them steam has superseded the use of sails as a motive power.

The well in which the fish or lobsters are placed is situated amidships at the bottom of the hold, extending from just forward of the main hatch nearly to the mainmast, and occupying about one-third of the length of the vessel. It is formed by two stout, water-tight bulkheads at either end, 4 or 5 feet high and about 5 inches thick, extending from keelson to deck and entirely across the vessel. Midway between these is usually another bulkhead, which assists in supporting the deck and divides the well into two compartments. Leading from the well to the deck is a funnel curb, about 2½ feet wide by 8 feet long at its upper end and 4 feet long at its lower end. The well has neither keelson nor ceiling, and the frames are usually the same distance apart as elsewhere in the vessel, but on some smacks they are twice as far apart, in order to permit the water to circulate freely and to facilitate dipping the fish from the well. About 300 auger-holes are bored in the bottom planking of the well, through which the sea water freely enters, and it is kept in circulation and constantly renewed by the motion of the boat. On the lobster smacks the auger-holes are generally 2 inches in diameter, whereas those on fish smacks are more frequently 1 inch. The vessels range from 12 to 60 tons and those using sails are either schooner or sloop rigged, though more frequently of the former type.

On the British coast a number of "dry-well" smacks, having an artificial circulation of water, are employed. In some of these there is a series of one or more lengths of perforated supply pipes arranged near the bottom of the well and connected at one end to a circulating pump operated by the main engine if on a steamer, or by a donkey engine or otherwise if on a sailing vessel, a two-way cock being on the pipes outside the well. The pipes, being at the bottom of the well, cause a continual circulation of water in an upward direction and thoroughly aerate the water as well as cause all the scum and refuse to rise to the top, whence, along with the used water, it escapes back into the sea through several bell mouthed overflow pipes, the lower ends of which pass through the vessel's bottom and are mounted so as to incline aft from the top, and thus allow the force from the forward movement of the vessel to suck them clear. The aeration of the water can thus be kept under perfect control and the well be readily emptied of water by pumping when it is desired to remove the fish.

The well-smacks running cod, sea bass, and tautog to the New York market, which fish off Sandy Hook and Long Island shore, have capacity for 8,000 to 20,000 pounds of fish each, depending on the time of the year and the length of the trip. Hand lines are employed for the most part and the fish are placed in the well as soon as taken from the water, the hook being carefully removed. Each vessel generally carries a small quantity of ice, with which to preserve such fish as may die, as well as the surplus that can not be placed in the well, this ice being carried in pens at either end of the well. The cod when caught in no great depth of water live in the well, under ordinary conditions, a week or more, but the sea bass and tautog are not so hardy and do not keep much more than half that long. The length of time which the cod will live depends also on the time they have been kept on the trawls, in case that form of apparatus is used. On arrival at Fulton Market the fish are removed from the well with long-handled dip nets, and placed in wooden cars, which are kept floating in the dock. For a description of these cars see next page.

Lobster smacks are employed mainly along the coast of Maine and Massachusetts, but there are a few at New York, Greenport, and New London. Up to a few years ago vessels of this type were used in bringing lobsters from Nova Scotia, but at present those shipments are made usually in barrels on regular commercial steamers. The lobster smacks are mostly old vessels which were formerly employed in the live fish trade before icing became the general practice; but many fine vessels are now coming into use, and at Portland, Maine, four steamers are engaged in this trade. About 60 well-smacks are now employed in transporting lobsters along the coast, running to Rockland, Portland, Boston, New London, New York, etc. Their capacity ranges from 3,000 to 16,000 lobsters, with an average of about 9,000 during cold weather and about two-thirds or half that number when the weather is warm. The loss in transit is small, rarely amounting to 2 per cent, unless the weather is calm or the loaded smack remains in still water very long, when the lobsters use up the air held in solution by the water and smother. These vessels are not so extensively employed as a few years ago, on account of the competition with steamer and railroad transportation, but they are yet an important factor in connection with the lobster trade.

The well-smacks until recently employed in the Gulf of Mexico red-snapper fishery were of the same type as those in use on the New England coast, indeed most of them were designed for the New England fisheries. At Key West a number of smaller sail craft, known locally as "smackees," are provided with wells. These boats average about 25 feet long, 8 feet wide, and 4 or 5 feet deep, with sharp bottom, the deep draft being necessary in order to submerge the hull sufficiently for the water to cover the fish in the well, which occupies about a quarter of the boat's length measured on the keel.

On account of the great depth from which red snappers and groupers are as a rule obtained, considerable difficulty was at first experienced in keeping them alive, the pressure of the water being so much less in the wells than at a depth of several fathoms that the air bladder would become greatly distended and the fish float belly up. To overcome this the fishermen adopted a practice of puncturing the air bladder as soon as the fish reaches the surface, forcing a hollow metal tube ⅛-inch in diameter into the side of the fish a little behind and just above the pectoral fin, thus relieving the air bladder of its extreme buoyancy so that the fish may control its movements in the well. Only those red snappers taken in less than 10 fathoms of water can be successfully held in the wells for a week or two; if caught in more than 10 fathoms they must be handled carefully, and if from over 20 fathoms they soon have a swollen surface, the eyes protruding and the scales becoming loosened and standing erect. For the purpose of holding the surplus fish when the well became overcrowded, some of the smack fishermen also carried two or three cars, about 8 feet long, 4 feet deep, and 4 feet wide, so constructed that they could be taken apart and stowed below deck. But, as before stated, the use of ice has almost entirely superseded the employment of well-smacks in the red-snapper fishery.

LIVE-CARS OR LIVE-BOXES.

The most usual method of keeping fishery products in captivity alive is by means of live-cars or live-boxes. These are employed in the market fishery of New York, the lobster fishery of the New England States, the catfish fishery of Louisiana, the seine fishery of the Gulf of Mexico, and in numerous other small fisheries along the coast and on the interior waters. Ordinarily they are plain wooden boxes, with open seams or numerous auger holes to permit a free circulation of water and yet not so large as to permit the escape of the fish, their size and shape conforming to the requirements of the fishes and the localities for which they are intended. The buoyancy of the material entering into their construction keeps them at the surface of the water, with little more than the upper portion exposed, this position being regulated if necessary by attaching floats or weights, as the case may require. When it is desirable to move them frequently from place to place they are made in the form of skiffs.

The live cars employed at Fulton Market, New York City, for retaining cod, sea bass, and tautog or blackfish brought in by the well-smacks, are of various sizes, but generally about 18 feet long, 12 feet wide, and from 2 to 3 feet deep, the depth being greater in the center than at the two ends. They are made of planking 1 inch thick and 6 inches wide, nailed to a rectangular frame of joist, with spaces of 1 or 2 inches between the planks to allow free circulation of water, and are without partitions on the inside and without barrels or other buoys. In the top of each are two pairs of doors, running the entire length of the car, but covering only about half the width, and which may be fastened with a padlock. The cars are moored in the dock at the rear of the market, and by means of tackle attached to the rear of each fish-house they are raised occasionally and rested on a platform or float running the entire length of Fulton Market, so that they may be cleaned and dried to prevent their becoming water-logged. They cost about $24 each and have capacity for 3,000 or 4,000 pounds of fish under ordinary conditions. No food is given the fish confined in the cars and the length of time during which they may be kept depends on the weather. If bottom ice forms, the tautog may all die in one night, but the cod are quite hardy. As soon as the fish are removed from the cars they are killed, and being much fresher and firmer they are sold at a higher price than that received for fish brought in packed in ice. These cars are used also for holding lobsters and green turtle alive.

At certain of the European fishing ports the retaining of live cod in floating cars is quite extensive. From Holdsworth's "Sea Fisheries" is obtained the following account of the business at Grimsby:

When the smacks arrive with their cargoes of live and dead fish at Grimsby, the cod in the well are taken out by means of long-handled landing nets, and are placed in wooden boxes or chests which are kept floating in the dock; there the fish are stored till wanted for the market. These cod chests are 7 feet long, 4 feet wide, and 2 feet deep; the bottom is made of stout battens placed a short distance apart, so that the water penetrates freely to the interior, as it does also between the planks of which the sides and ends are built up. The top is wholly planked over, except in the center, where there is an oblong opening for putting in and taking out fish. This opening is closed by a cover when the chest is in the water. Two ropes or chains are fixed in the ends of each chest for convenience in moving it about and hoisting it out of the water. About 40 good-sized cod, or nearly 100 smaller ones, may be put into one of these chests, and will live there without much deterioration for about a fortnight. There are usually as many as 400 of these chests in the Grimsby fish-dock, sometimes all in use and containing from 15,000 to 20,000 live cod. Every day during the cod season a remarkable scene is presented here, and the same thing occurs at Harwich, although on a smaller scale, Grimsby and Harwich being the two ports where the live cod are stored. A certain number of fish being wanted for market, the salesmen make their preparations accordingly, and the cod are taken out of the chests and killed. I say killed, because the fish are not merely taken out of the water and allowed to die, but they are dispatched in a very summary manner. A chest of cod is brought alongside an old hulk kept for the purpose, and moored in the dock close to the market place; tackles from a couple of davits are then hooked on to the handles, and the chest is hoisted up till nearly clear of the water, which drains through the bottom and leaves the fish dry. The cover is then taken off, and a man gets into the opening and takes out the fish, seizing them by the head and tail. As may be supposed, the commotion among 50 or 60 cod just out of the water is very great, and it is often a work of difficulty to get a good hold of the fish; but, one after another, they are lifted out and thrown up to the deck of the hulk, when they come into the hands of another man, who acts as executioner; he grasps the fish tightly behind the head with his left hand, holds it firmly on the deck, and, giving a few heavy blows on the nose with a short club, kills it at once.

It is sometimes as much as can be done to hold down a large and lively fish on the slippery deck while giving it the coup de grâce; but the work is generally skillfully performed, and the dead fish rapidly accumulate into a large heap, whence they are taken to the adjoining quay to be packed in bulk in the railway trucks waiting close by to receive them. Each truck will hold about twelve score of good-sized fish, or a proportionately larger number of smaller ones. The fish thus killed and packed reached Billingsgate in time for the early market next morning, and are known in the trade by the name of "live cod," the manner in which they are killed affecting the muscles of the fish in some way that enables the crimping process to be carried out successfully some hours after the fish have been taken out of the water. These cod command a high price, and are looked upon as essentially "West End" fish. There is, of course, a great advantage gained by thus storing the cod alive, for not only is the market more regularly supplied than would otherwise be the case, owing to small catches during bad weather, or delays from calms or adverse winds, but the fish themselves also come into the hands of the fishmongers in a fresher state than almost any other kinds supplied to them.

In connection with its hatching operations at Woods Hole, Mass., the U. S. Fish Commission retains live cod in cars, and for protection in stormy weather these are sheltered in an inclosure. The method is as follows:

The fish are taken with hand lines fished from the deck while the vessel is drifting in water from 10 to 40 fathoms deep. Those taken in the shoaler water are preferable to those coming from deep water, as the change to the shallow cars in which they are held at the station is less pronounced. Great care is exercised in catching the fish, for when hastily hauled up from deep water they are very liable to be "poke-blown"; that is, they have their stomachs turned inside out through the mouth. When drawn in with moderate speed, they become adapted to the gradually diminishing pressure and do not suffer injury. It is also important in unhooking the fish not to injure its mouth any more than is absolutely necessary, as the wound caused by the hook frequently spreads and forms a large sore and eventually kills the fish. All the vessels which collect cod for the station are provided with wells, in which the fish are placed and held while in transit. When a vessel arrives at the station with cod, the fish are immediately transferred with dip nets from the well to live-cars 16 feet long, 6 feet wide, and 5 feet deep, which are constructed of wood and divided into two compartments by a crosswise partition. As the fish obtained from smacks are paid for by the pound, it is customary to weigh about 40 per cent of each load and estimate the total weight by the average of those weighed. While being weighed, the cod are also counted, about 500 being put in each car. The cars are moored in the middle of a pool or basin protected on all sides by a wharf, which breaks the force of the sea in stormy weather and affords a sheltered place for handling the fish and taking the eggs. Cod take little or no food when spawning. The impounded brood fish are often tempted with fresh fish and with fresh and salted clams. (Report U. S. Commissioner of Fish and Fisheries for 1897, pp. 200-201.)

In the catfish trade centered at Morgan City and Melville, La., very substantial live-cars are used in transporting the catch from the fishing-grounds to the markets. These are built in the shape of a flat-bottomed skiff, sharp at each end, the sides, top, and bottom being formed of slats, with space between each slat for the free circulation of water. They range in length from 18 to 30 feet and about 5 feet in width. At each end there is a water-tight compartment with about 40 gallons capacity, and by emptying or filling these compartments with water the buoyancy of the car may be regulated.

Since the cars are usually towed by steam tugs at a speed of 6 or 8 miles per hour, the determination of the proper buoyancy at either end suitable for towing requires considerable judgment and experience.

These cars are divided by a slat partition into two or more compartments, so that the fish will not all crowd together. Their capacity is from 5 to 10 tons of fish, dependent on the temperature and condition of the water. During warm weather, or when there is considerable sediment in the water, the tugs usually carry ice in which the fish are packed in preference to carrying them in the live-cars.

The fishermen catching hogfish along the coasts of Virginia and North Carolina usually transport them in live-cars to the marketing ports, and the same is true in a number of other minor fisheries of the coast. In the sea bass and tautog fisheries prosecuted on the southern coast of New England the fishermen occasionally use boat-shaped cars made of wood, sharp at both ends, with auger-holes in sides and bottom and with top covered with hinged lid. A common size is 5 feet long on top, 3 feet long on the bottom, which is flat, and 2 feet wide on top at the middle.

Live-boxes are generally employed in the eel fishery of Connecticut, Long Island, and other places on the Atlantic coast, but these conform to no established shape or size, suiting the convenience and needs of the individual fisherman. Several of the catfish fishermen of Philadelphia retain their catch for several days or even weeks by putting them in large boxes lined with tin, which are placed in their yards and kept covered over, the water being changed frequently.

In this connection it may be well to describe the cars used in the Penobscot by the United States Fish Commission in transferring live salmon from the fishermen's weirs to the retaining ponds, preparatory to stripping them of spawn for hatching purposes:

The car employed is made from the common dory, divided transversely into three compartments. The central one, which is much the larger, is occupied by the fish, and is smoothly lined with thin boards and covered with a net to prevent the fish jumping out or being lost by the car capsizing, which sometimes occurs, while to guard them from fright and the rays of the sun a canvas cover is drawn over all.

The first cars of this form constructed had iron gratings to separate the central from the forward and after compartments, the water being admitted through the forward and discharged through the after compartment, but this was objectionable because the salmon were constantly seeking to escape through the forward grating, and often injured themselves by rushing against it. Smooth wooden gratings were afterwards used, and for many years cars were employed in which the compartments were separated by tight board partitions, the openings for the circulation of water communicating through the sides of the boat directly with the fish compartment, and being, of course, grated. This was very satisfactory, but when it was found desirable and practicable to use ice in transportation the forward compartment became the ice room, and it was necessary to perforate the partition again to admit the cold water to the fish. Finally, stout wooden blanket cloth was substituted in the partitions, with eyelet holes wrought in to afford passage to the water. This is the form now in use, in which the water is admitted through openings in the sides to the ice room, from which it passes through the fish room to the after room, whence it is discharged. The car is ballasted so that the rail is just above water, or, in case of an unusually large load of fish, a little below it. All the openings communicating with the outside are controlled by slides, which can be closed so as to let the car swim high and light when it is towed empty.

To avoid injury to the fish in transferring them to the cars, fine minnow dip nets, lined with woolen flannel of open texture, are used. The bow on which the net is hung is 22 inches in diameter, and to secure a net of ample width three ordinary nets, 36 inches in depth, are cut open down one side quite to the bottom, and then sewed together, giving thus three times the ordinary breadth without increasing the depth.

The collection of salmon is begun each season usually from the 20th of May to the 1st of June, but as the maximum temperature that the fish fresh from the weirs will endure is about 75° F., the temperature of the water through which the cars are towed must be taken into consideration, and the collection not be postponed until too late in the season. If the collection is prolonged, this difficulty is obviated by using ice, as it has been found that by moderating the volume of water passing through the car and introducing it all through the ice compartment it is possible to keep a uniform temperature in the compartment in which the fish are held several degrees below that of the water in the river, thereby insuring the safe transfer of the salmon. (Report U. S. Commissioner of Fish and Fisheries for 1897, pp. 32–33.)

The live cars used by the lobster dealers on the New England coast are usually substantially constructed, of large size, and divided into compartments. Those at Portland, Me., are mostly 30 feet long, 12 feet wide, and 3 feet deep, with capacity for 2,000 to 3,000 lobsters. The framework consists of six rectangular frames, 6 feet apart, to which are nailed boards 6 inches wide and 1 inch thick, forming the top, bottom, sides, and ends, with spaces of 1 to 2 inches between the adjacent boards. The cars are thus divided into five transverse compartments, each of which is provided with two large doors entering from the top, one door on each side of the middle line of the car. The cost of each approximates $60, and they last four or five years. At Portland there are about sixty of these cars, providing storage capacity for 150,000 live lobsters, which may be retained for three or four weeks under favorable conditions.

The usual size of the lobster cars employed at Boston is 28 feet long, 14 feet wide, and 5 feet deep, divided into four compartments, each of which holds from 500 to 800 lobsters, according to the season. The compartments are separated from each other by vertical lathes, and each has two doors opening from the top. Some dealers omit two or three of the middle lathes in each partition between the compartments, so that when the doors in the two middle ones are opened the light causes the active and more healthy lobsters to scurry into the end compartments, where, huddled closely together, they are more easily removed with a dip net. The weaker lobsters, being less active, remain behind, and, thus separated from the stronger ones, may be removed as desired. During the first year after its construction the buoyancy of its material keeps the car afloat with the top slightly above the surface. But as it becomes water-soaked it is necessary to buoy it, which is accomplished by placing an empty water-tight barrel within the car at each corner. Small marine ways are usually built adjacent to the cars for convenience in raising them above the surface of the water. The cars cost $90 each. They last about five years only, their period of usefulness being shortened by the destructiveness of the teredo. There are 65 of them in Boston, with an aggregate carrying capacity of about 170,000 lobsters.

At Friendship and Tremont, in Maine, lobsters are retained in cars constructed on a plan invented and patented by J. R. Burns, of Friendship, and differing from the usual type in being divided horizontally into separate compartments, each about a foot in height, thus preventing the lobsters from crowding and killing each other by their own weight. Each compartment is provided with convenient openings at the sides, so that lobsters and food can be introduced as desired. The cars are about 35 feet long, 18 feet wide, and 6 feet deep, with capacity for 5,000 lobsters each.

In New York the market floats already described as being employed in connection with the live-fish trade are also used for retaining lobsters. The aggregate storage capacity of the floats at New York probably does not exceed 25,000 lobsters.

The cars used by the lobster fishermen of the New England coast are generally much smaller and more rudely constructed than those of the dealers. It is desirable to have them small, because of the convenience in removing the lobsters by hoisting the cars rather than by bailing; but some are so large that bailing is necessary. In general their capacity ranges from 100 to 1,000 lobsters, and entrance is made through a door on the top. At Woods Hole, Mass., the cars are about 6 feet long, 4 feet wide, and 3 feet deep. At No Man's Land, Mass., the average size is about 10 feet long, 5 feet wide, and 3 feet deep, and some of them are constructed for breaking the force of the waves that beat against them, having the top and bottom converging toward the ends, which are somewhat pointed. Old dories provided with a cover and with numerous holes bored in the sides and bottom are frequently employed, but slat-work boxes are the most common.

While the size and form of the live cars or boxes are largely matters of local fancy and convenience, it is important that they be of sufficient capacity to hold the lobsters without crowding. In estimating the capacity of live cars several modifying conditions must be considered, such as the roughness of the water, temperature, shade, etc. In localities where the water is still and quiet, fewer lobsters should be put in a car of definite size than in more exposed localities, because lobsters must have air as well as water. When the water is still the air is quickly exhausted and agitation of the water is necessary to replenish it. More lobsters can be carried in a given space during cool weather than when it is warm. The number that can profitably be put in a car depends also on the length of time they will remain there. In general, 150 lobsters to each 100 cubic feet of space is most satisfactory, although sometimes 300 and even more are placed in 100 cubic feet. In shallow cars a greater number of lobsters can be carried per 100 cubic feet than in deep ones. When given sufficient room, lobsters may be kept alive in these inclosures for several days or weeks, while awaiting the arrival of the market boat or while holding them for better prices. If the length of the confinement extends beyond a week or two it is desirable to feed the lobsters, otherwise they will eat each other. Any refuse fish which is not very oily is used for food. It is not advisable to confine them in live cars and feed them for the purpose of increasing the weight, and unless they are being held for a better market price the sooner they are removed from the car the better. The practice of plugging the claws of lobsters has been almost entirely abandoned.

OVERLAND TRANSPORTATION OF LIVE FISH.

Live fish are rarely shipped overland in the United States for commercial purposes, owing to the expense and also to the difficulty in keeping the water properly aerated and at the right temperature; but in connection with its work of stocking streams, etc., the United States Fish Commission is almost constantly engaged in work of that nature, adult fish as well as fry being carried in specially prepared tank cars on trips that last sometimes a week or more. The best type of these cars is described as follows in the report of the Commissioner for 1898:

The dimensions of car No. 3 as rebuilt are as follows: Length of body, 60 feet; total length from end of platform to end of platform, 67 feet 10 inches; width, 9½ feet; height from top of rail to top of roof, 13 feet 8 inches. The frame of the car is so braced as to permit of the two large doors in the center extending from floor to roof. This feature very materially simplifies loading and unloading. The interior of the car is finished in ash, and in one end is an office, an ice-box of 1½ tons capacity, and a pressure tank holding 500 gallons of water; at the other end are the boiler room and kitchen. The boiler room is equipped with a 5-horsepower boiler, circulating water pump, and air and feed pump. The tanks and cans used in transporting fish are carried in two compartments running along the sides of the car between the office and boiler room. They are 30 feet long, 3 feet wide, and 25 inches deep. Under the car, between the trucks, is a reservoir tank holding 600 gallons of water, and from which water is pumped into the pressure tank near the office; it then passes from this tank to the fish cans and tanks, and then back to the reservoir. In the middle of the car, over the compartments referred to, are four berths and several lockers for the use of the crew. The office also contains two berths, a writing desk, and a typewriter. These cars are fully equipped with all modern improvements in the way of brakes, couplers, signal whistles, etc., and have Pullman trucks and 33 inch Allen paper wheels. With the large water capacity provided, they are capable of carrying much greater loads of fish than ever before.

In transporting fresh-water species both water and air circulation are used, but with salt-water species the salt water is usually kept aerated by circulation only as it is not generally practicable to provide for a change of water. When the temperature is high, ice is sometimes packed about the transportation tanks to keep them cool, and in extreme cases a can filled with ice is placed in the water. In this manner marine species have been carried successfully for six days or more.

In the above-described cars the carrying capacity is to some extent sacrificed for the comfort of the crew, since they live on the cars throughout the year. Also the fish must not only reach their destination alive, but in a vigorous, healthy condition, so that they may live and be used in reproducing. Neither of these conditions is essential in transporting live fish to market, consequently a greater carrying capacity could be secured in cars designed especially for that trade.

A method of operating the air-pump by means of the rotary motion of the car axles was attempted on the U. S. Fish Commission transportation cars. The experiment is thus described on page 241 of the report for 1897:

An arrangement was adopted to furnish power for the pump and an air-blower by means of a friction wheel placed on the truck at one end of the car. This wheel was attached near one end to the top of the truck, so that it rested on the tread of the car wheel and was held there by two spiral springs. When not in use it could be elevated above the car wheel by a lever operated from inside the car. Power was transmitted from the friction wheel by means of a countershaft and rubber belting. The friction wheel gave a great deal of trouble, however, as it was impossible to make it strong enough to stand the wear to which it was subjected. As the action of the truck springs while the car was in motion moved the truck frame up and down, sometimes 3 to 5 inches, the friction wheel would be jolted out of position, and so uncertain was its operation that it could not be relied upon, and the pump and blower had to be worked by hand.

As a general rule fish will carry best in water of a low temperature. Cold water absorbs more air than warm; it also lessens the activity of the fish, causing them to consume less oxygen, and it retards decomposition in the organic substances contained in the water and the consequent generation of noxious gases. The lowering of the temperature therefore offers a threefold advantage. Whenever practicable the fish should be kept in confinement without food for a day or two before being transported, so that there may be no danger of the water being made impure by excrements of the fish.

In Europe considerable attention has been given to transporting fish alive. Well-smacks are used in the North Sea fisheries prosecuted by Germany, Holland, and England. Live cars are employed by many of the shore fishermen, and in many of the fish-markets both fresh-water and marine species are kept alive in tanks. The difficulty of keeping sea fish alive when natural salt water can not be obtained is met by the use of artificial salt water. But the most interesting feature of the European fish marketing is the overland transportation of live fish.

In Germany fresh-water species are transported alive in barrels about three-fourths full of water, the quantity of fish to each barrel depending on the variety, the length of the journey, and the season of the year. During the journey the water in the barrel is in almost constant motion, presenting considerable surface to the air, so that during a short distance sufficient oxygen is in this manner introduced into the water. But if longer journeys are made air must be introduced, which is accomplished by filling a sprinkler with water and squirting this water into the barrel with considerable force from a short distance, or the water is agitated by a vertical paddle-wheel fastened on the upper part of the vessel and separated from the fish by a perforated wall. Some of the barrels are provided with a tube running almost to the bottom of the barrel, the lower end containing many openings, and through this tube air is forced by means of a bellows on the outside. The last method is preferred, for by its use the barrel may be filled with water and fish, the carbonic acid is driven off, and agitation of the water is avoided. In the manner above described fish can be kept alive for a considerable period in a quantity of water weighing much less than their combined weight.

In 1881 a company was organized in Germany for the wholesale transportation of fresh salt-water fish from Cuxhaven, on the border of the North Sea, to Berlin, in specially constructed cars. The form of car adopted was invented and patented in Germany by Arno Gustav Pachaly, a Bohemian.

The following description is from the German letters patent dated March 20, 1880:

The transporting vessel is a railroad car, which can be taken off the wheels, the walls of which are double, the intervening space being filled with nonconductors of heat. Inside the car, and resting on the double floor, there is a shallow tank of forged iron with a vaulted roof, in which is placed the live fish with a quantity of fresh sea water. Along the inside walls of the car are shelves for storing the dead fish, and ice-boxes attached to the ceiling serve to keep the air cool. With a view to supplying the live fish in the tank with the necessary oxygen, air is led by means of pipes from the top of the car into the ice-boxes in which it is cooled and then by means of an air-pump it is forced into the fish tank. This air-pump is connected by means of a belt with one of the axles of the car, so that the necessary power may be obtained while the car is in motion, and in order to protect the fish from suffocation during the stoppages the air pump is so arranged that it can be operated also by means of a crank. To prevent violent motion of the water the air above the water in the tank is kept at a slight pressure, this being regulated by a suitable escape valve in the roof of the tank.

SHIPPING LIVE LOBSTERS.

In shipping lobsters alive well-smacks are employed to a great extent where the transportation is in large quantities from one part of the coast to another, but much of the coastal shipments, as well as the great bulk of those overland, are made in barrels. Flour barrels holding about 140 pounds, or sugar barrels with 185 pounds capacity, are employed, in the bottoms of which several holes are bored to afford drainage. In placing the lobsters in the barrels, each lobster is seized by the carapax, the tail is bent up under the body, and it is placed in the barrel with the back upper-most, being packed quickly and snugly together, so that they can not move from the position in which they are placed. Unless the weather is cold a long, narrow block of ice, weighing from 20 to 40 pounds, is placed in the center, its length following the axis of the barrel. On top of the lobsters a handful of seaweed is placed, and this is covered with 5 to 20 pounds of crushed ice, and the whole is inclosed by sacking secured under the upper hoop of the barrel. Packed in this way, the lobsters readily survive a trip lasting three or four days. Some dealers have tried separating the lobsters from the ice, using for this purpose a long, narrow box, divided transversely into three compartments, of which the middle is much the largest, and in this the lobsters are placed, while the ice is put in the two smaller compartments: but lobsters do better when in contact with the ice, the moisture appearing to be necessary for their preservation.

The United States Fish Commission has successfully carried live lobsters in its transportation cars for distances upward of 3,000 miles. The method pursued is thus described on pp. 243–244 of the report for 1897:

Large, mature lobsters, on long trips, are packed in seaweed in wooden trays about 6 inches high and of a size convenient for handling. Strips of wood attached to the bottom of trays have open spaces between them to allow air circulation. About 2 inches of seaweed are spread on the bottom of the tray and the lobsters placed on it with their claws toward the outer ends, so that they can not injure each other, and the trays are then filled with seaweed. They are packed in the refrigerator compartments, and the temperature of the air is kept, if possible, at from 40 to 48 F. A supply of salt water, filtered through cotton, is taken along, and the lobsters are sprinkled with it three or four times a day, and they are also daily overhauled and repacked. If the desired temperature is maintained, 50 to 60 per cent can be carried for five or six days.

Attempts have been made to ship live lobsters in sea water by having a water tank with a series of shelves either communicating or separate, with supply and discharge pipes connecting with the shelves, so that the lobsters on each shelf may be kept supplied with fresh sea water. This apparatus was intended especially for transporting lobsters on shipboard to England, but it has not been used to any great extent.

The following article from the *Canadian Gazette*, of London, contains an account of the experiments with it:

The Canadian lobster has long been well known and appreciated in England, but only in its preserved state, packed in the tins familiar to all housekeepers. A successful attempt has just been made to import live lobsters from Canada, where they are abundant and cheap, to England, where they are so dear as to render them a positive luxury. Many attempts have been made at different times to land live Canadian lobsters in England, but none of them had proved successful, owing to various causes too numerous to explain here. The idea was, however, too good, too tempting, to be definitely abandoned, and experiments were constantly being made, though with but little success. Finally Messrs. Arthur and Harold McGray instituted careful inquiries in the principal lobster districts, the result of which led them to the conclusion that the methods adopted by previous shippers had been defective, owing to their ignorance of the habits and requirements of the lobster. These shippers had simply placed the fish in large tubs, renewing the salt water at frequent intervals. This was clearly insufficient, for the lobsters invariably died within 12 or 15 hours. Having concluded their inquiries and carefully tabulated the information they had obtained, Messrs. McGray commenced to experiment with a system entirely different, devised by themselves. This improved apparatus, which appears simple in itself, is the outcome of patient observation and study of the habits of the lobster at various points along the coast. It enables the crustaceans to continue while in transport an almost identical mode of life to that led by them at the bottom of the sea. This system constituted the inventor's secret, which we cannot of course divulge at the present moment. They commenced with ten lobsters, which they placed in their improved receptacle and contrived to keep them alive for 18 hours. This was a decided improvement on the results previously obtained by other merchants. Thus encouraged, they continued their experiments with successive series of lobsters. In the course of the summer of 1891 they succeeded in keeping them alive 5, 8, 11, 13, and ultimately 18 days. These experiments, diversified by innumerable incidents, trials, failures, and partial successes, were conducted on board a light-ship stationed off Barrington, with water always taken from the bay and naturally of about the same temperature. An important point was thus established—lobsters could be kept alive for 18 days on board a stationary ship.

The question then arose, would similar lobsters live the same length of time on board a ship crossing the Atlantic, and in water constantly changing in temperature? Messrs. McGray were quite convinced that they would. They, therefore, arranged to ship 50 lobsters by the steamship *Historian*, running from Halifax, Nova Scotia, to London direct. The passage was expected to occupy 14 days. This was more than sufficient to thoroughly test the system, seeing that steamers are available which make the passage in 10 days. The ship left Halifax at 8 a. m. on Thursday, December 10, Mr. Harold McGray being on board to personally conduct the experiments. The lobsters were shipped under rather unfavorable circumstances, they having then been out of water for 24 hours. The losses during the voyage were as follows: On the first day 2 lobsters died; on the fifth day, 4; on the sixth day, 1; on the seventh day, 3; on the eighth day, 1; on the ninth day, 1; on the twelfth day, 2.

The fifth day a receptacle containing 15 lobsters was swept overboard during a southwest hurricane. The first 2 deaths were due to the unsatisfactory condition of the fish when shipped; the next 4 were killed by the rapid change in temperature during the passage across the Gulf stream; 2 died from injuries inflicted by larger and stronger ones, while the remainder died from some unknown cause.

On leaving Halifax the temperature of the water was 44°, and this was maintained for 4 days. On the banks of Newfoundland it varied from 45° to 48°, while, on arriving in the Gulf stream, it suddenly rose to 65°. Mr. McGray was naturally anxious to ascertain the effect produced on the crustaceans by this rapid rise in temperature. Four of them succumbed, as we have said; but the rest remained in good condition.

Strange to say, the cold air and the warm water exercise an equally fatal effect on these delicate fish, accustomed to live in depths where the air never penetrates, and where the water never rises above a certain temperature. Another curious point was that they traveled the entire distance—2,800 miles—without requiring anything in the shape of food. When at the bottom of the ocean they eat fish, and when brought to the surface to be kept for a certain time they can be fed on oatmeal. They would, of course, eat fish, but it has been found that they fight for this food like hungry wolves, biting and seriously injuring one another. To avoid all possible risk Mr. McGray decided to give them nothing to eat and found that they still remained in good condition.

Up to the time of the arrival of the shipment in the Victoria docks, at noon on December 26, everything had come up to the expectations of the exporters. Unfortunately, however, they reached London just at the time when, owing to the Christmas holidays, the markets were closed for 3 days. They had consequently to be kept on the ship for nearly 2 days, until the morning of Monday, December 28, and the water in the dock had to be used in the endeavor to keep them alive. That dock water, helped by the fog, killed all but four. It will, however, be admitted that these quite exceptional circumstances do not detract from the value of the experiment, as showing that live lobsters can be brought to this country in a marketable state, and Mr. McGray is confident, from the experience he has gained, that the next shipment will establish beyond doubt the feasibility of a successful and profitable trade.

The practicability of the transport of live lobsters having been thus far demonstrated, the promoters will later on arrange for the acquisition of a 15-knot boat specially fitted with the necessary apparatus for the conveyance of live lobsters in large quantities across the Atlantic. This will enable them to supply the markets of London and Paris with first-class lobsters delivered alive in these cities at less than half the price now paid for English lobsters of equal quality.

Mr. Adolph Nielsen, fishery expert connected with the Newfoundland Fisheries Commission, thus discusses the practicability of shipping live lobsters to Europe:

The exportation of lobsters alive to England and the European Continent is a matter which has occupied my attention very much, and is worthy of the Fishery Commission's greatest attention, because if it could be carried out successfully it would mean a large increase in the value of the lobster fishery. The greatest difficulties to be overcome in carrying lobsters alive in large quantities for a long distance is to prevent them from crowding together on top of each other if shipped in bulk, in which way a large number of them suffocate, and to keep them alive for any length of time in hot weather in the summer, especially in water of a high temperature. Several experiments have been made in the United States with shipments of lobsters alive in the hot season in vessels built for the purpose, fitted out either with wells or after other plans, but as far as I am aware the attempt has not yet been successful. I saw a few years ago a steamer fitted out to carry lobsters alive from Nova Scotia to the States. She had her hold divided into compartments, which were intended to be filled with the live stock. Along the keelson was laid a pipe through which water forced in from the bow. This pipe was furnished with a valve so that the water could be shut off and let in according to wish. From this main supply pipe smaller perforated pipes were laid around the bottom of each compartment. The idea of this system was to obtain a strong current of water upwards, which it was thought would keep the lobsters from crowding together on top of each other and smothering. I was informed that this system worked well enough in a cold season, but noticed myself that it was condemned in the summer months, and that the lobsters were packed in boxes with ice and shipped away to the States in this way in the same steamer. The difficulties in keeping the lobsters from crowding together and smothering, and in keeping the water cold in a vessel while crossing the Atlantic, I think could easily be overcome, and I even think it would be sufficient to bring the temperature of the water down while crossing the Gulf Stream in the summer months.

My plan would be to divide the hold of the vessel into compartments and have each compartment again floored over with boards 3 to 4 inches apart through its whole height. These boards could be put down and fastened according as the vessel was loaded, and taken up according as it was discharged. In this way the difficulty of keeping the lobsters from crowding on top of each other would be overcome. The height of water in the compartments could be regulated according to wish while loading or discharging. By means of refrigerators the temperature of the water could be brought down at a little expense whenever it was found too high. In this way I am of opinion that lobsters could be kept alive in good condition for a considerable length of time. Steamers would be preferable to sailing vessels. The collection of lobsters could easily be arranged by having fixed stations in a bay, at which the vessels could call and take in their cargoes, and I am sure there would not be much trouble, nor would it take a long time to secure full loads when everything was well arranged. * * * With the great demand and high prices paid for lobsters alive in England and on the Continent, there is the best reason to anticipate that a large and profitable business could be done in carrying these crustaceans across the Atlantic alive. When it can pay English people to send their vessels up to the western coast of Norway, where the lobsters are far from plentiful and where only a limited quantity can be secured, and purchase there at a high figure and carry them alive to England every year, it is reasonable to presume that it would pay very much better to ship them from Newfoundland, where they can be secured in much larger quantities and at a very small cost, if the difficulty in bringing them across the Atlantic alive can be overcome. (Report of Newfoundland Fisheries Commission for 1890, pp. 54-56.)

While lobsters are generally shipped alive, yet some are first boiled and then cooled and placed in barrels or boxes, and if the weather be warm some ice is added. Boiling before shipment is applied to about one eighth of the lobsters handled on the United States coast. They are boiled in salt water in a covered box, to which steam is admitted for 20 or 30 minutes, the temperature being about 250° F. On removal they are carefully folded and placed in piles like cord wood for cooling, when they are packed in a manner similar to live lobsters. They will keep a week or longer when well iced. Only live lobsters are boiled, for after death the muscles so relax that the fibers become short and the meat crumbles, not having sufficient tenacity to hold together, and the tail bends readily upon slight pressure. Lobsters that die before being cooked are so much loss, since they are not then suitable for food markets.

SHIPPING LIVE OYSTERS AND CLAMS.

While the great bulk of oysters and clams produced in this country are opened before being marketed, yet there is an extensive trade in these mollusks in the shell, not only at markets near the source of supply, but at points quite inland, and even in foreign countries. When out of their native element, oysters and clams will ordinarily live only a few days, and in order to retain them alive during shipment and in storage prior to consumption it is necessary to keep them cool and to prevent the loss of liquor from the shell. With care they may be retained in this manner for months. Half a century ago it was customary with many families in Connecticut and New York to lay in a supply of oysters every fall for use during the winter. Piled up in some cool place, usually in the cellar, with the deep shell downwards and between layers of seaweed, they would live sometimes for three or four months. At present during cold weather it is not unusual for them to remain in bulk in the holds of vessels for two or three weeks without great deterioration of quality, and they are at times kept as long in the holds of the oyster scows in the New York market, or in the cellars of other wholesale depots.

At several points along the Atlantic coast, and especially at Franklin City, Va., very convenient floats for storing oysters or clams are so arranged in the water adjacent to the market-houses that they may be raised or lowered by means of windlasses. The most convenient size of the floats is 20 or 25 feet long, 8 or 10 feet wide, and 2 or 3 feet deep, the sides and bottom being of strong slats. Four piles are driven into the ground, two at each end of a float, and on these rest the windlasses for raising the float when it is desirable to remove the oysters.

An interesting feature in connection with the marketing of oysters is practiced in Connecticut, New York, New Jersey, Virginia, and a few other localities, viz, the "drinking" or "freshening" of oysters. The oysters are removed from the reefs or planting-grounds and placed in floats or on private areas near the mouth of a small stream of fresh water. Here they at once eject the mud and other impurities within the shells and clinging to the edges of the mantle and gills, and imbibe a large quantity of fresh water, improving the color of the flesh, making it a purer white and bloating it into an appearance of fatness. From 10 to 36 hours is the usual length of time the oysters remain in fresh water before being marketed. The same result is accomplished in some places with the aid of platforms of rough planking set in the river bank, on which oysters are thrown at high tide and are left bare by the receding tide; a sluice-gate is then opened and fresh water is allowed to flow over them.

A more elaborate affair is constructed in the following manner: The shore or bank is excavated and piles are driven until a floor can be laid at a suitable level below high-water mark. A tight shed is built over this, and on one side a canal is dug, into which a boat may run and its cargo be easily shoveled through large openings in the side of the shed upon the floor within. As the tide recedes it leaves the oysters upon the platform within the shed nearly bare, a depth of 8 or 10 inches of water being retained by a footboard at the seaward end of the shed. By an arrangement of sluices the fresh water is then admitted and the freshening begun, and the bulk of salt and of fresh water can be so proportioned as to impart the degree of freshness desired. At a height of 7 or 8 feet above the oyster platform or pen is another platform or garret where barrels, baskets, boat gear, and other small property can be safely stowed.

When oysters are removed from bulk, and subjected to the varying conditions met with in transportation, greater care must be taken, especially to prevent loss of the liquor, and, secondarily, to maintain a moderately low temperature.

In shipping short distances, merely placing the mollusks in barrels or in bulk is sufficient if the weather is cool. For longer shipments, or in warm weather, they must be packed more carefully, and some shippers place each oyster with the deep or concave shell underneath and press the head of the barrel down tightly. Refrigerator cars are used to some extent during both warm and cold weather, and when the destination is reached, if not intended for immediate consumption, oysters should be kept at a temperature between 35° and 45°. If they become frozen they should be thawed gradually in a cool place. Oysters will not freeze as readily as clams; and oysters and clams in transit during a snowstorm do not freeze as readily as when the weather is clear and a stiff wind prevails.

The European trade in American oysters depends on shipments of live oysters. This began in 1861, and a trade has been built up amounting to about 100,000 barrels annually, small East River and Long Island oysters being selected, averaging 1,200 to 2,000 to the barrel. The oysters are packed as snugly as possible in the barrels, sometimes with the more concave shell underneath, to prevent escape of liquor, and all are pressed down tightly by the cover, to keep the shells of the oysters closed. In shipping, the barrels are stowed, head up, in some part of the steamer where they may keep cool, and two or three weeks frequently elapse from the time of gathering them until their bedding or consumption in England.

A large proportion of the oyster trade on the Pacific coast depends on the transportation from New York of seed oysters, running from 2,000 to 7,000 to the barrel. These oysters are transported during the spring and fall, carefully packed in barrels, in carload lots, and are usually two or three weeks on the road, being carried on fast freight trains. Unless the weather conditions be unusual they survive the journey with small loss, usually about 10 per cent. It is not considered injurious if the liquid about the oysters freezes, provided the mollusk itself does not freeze. In illustration of the vitality of these small oysters it is stated that several years ago, in a shipment of several carloads, one car was missent through some blunder, and on reaching San Francisco, after being two months on the way, the percentage of loss among the oysters, which were already partly frozen, was but little more than ordinary. The cost of the seed at New York is about $3.50 per barrel, and the transportation charges are about $5 per barrel. During some years as many as 100 carloads, of 85 to 95 barrels each, are planted in Pacific coast waters, principally in San Francisco Bay.

In 1882 a patent* was issued for a somewhat unique method of preparing oysters and other mollusks for long shipments. It consists in binding the shells firmly together, while the mollusk is fresh and alive, by means of a wire or wires made to embrace the shells between which the animal is contained, the ends of the wire being secured by being twisted. It is claimed that by this process the natural juices are retained and the deterioration in quality which ensues upon their evaporation is prevented.

Prof. John A. Ryder is quoted as indorsing the value of the method as follows:

I have examined and had in my possession a number of wired oysters, and I am satisfied that the oyster can be preserved, when the shells are thus wired, for a considerable length of time. I have carefully examined oysters, which I am satisfied have been wired for 60 days, and I find that their vitality is fully preserved and the oyster in no way deteriorated in quality or flavor. I think the process of preserving oysters by placing a wire around them is a practically useful process, and, in my opinion, would lead to the transportation of oysters to distant points as an article of commerce, when it would otherwise be impossible to transport them alive in the shell.

The method was employed on a small scale in Philadelphia for several years, and in 1888 a stock company was formed and a plant established at Cape Charles, Va. At first the work of wiring was done with pliers, but in 1890 special machines were introduced, by means of which one man has been known to wire 48 oysters in a minute, but the average is much less. The shipments extended quite over the country in an experimental way, many being sold on the Pacific coast. A few oysters are yet shipped in this manner, but the practice has not come into general use.

---

* No. 265255, dated October 3, 1882.

A number of other methods have been proposed for fastening the shells of oysters together, such as inclosing them in a batter of plaster of paris or similar material,* securing a rivet of soft iron through the nib or bill,† inserting a plug of hard wood or other suitable material between the shells immediately in the rear of the hinge,‡ etc.

A railroad car for the special transportation of live oysters, invented by Mr. A. E. Stilwell, of Kansas City, Mo., has been used to some extent during the last year or two on the Kansas City, Port Arthur and Gulf Railroad. The interior of the car is 34 feet long, 8 feet wide, and 4 feet high, the space being divided into four compartments, each of which has two ventilators in the top through which the oysters are loaded, and two unloading spaces in the side. The floor and sides of the car are constructed of 3-inch white pine, calked and pitched in the manner of ships, so as to make the compartments water-tight. These compartments are first nearly filled with oysters, and then sufficient sea water is added to cover them.

SHIPPING LIVE CRABS.

Hard crabs require little care in packing and shipping. They are placed, back up, in barrels or boxes, usually without ice if the weather be cool, and covered with cloth. Little mortality occurs if their destination be reached in two or three days. But during warm weather a block of ice, weighing 5 to 15 pounds, is placed in the top of the barrel and separated from the crabs by a double handful of wet seaweed. The receipts of hard crabs in the markets are quite extensive, the supply being obtained at various points along the coast, but principally from the shores of Chesapeake Bay. From that bay about 75,000 barrels are marketed annually, each barrel holding from 200 to 300 crabs, which weigh about 75 pounds.

Much greater care is required in handling soft-shell crabs. This industry originated about 1873, and it has reached its greatest development at Crisfield, Md., but supplies are received also from New Jersey, Virginia, North Carolina, Louisiana, and various other points. The soft-shell and "peeler" crabs are caught together, the former being shipped at once and the latter impounded until after the shedding process and then shipped to market. The present trade amounts to about 16,000,000 crabs annually, worth 2 or 3 cents at the fishing port, and from 3 to 10 cents each in the retail markets. The peeler crabs are impounded in floats made of light plank and scantling, with plain board bottoms and latticed sides. The size of the floats varies somewhat, but most of those at Crisfield are 20 feet long, 3 to 5 feet wide, and 15 inches deep, with a projecting ledge at half their height corresponding to their water line. The average value is about $2, with a capacity for 300 or 400 crabs each. They are frequently inclosed by a board fence, which serves as a breakwater. The floats are visited three or four times daily, and the crabs that have shed since the last visit are taken out and at once marketed.

The following, in reference to the live-crab business, taken from an article by Hugh M. Smith, in the Bulletin of the United States Fish Commission for 1889, is of interest in this connection:

The one factor which, more than any other, tends to reduce the profits of the shippers and indirectly the receipts of the fishermen, is the high death-rate among the impounded crabs. Owing to the injuries which many crabs receive in being caught and handled, and, in a measure, to the severity of the shedding process, a comparatively large number of crabs die after being purchased by the dealers,

* See Letters Patent No. 431212, dated July 1, 1890.
† See Letters Patent No. 453114, dated May 26, 1891.
‡ See Letters Patent No. 459220, dated September 8, 1891.

and are a total loss. As an illustration of the uncertainty of the business and of the risks which the dealers have to run at times, it may be stated that of 3,200 crabs purchased by a firm one day in July, 1888, no less than 3,000 died before shipment. This, of course, is an unusually great loss, and is not to be taken as a basis, although the individual dealers estimate their losses at from 10 to 30 per cent, and even as high as 50 per cent during certain periods. A few crabs die after leaving the hands of the shippers on the way to their destination, but this element of loss is being overcome by greater care and experience in packing the crabs prior to shipping them. A comparison of the total catch with the aggregate shipments for 1888 gives a difference of 628,706 crabs, with a market value of about $23,600, which figures represent the mortality and consequent losses. The death rate in 1887 was even higher than in 1888, being 21 per cent, as against 16 per cent in the latter year. It is impossible to determine with accuracy the number of crabs which die during shipment to market. There seems to be no remedy for this state of affairs. Although the season of 1888 showed a small but gratifying improvement over the previous year, it can hardly be hoped that the mortality will ever be reduced below a somewhat high limit, owing to the methods of capture and handling, and to the normal vicissitudes of the molting process, increased as they are by the unnatural surroundings and conditions to which the crabs are subjected.

The crabs are shipped to market in crates or boxes. The crates used in the Chesapeake region are about 4 feet long, 18 to 24 inches deep, and the same in width, and are provided with closely fitting trays, in which the crabs are carefully packed side by side, with their legs well folded up and their bodies lying obliquely, so that the moisture may not run from their mouths, in rows between layers of cold seaweed, on which finely crushed ice is sometimes placed. The capacity of each crate is from 8 to 10 dozen, and as the crabs possess little tendency to move when once packed in position, they remain quiescent for a long time. The principal markets for soft-shell crabs are New York, Philadelphia, Baltimore, and Washington, but the demand from the interior is increasing. In some localities the crabs are carefully placed in stout boxes in rows and tiers or layers separated with cold, moist seaweed, and with crushed ice in the top of the box over all, the entire contents being so arranged that the respective positions of the crabs can not be disturbed.

TERRAPIN AND TURTLES.

Among fishery products that are nearly always marketed alive are the various species of edible terrapin and turtles. These reptiles are remarkable for their tenacity of life; with very little care they may be retained alive for six months or more. In the Middle Atlantic States terrapin caught in summer or fall are usually placed in dark inclosures, as in cellars, with a quantity of seaweed or grass, into which they may burrow, and without food or water they are kept in excellent condition until the following spring.

It may be remarked incidentally that terrapin and certain kinds of fresh-water fish, as catfish and pike, may be frozen alive in a block of ice, and kept there for several days at least, and on thawing the ice the animals are found to be unharmed. I am not aware that experiments have been made to determine how long they will live under these circumstances or the lowest temperature they will stand.

There are numerous inclosures along the Atlantic coast where terrapin are confined throughout the year for growing and breeding purposes, but this interesting feature of our fisheries is scarcely within the scope of the present chapter.

REFRIGERATION, OR PRESERVATION BY LOW TEMPERATURE.

The temperature of fish, unlike that of mammals and other warm-blooded animals, corresponds to that of the medium in which they live. The atmosphere during the day being usually warmer than the seas and rivers the temperature of fish is generally increased on their removal therefrom and their consequent death, whereas in case of most land animals death usually results in a decrease in the temperature. This increase in temperature, together with the delicacy of the texture of flesh and the very large number of bacteria in the atmosphere to which the flesh is unaccustomed, makes fish extremely susceptible to putrefaction soon after life is extinct, especially if there be considerable moisture in the stomach cavities. In order to overcome this tendency it is important, in case fish are to be used fresh, that the temperature be kept at a low point while they are awaiting consumption. As the markets are generally situated at some distance from the sources of supply, preservation for a short time is a necessity, and for this purpose the application of low temperature is so general that it is almost coextensive with the fresh-fish trade in this country.

The importance of this method of preserving fish is not readily overestimated. It has resulted in a wonderful development of the Gulf and South Atlantic fisheries; and, indeed, without its agency the fishery resources of those regions would be of comparatively little value. It has enlarged and widened the general fishery trade so extensively that at present salmon fresh from the Columbia River, halibut from Alaskan waters, and oysters from Chesapeake Bay and Long Island Sound, are sold throughout the United States and in foreign countries. and numerous other fishery products are marketed thousands of miles from the source of supply, and for weeks after their capture, in condition not dissimilar to that when removed from the water.

It is only within the last half century that much attention has been given to the fresh-fish trade. Prior to 1830 it was of very limited extent, being confined during the warm months to a retail business in the towns near the fishing ports, while in winter the fish were frozen naturally and transported to the neighboring markets, the business being largely in the control of peddlers. Following the introduction of ice, about 1830, the handling of fresh fish developed more extensively than any other branch of the fishery industries, and at present the quantity of fish marketed fresh in the United States is much greater than the quantity placed on the market in all other conditions. The increase in this trade is one of the most noticeable features in connection with the fishery industries. An important factor in developing and maintaining it is the improvement in transportation facilities—not only on shore but also in bringing fish from the sea to the fishery ports, the improvements in railway traffic, and the addition of fast types of vessels. This feature of the trade, however, is scarcely within the scope of the present paper, as it does not tend to preserve the product, although it serves better than methods of preservation.

The processes generally employed for retarding putrefaction in fish by low temperature are (1) simple cooling with block or crushed ice, (2) open-air refrigeration during cold weather, and (3) artificial freezing and subsequent cold storage. The first process is employed quite generally throughout all countries in which ice is obtainable, in transporting the fish from the source of supply to the wholesale markets and thence to the retail stands, and in preserving the fish while awaiting immediate sale. The second is used principally in the winter herring and smelt trade with the British North American Provinces, and the third in preparing them fresh for very long ship ments and in storing them for several weeks or months to await a better market. All of these methods are of comparatively recent development, the oldest in general use—cooling with ice—being used commercially only about 70 years, and doubtless none of them have yet reached their highest development.

COOLING FISH WITH ICE.

Probably half of the fishery food-products in the United States are preserved in ice for transportation to the markets and in holding them for immediate sale, and this is also true with respect to England and possibly some other countries of northern Europe. The process does not result in freezing the fish, the resulting temperature being never less than $32°$ F. The ice adds greatly to the expense, especially in transportation, and confines the process to preserving fish for brief periods of time.

The use of ice for preserving fish in the United States began in 1838, when a Gloucester smack is reported to have carried ice with which to preserve the halibut dying in the well or killed before being placed there. For a number of years there was a strong prejudice against iced fish, almost equal to the present opposition to frozen fish, and it was not until 1845 that it became common for vessels to carry ice as a preservative. Care was at first taken that the ice be kept separate from the fish, being placed in a corner of the hold. It was soon found, however, that stowing the fish in crushed ice did not materially injure them, and this method was soon in general use and largely superseded the trade in live fish north of Cape Cod.

For many years after ice was introduced in the vessel fisheries it was still thought inadvisable to ship iced fish inland, and not until 1858 could New England dealers be induced to experiment in sending them as far as New York City, but as the experiment was successful a large trade was quickly developed, and iced fish are now shipped to all parts of the United States.

The usual method of applying the ice is to crush it and mix it with the fish in successive layers of ice and fish. The process requires no great skill, yet there must be a good knowledge as to the quantity of ice necessary, the most economical size of the pieces, the convenient form of the receptacle, and the manner of packing—all of which depend on the kind and quantity of fish, the length of time for which they are to be held, and the temperature of the atmosphere. Fresh fish should have the very best of care in handling at the originating point, be promptly and thoroughly chilled, and so placed in the shipping box or barrel that bruising and the possibility of an increase in temperature are reduced to a minimum. It is advisable that fish be killed immediately after capture, as this prevents their thrashing about and bruising themselves, and they remain firmer and bear shipment better than those allowed to die slowly. Bleeding the fish is very frequently advantageous, but it is rarely done unless the fish are to be dressed. The practice of piling freshly-caught fish en masse, one upon the other to the depth of several feet, is extremely objectionable, especially when the weather is warm; since it unnecessarily bruises and heats the fish, causing putrefaction to set in much earlier than would otherwise be the case. Absolute cleanliness is essential at every stage in handling fish, care being taken to keep the market houses and the shipping boxes or barrels free from every particle of putrefying refuse, otherwise the fish will become infected with bacteria already developed and natural putrefaction will be thereby accelerated by several hours. To secure the highest degree of cleanliness all stationary storage compartments should have metal linings, since the wooden walls and floors of the compartments furnish a lodgment extremely favorable to the developed bacteria.

Much difference exists as to the dressing of fish before shipment to the wholesale markets, but in general it is best to ship them round, or just as they come from the water. The choice grades of fish should never be eviscerated or beheaded when intended for the fresh trade; but cod, haddock, bluefish, lake trout, and all large fish, such as halibut, sturgeon, etc., are usually dressed before delivery. In every case putrefaction would be retarded longer if the viscera were removed; but the round, plump appearance of the fish is thereby impaired, and in case of certain species, as shad and herring, the roes, which are highly prized, would thereby become wasted. Dressing the fish also decreases the weight 15 or 20 per cent, and sometimes even more, and a correspondingly higher price is expected for the dressed fish than for the round. The practice varies not only in different localities, but in the same locality at different seasons of the year. Mackerel caught in the vessel fishery between June 1 and October 1 are usually dressed by drawing the viscera out through the gill-openings, whereas those taken during the spring and fall are generally iced round. The whitefish received at Detroit and some other lake ports are commonly split and eviscerated, except that Lake Erie whitefish are nearly always sold round. These are caught mostly in the fall, when they are full of spawn, and if eviscerated they would decrease about 24 per cent in weight, and, moreover, many customers desire the spawn. The yellow pike are received round mostly, but those coming from the Dominion of Canada are usually dressed, in order to lessen the import duties.

The importance of the careful handling of fish and their arrangement in the shipping boxes is scarcely appreciated by the majority of the fishermen. In discussing this subject in the National Fishery Congress, at Tampa, in January, 1898, Mr. E. G. Blackford, of New York, said:

As an example of the increased returns to the shippers from careful handling, I call attention to the fact that certain shipments of shad, going to the New York market from North Carolina, bring from 25 per cent to 40 per cent more than other shad from the same locality. For instance, a certain shipper from Albemarle Sound, North Carolina, pursues the following method: His shad are carefully taken from the nets and placed in a cold room until thoroughly chilled, then packed in boxes; first a layer of fine ice, broken into lumps no larger than chestnuts, is placed in the bottom of the box; then the shad are placed in rows, lying on their backs, making a complete layer on the ice; then a layer of fine ice is spread over the bellies of the shad, and on this layer is another row of shad; all the shad are packed in a similar way; then the top of the box is filled with fine ice and the cover nailed securely on. These shad reach the New York market in a perfect condition, and so well known has this shipper's mark or brand become that buyers are always on the lookout for this particular brand, and these shad are the first sold and bring the highest prices. On arriving in New York, the fish have not moved from their position in the box, the ice is still intact, and on opening the box we find all the fish to be in a perfect condition, each scale undisturbed, and the whole presenting the appearance of a glistening jewel just taken from a casket. (Bull. U. S. Fish Commission, 1897, pp. 157–158.)

No matter what kind of fish are shipped, they should be thoroughly chilled before being placed in the shipping box or barrel, whenever the time will admit. It too frequently happens that fishermen place the fish in shipping packages immediately after their capture or after they have lain in the boat several hours exposed to the heat of the sun. Having to contend with the warmth of the fish as well as the atmospheric heat, the ice packed with the fish melts very rapidly, whereas if the fish had been chilled before being packed they would carry for a much longer time, and less ice would be necessary in the shipping package, resulting in reduced transportation charges. Cooling the fish is generally best accomplished by laying them thinly on a clean floor or platform in the fish-house and spreading finely chiseled ice over them; but if the air is unusually warm they should be cooled in a suitably insulated ice-box.

It is of prime importance that the temperature of the fish be reduced as soon as practicable after they are caught. The sooner they are placed with the ice after coming out of the water the longer the fish will carry and the better their condition on reaching the consumer. On the death of the fish the tissues relax, and offer a favorable lodgment for bacteria, whereas the application of ice as soon as the fish are removed from the water hardens the tissues and counteracts the development of bacteria.

Mr. E. Le Clair, of the Baltimore Packing and Cold Storage Company, of Minneapolis, Minn., in writing on this subject, states:

While at Lake of the Woods four years ago, during the month of July, when the weather was warm, the writer took two boxes in the boat; one of the boxes was filled with fish as soon as they were taken out of the nets, without ice; the other contained about the same quantity of the same kind of fish, which were immediately iced, while they were yet alive, and a test was made as to the keeping qualities of the two. As soon as the boat reached the fish-house where the fish were dressed, the un-iced fish were immediately iced after having the inwards and gills removed, and the fish that had been iced in the box were also dressed and treated in the same way as the other un-iced fish were. The fish that were not iced when taken out of the nets became unfit for human food in six days, and the fish that had been promptly iced were kept in the shanty for two weeks and then shipped from Lake of the Woods here, a distance of 600 miles, in a refrigerator car, the boxes marked; and when the fish arrived they were found to be in good condition, and we reshipped the same fish to Butte, Mont., and never had any complaint of them. The time that elapsed from the time that the fish were taken out of the water until they were iced in the fish-house was two hours and ten minutes, but the weather and water were warm. It therefore is evident that the greater care exercised at the originating point as to the proper icing of fish, the better will be the result.

The quantity of ice used in shipping fish depends on the size of the package and the season of the year. During the summer months, for a shipment covering one or two days, 50 pounds of ice is generally required to each 100-pound box of fish, more in proportion being necessary for smaller packages and less for larger ones. This ice should be crushed quite fine, so as to completely surround the fish and yet not bruise them. It is generally better to "chisel" the ice than to crush it, especially for packing among the layers of fish, since crushed ice is generally somewhat coarse. "Chiseling" consists in planing the ice from a large block by means of a long-handled chisel, the face or edge of which is formed by three or four thin, sharp teeth.

The most usual forms of shipping packages on the Atlantic coast are the flour barrel, with 200 pounds capacity, and boxes holding 400 or 500 pounds of fish. The 450-pound box, so popular in the Boston wholesale trade, measures 42 inches long, 24 inches wide, and 18 inches deep, and the 500-pound box is 48 inches long, 26 inches wide, and 18 inches deep. Auger-holes in the bottom of the barrels and the edge-cracks in the boxes suffice for drainage of the water resulting from the melting ice. After placing a shovelful of crushed ice in the bottom of the barrel, about 50 pounds of fish are put in, followed by succeeding layers of ice and fish, with a top layer of two or three shovelfuls of ice. In packing in boxes a layer of crushed ice is placed in the bottom, another in the middle, and a third on top, the fish being between, with their heads toward the ends of the box. During cold weather, or when shallow boxes are used, the middle layer of ice may be omitted.

The arrangement of the fish in the boxes differs according to the species and the individual ideas of the shippers. The common practice is to place the fish on their backs if round, and on their bellies if dressed; but sometimes, in case of small fish, two layers are placed together, backs to backs and bellies to the ice. Cheap fish, such as cod, haddock, etc., and other large split fish are usually not arranged in any particular manner, but are permitted to lie in the barrel as they fall, while shad and other delicate species must be carefully placed. Small or medium-sized fish are rarely eviscerated before being marketed, as they sell much better in the round.

If the fish are thoroughly chilled as soon as practicable after removal from the water and carefully placed in the shipping package, so that they will not be bruised or the temperature rise above 40° or 45° F., they will generally keep in good condition for two weeks or more.

Natural ice, being usually less cold, is generally more desirable for shipping fish than artificial ice, since it gives off its coldness quickly and the moisture coming in contact with the fish acts as a good conductor, and the fish are more rapidly cooled. But as artificial ice lasts longer, it is better for long-distance shipments, provided the fish have been thoroughly chilled before being placed in the shipping-box.

Among the most effective devices for holding and shipping fresh fish cooled by means of ice is the shipping-car used in the trade on the Great Lakes. This consists of a large box mounted on a four-wheeled iron truck, the size of the boxes ranging from 4 feet long, 2½ feet wide, and 28 inches deep, to 7 feet long, 4 feet wide, and 4 feet deep, with a corresponding capacity of from 800 to 3,000 pounds of fish. The walls of the box are made double, of tongued-and-grooved boards, with an intervening air space of 2 or 3 inches. The truck wheels range from 9 to 12 inches in diameter. A layer of ice is placed in the bottom of the box and then one or two layers of fish, succeeded by alternate layers of ice and fish, the latter, in case of two layers together, being placed backs to backs and bellies to the ice, and the whole covered by a layer of ice. Some of the large boxes were formerly so constructed that the upper half of one of the sides might be let down to facilitate placing the fish near the bottom of the box. But because of the tendency of this hinged side to become loose, that form of box is no longer used, and fish are placed in the bottom of the box by tilting it down on one edge, the side forming an angle of 45° with the floor and resting on a triangular frame or horse.

The cars average in value about $25 each, and about 2,000 are employed on the lakes. They are used in transporting the fish from the receiving ports to the large wholesale markets and for distributing them to the various inland dealers. When emptied, the cars are returned to the shippers.

The foregoing represents the general methods employed in icing fresh fish at the various marketing centers. In addition to these there are special methods of handling certain important varieties of marine products which experience has proven to be of value, such as icing cod, haddock, halibut, mackerel, shad, oysters, etc.

ICING COD, HADDOCK, AND BLUEFISH.

During the last thirty years cod and haddock have largely increased the fresh-fish trade. Prior to 1860 haddock was very little esteemed, but at present it is the most important fish in the fresh-fish markets of the New England States and one of the most important in the United States. During 1889 the quantity of fresh haddock received at New England ports aggregated 41,155,481 pounds, and of cod, 30,168,643 pounds, nearly all being the product of the vessel fisheries.

In the fresh cod and haddock fisheries the hold of each vessel is generally fitted up with twelve to fifteen pens, each about 6 or 8 feet long and 4 or 5 feet wide, with capacity for 1 or 2 tons of fish and the necessary ice. When the vessel is making a long trip, from 10 to 20 tons of ice are carried, but during cold weather this quantity is much reduced. When the fish are received on the vessel the men dress them, seizing each fish by grasping it about the eye or some part of the head with the left hand and ripping it down the throat, removing the viscera, which is thrown overboard, while the liver and roes are placed in barrels. The fish are washed in tubs or by pouring buckets of water over them as they lie on the deck, and are then ready for icing in the pens. A layer of block ice is placed at the bottom of each pen, next a layer of fish, backs up, and, sometimes, when the weather is warm, the abdominal cavity of the fish is filled with fine ice. A layer of ice is placed over the fish and about the ends and sides, and successive layers of fish and ice are added with a layer of ice on top of all, the ice being chiseled or planed with a sharp dentated chisel attached to a long handle. The care taken in icing the fish conforms to the probable time that will elapse before the schooner reaches port; when the weather is very cold and the wind favorable for a quick run to market the quantity of ice used is largely reduced or it is even dispensed with altogether.

The method of icing bluefish on the New York market vessels differs little from the process applied to cod and haddock. The vessels carry 15 to 25 tons of ice each in the pens, whence it is removed as the fish are stored. Immediately on landing on the vessel's deck the fish are split from the pectoral fin to the vent, the viscera removed and the stomach cavities washed thoroughly. At the bottom of each pen is placed a layer of block ice, 6 or 8 inches thick, covered with a thin layer of chiseled ice. On this is placed a layer of bluefish, backs upward, and inclined slightly on the side, so that all moisture may run from the stomach cavity. This is succeeded by alternate layers of chiseled ice and bluefish until the pen is nearly full, the whole being covered with a layer of crushed ice.

ICING HALIBUT.

The fresh-halibut industry is one of the most extensive branches of fishery trade depending almost entirely on the use of ice. On the Atlantic coast alone about 10,000,000 pounds of fresh halibut are handled annually, and there is also a considerable trade on the Pacific coast. In the early years of the halibut fishery, on Georges and other neighboring banks, no ice was used, the fish being simply eviscerated and placed in heaps in the hold. Later, in order to prevent a bruised and compressed appearance in those in the lower part of the heap, the fishermen suspended the fish by the tails in the hold, this being practiced as late as 1846 in tight-bottomed vessels.

From 1835 to 1850 well-smacks were employed, and though they were quite popular with New London and Greenport fishermen, they were not generally used on the Massachusetts coast. About 1840 a small quantity of ice was generally carried for icing the fish accidentally killed, and with the development of the trawl fishery the use of ice became general, both on smacks and on the tight-bottom vessels. At first the ice was not placed among the fish, being carried for the purpose of cooling the hold, but about 1846 it became customary to crush it and mix it with the fish.

The New England halibut vessels now usually carry from 15 to 30 tons of ice stored in pens in the hold of the vessel, similar to those used in the haddock fishery, and the present method of handling and icing the halibut is as follows: In dressing, each halibut is grasped by the gills with the left hand, the head raised from the deck, and with quick strokes of the knife the gills are separated from the head and from the napes. The fish is then ripped down the belly, and the gills and viscera are removed with the left hand. A second operator takes the halibut and with his bare right hand removes the ovaries from their cavities and the blood from the backbone, pressing his thumb along each side of the backbone to express the blood therefrom. The fish is then passed to the scrub gang, composed usually of three men. One of these, hooking an iron gaff into the head of the fish and another gaff into one side of the nape, holds the fish up and open, while a second workman, with a hickory or oak broom, scrubs off all loose blood, slime, etc., from the spawn cavities and the backbone. During this process a third man souses water into the fish, completely rinsing it out.

The halibut is next passed into the vessel's hold, which is divided by permanent compartments into 10 or 12 pens or bins, half on either side of the vessel, with an alleyway in the middle. A layer of ice is placed in the bottom of the bin, and on this is placed the halibut in rows or tiers, with heads toward the front and back of the bins and tails overlapping in the middle. The abdominal cavity of each fish is filled with finely chiseled ice and the fish so placed in the bin that the ice will not spill from the cavity. No ice is placed between the fish, as in case of fresh cod and bluefish, because of its tendency to bruise the sides of the halibut. As each bin is filled, slide boards are placed at the front and a quantity of ice is put about the heads of the fish, both in the back and front of the bin, and on top of the fish is a layer 6 to 12 inches deep. The amount used varies, of course, according to the temperature, much more being required during warm weather than when the temperature is low. When packed in this manner, halibut will keep in good condition for three or four weeks.

On arriving in port the halibut are removed from the hold of the vessel to the fish-house and assorted, the "white," "gray," and "seconds" or "poor" halibut being kept separate. The "white" halibut are those having the under side pure white, the "gray" are more or less tinged with gray or drab in the same place, while the "poor" halibut or "seconds" are such as are slightly tainted in the vicinity of the abdominal cavity. The "white" halibut ranges from 3 to 15 cents per pound out of the vessel, the "gray" halibut is generally about two-thirds that allowed for white, and "poor" halibut sells for about 1 cent per pound. The distinction between "white" and "gray" halibut was made first in 1848, and while fishermen receive much less for the latter, little distinction is made in the retail trade between the two, both selling at nearly the same price, and it is impossible to distinguish them separately when cooked. After assortment the fish are weighed and 14 per cent is deducted as the weight of the heads to obtain the basis for settlement with the fishermen. The heads are cut off and sold to oil and fertilizer factories for about $1 per 100 pounds. The fish are then placed mostly in stout pine boxes containing each about 450 pounds, and also in 100-pound boxes and 200-pound barrels. In case a halibut is too long to go in the box the tail may be cut off; otherwise the tail and fins remain on. As the fish are placed in the box, the abdominal cavities are filled with crushed ice, and in summer time the fish are surrounded with chiseled ice. The cover is nailed down, the box and its contents weighed and shipped to its destination.*

ICE IN FRESH-MACKEREL FISHERY.

It is customary for vessels engaging in the fresh-mackerel fishery, especially in the spring southern fishery, to carry from 5 to 20 tons of ice, according to the size of the vessel and the time of the year. Usually from June 1 to September 30 the fish are gibbed, the gills and viscera being drawn out together through the gill-openings, but during the spring and fall the fish are iced round. The fish are stowed away as soon as practicable after being caught, being packed in bins in the hold of the vessel similar to those in the fresh cod and haddock vessels, except that each bin is divided horizontally by movable platforms into two or three parts or shelves to prevent crushing the lower layers of fish, as would be the case were they the full depth of the haddock bins. A thick layer of ice is placed in the bottom of the bin, and this is followed by alternate layers of fish and fine ice, the topmost layer being of ice.

The extent of icing depends, of course, on the length of time that will probably elapse before the fish are placed on the market. Ice-grinding machines are no longer carried on any fishing vessels, since they take up much room and the ice may be chiseled more quickly than ground. Those mackerel caught when the hold is full or just before leaving the fishing-ground for market are sometimes placed with crushed ice in barrels on the deck of the vessel. Each vessel also carries barrels for salting such fish as are not to be carried to market fresh.

Vessels engaged in the salt-mackerel fishery occasionally take fresh mackerel into port by placing them, after being gibbed, in barrels of water, but this is practiced only to a limited extent.

ICING SHAD.

Few species of fish show greater increase in selling price as the result of careful handling and icing than the shad, yet in the marketing of few valuable species is greater carelessness shown by the average shipper. It frequently happens that the fish immediately from the water are carelessly placed in the shipping boxes or barrels with insufficient ice, much of which is melted in cooling the fish before the package starts on its journey. The package then being not quite full, the contents move from their respective positions in handling, resulting in loosening the scales and bruising the surface of the fish considerably. The shad trade is very large, the quantity marketed annually on the Atlantic coast of the United States approximating 11,000,000 in number, or 52,500,000 pounds, nearly all of which are used fresh.

The following is the best method of handling fresh shad: If practicable the fish-house should be raised 2 or 3 feet above the water or the shore, so that the wind may freely circulate and cool the floor. The fish ought to be handled carefully, bruises

---

* See the Fishery Industries of the United States, sec. v, vol. 1, pp. 21-22.

being avoided as much as practicable, protected from the heat of the sun, laid on the floor as received from the boats, the heads of each row resting on the tails of each preceding row, and a layer of crushed ice 1 or 2 inches thick spread over the whole. There the fish remain until it is necessary to place them in the shipping packages, which may be boxes or ordinary sugar-barrels. The former are preferred if satisfactory shipping rates are secured, but because of more favorable shipping rates barrels are more frequently employed. A layer of crushed ice is put in the bottom, on which the shad are placed on their backs, with the heads at the ends if boxes are used. This is followed by a layer of fine ice, succeeded by similar layers of fish and ice, with a double quantity of ice at the top, the boxes being covered with boards securely nailed and the barrels with bagging fastened under the top hoop.

ICING OYSTERS.

Ice is very generally employed in preserving and shipping shucked oysters to the retail trade. During cold weather only a small quantity is used, but the ice bill of the oyster-dealers throughout the season is a considerable item. In using ice for this purpose it was formerly considered important to keep it separate from the oysters—generally by having a separate ice chamber in the package or keg containing the oysters. In some packages the chamber occupied one side of the package, or it filled a space entirely surrounding the oysters, but usually the ice was in the central part, the oysters filling the annular space about the chamber. A variety of packages have been invented for this purpose, differing particularly in the manner of affording separate access to the two chambers. Many of these have been patented, and for further description of them reference may be made to the following United States letters patent:

No. of patent.	Date.	In favor of—	No. of patent.	Date.	In favor of—
103551	May 31, 1870	Alfred Booth, Chicago, Ill.	240281	Apr. 19, 1881	Alvin Squires, Hartford, Conn.
111722	Feb. 14, 1871	Do.	250107	Nov. 29, 1881	C. A. Sheridan, Detroit, Mich.
135083	Jan. 21, 1873	L. R. Comstock, Baltimore, Md.	255017	Mar. 14, 1882	C. L. Pond, Buffalo, N. Y.
142704	Sept. 9, 1873	James C. Jones, New York.	265137	Sept. 26 1882	Do.
158089	Dec. 22, 1874	Do.	300061	June 10, 1884	S. L. Frazer, Toledo, Ohio.
190353	May 1, 1877	Do.	300476	June 17, 1884	O. P. Johnson, Washington, D. C.
199569	Jan. 22, 1878	James d. Phillips, Norfolk, Va.	374119	Nov. 29, 1887	M. U. Dotson, Baltimore, Md.
209189	Oct. 22, 1878	Frank Pfeiffer, Norfolk, Va.	405488	June 18, 1889	John P. Kuhn, Alton, Ill.
240143	Apr. 12, 1881	O. P. Johnson, Cambridge, Md.	438391	Oct. 14, 1890	John T. Store, Baltimore, Md.

A refrigerator oyster-shipping package, patented by a leading oyster-dealer, and formerly used to a considerable extent, is constructed as follows:

A sheet-metal can is employed with a capacity of 20 gallons, flanged edges projecting from the top ends of the can. In the center of the can, extending from top to bottom, is formed a rectangular ice chamber which is opened at the top and has four sides exposed within the body and to the contents of the can. This chamber is closed by means of a wooden cover of suitable dimensions to fit snugly within the flanged edges of the end of the can over the opening into the ice chamber and against the screw-cap, and projecting a little beyond the ends of the flanges, and secured thereto so as to be easily removed. At the opposite end of the can a similar wooden cover or guard is fastened, these covers also serving to protect the ends of the can from injury during transportation. The can is placed in a wooden case or enveloped with wooden covering to protect it during shipment and to more effectually exclude heat from the contents of the can. The oysters are placed in the annular space about the ice chamber, this space holding about 15 gallons; the ice chamber is filled with ice and the covers fastened, when it is ready for transportation. (See Letters Patent No. 111722, dated February 14, 1871.)

The use of these intricate forms of shipping packages was abandoned several years ago, and at present the trade throughout the country uses ordinary package-tubs in various sizes, ranging in capacity from 1 to 10 gallons, the tubs being returned to the wholesalers as soon as the oysters are sold. The covers are loosely fitted on top and fastened by tacking small tin clasps to the tops and sides. Handles are provided at the sides of the tubs for convenience in transfer. The oysters are chilled with chunks of ice before being packed for shipment, and when placed in the shipping-tub a block of suitable size to last through the journey is added. During cold weather ice is sometimes omitted, but it is poor economy to stint in its use. The size of the tubs should be adapted to the quantity of oysters shipped, so that each tub may be quite full, to prevent the agitation or slushing of the oysters. In order to prevent the cutting and bruising of the oysters by the block of ice while the tub is being handled and in transit, a flexible pocket of cotton cloth, muslin, or other cheap texture is occasionally used by some shippers, the block of ice, of a size suited to the oyster package, being placed in this pocket and the whole suspended rigidly from the sides of the tub.*

The oyster tubs are generally shipped in refrigerator cars, these leaving the principal oyster markets regularly during certain days of the week. When the weather is very cold, the refrigerator car is a protection against the oysters freezing. Fat oysters will not freeze as quickly as thin ones, as the latter contain more water. But freezing does not greatly injure shucked oysters when mixed with their own liquor, provided they are consumed soon after thawing.

In many localities, especially along the Gulf coast and through the West, a practice prevails of shipping opened oysters in hermetically sealed square cans, containing from 25 to 100 oysters, these cans being then placed in boxes with the tops and sometimes the sides in contact with crushed ice. This method is not so general as it was several years ago, on account of the extra expense incurred, and the condition of the oysters shipped in bulk is generally about as satisfactory.

REFRIGERATOR CARS.

The large inland trade in fresh fish and the liability of frozen fish to rapid decay when subjected to a higher temperature have resulted in an extensive use of refrigerator cars for transportation purposes. The refrigerator car is little more than 30 years old, the first American patent being issued on November 26, 1867, to J. B. Sutherland. His claim covered a car with double walls, roof, and floor, with ice chests at each extremity, closed by hanging flaps, and having spaces so arranged as to produce a constant circulation of air in the car. The air was admitted at the top of the car and passed down through the ice chamber, and entered the room near the bottom at a low temperature. In March, 1868, George K. Wood, of Morristown, N. J., brought out a car with a plurality of metallic chambers for the respective reception of way and through freight, with an ice chamber above; while the car of A. L. McCrea, of Chicago (March, 1869), had interior movable sections. Numerous other patents followed in quick succession, scarcely any of which embody the features of those now in extensive use.

The following description of one of the most practical of the forms of refrigerator car in general use at present gives an idea of their construction:

The ice and salt receptacles are four galvanized-iron cans strongly jacketed at each end of the car, extending from the roof to within 6 or 8 inches of the floor, and under them is a pan to catch the drip, the overflow escaping through an air-tight trap. The walls, roof, and bottom are 7 inches

---

* See Letters Patent No. 438391, dated October 14, 1890.

in thickness, made with a dead-air space and three 1-inch layers of hair felt, the joints of the doors being padded. No air enters the car when closed, and there is no provision for circulation of air within the cooling chamber. Each car carries about 6,000 pounds of crushed ice mixed with about 600 pounds of fine rock salt, which is entered at the top and tapped down in the cans, after which the covers are put on and the roof holes closed. In eight or ten hours the receiving room of the car has become chilled, when additional ice and salt is added and the car is ready for the freight.

The longest transportation of fresh fish in this country is the sending of salmon from Columbia River to the Atlantic coast, requiring five or six days. The methods are thus described by Messrs. Senfert Bros. Co., of The Dalles, Oregon:

We ship all our fresh salmon by express for New York and all points east in boxes 44 inches long, 20 inches wide, and 12 inches deep. We put in each box 175 pounds of fish undressed, or just as they leave the water, and 75 pounds of crushed ice in each box. The express company refills these boxes daily at certain icing stations along the line, and makes no extra charge on these icings, that being all included in the express charge of 8½ cents per pound on the net weight of the fish to Chicago, or 10½ cents to New York, or 7½ cents to the river or Union depot, Council Bluffs or St. Paul, Minn. In shipping carload lots we put 150 pounds in each box, fill the box with ice, and load 12 tons of fish in a car. We use about 8 tons of ice, and these cars are not opened or re-iced until they reach New York, by passenger train service to Chicago and fast freight from Chicago to New York over Erie Railway, on Wells Fargo express trains, 30 hours' time. These cars reach New York in 5½ days from this river.

The shipment of fresh salmon in carload lots across the continent began in 1884, during which year eight carloads of fresh salmon were sent east, all arriving in good condition. On account of the high rate for freight service in refrigerator cars the profits were so small that further shipments were postponed until a reduction in rates was made in 1890.

FREEZING FISH IN THE OPEN AIR.

In cold countries the freezing of fish in the open air during cold weather is a natural and doubtless one of the oldest forms of preservation. In the northern portions of Europe and America fish are frequently preserved in this manner. Prior to the use of ice in the United States it was not unusual during the winter and early spring for dealers to take fish frozen by natural cold from Boston or New York 200 or 300 miles inland. But the uncertainty of depending on continued cold weather, and the advent of the use of ice and quick transportation, have resulted in an abandonment of that trade.

There is yet a very extensive trade in frozen smelt during the winter, especially in December and January. These fish are frozen in Maine and the British Provinces, boxed and shipped by steamer or rail to Boston or New York, whence they are supplied to the retail trade. During the season of 1897, 82,306 boxes, each holding an average of 25 pounds of smelt, were received in Boston. Most of these come from the British Provinces, being admitted free of duty, and they are sold from 2 to 8 cents per pound, averaging perhaps 4 cents per pound, wholesale.

FROZEN-HERRING INDUSTRY.

The most important industry depending on open-air refrigeration is the freezing of herring on the Newfoundland and New Brunswick coasts for the United States markets. This is scarcely more complicated than the usual method of packing in crushed ice, and not by any means so intricate as the process of mechanical or chemical refrigeration now employed in the large marketing centers of the United States. It originated during the winter of 1854–55, the immediate object being to supply the vessels engaged in the Georges Bank cod fishery with bait. At present those vessels, as well as those employed in the bluefish fishery, depend almost entirely on this form of bait during the winter and early spring, and in addition large quantities of frozen herring are used for food.

The process of freezing is as follows: When the vessel has been moored in some cove convenient to the fishing-ground, the ballast is thrown overboard, the hold sheathed up around the sides with spruce boughs, and a platform built in the bottom of the hold several inches above the keelson. A bulkhead 6 inches thick, with the space in the middle filled with sawdust, is placed across the forward part of the hold to separate it from the forecastle. Sometimes the fish are frozen ashore by the natives, a clear, gravelly beach above high-water mark or a surface of crushed snow and ice being chosen. But usually the freezing is done on a large scaffolding on the deck of the vessel. This scaffolding is generally about 100 feet long by 25 feet wide and is built of rough boards, most of which are purchased at Nova Scotia points on the way to Newfoundland, they being obtainable much cheaper there than in Gloucester. The quantity of fish placed on the scaffold varies according to the weather.

When the temperature is little below the freezing point, the fish must be spread very thin, in order that those underneath may be thoroughly frozen; but, with a lower temperature, the fish can be heaped together to the depth of a foot or more, though in such cases it is necessary to turn them every few hours. A constant watch must be kept to guard against loss from a sudden rise in temperature or a storm of rain or snow. The watch usually turns the fish with a wooden shovel or stirs them with his feet every few hours, and during a snow storm it becomes necessary to work constantly among them to keep them from being covered up. Should the weather become so warm that the fish would be thawed by exposure, it is necessary to place them in piles and cover them with canvas or other material, again exposing them when the temperature has fallen sufficiently. The usual method of ascertaining whether a herring is sufficiently frozen is by breaking. If the fish bends at all it is not thoroughly frozen; but if it breaks short, like a dry stick, it is ready to be stowed in bulk.

The herring are roughly shoveled in the vessel, the hold, and sometimes even the cabin, being filled, the crew in the latter case living in the forecastle on the homeward passage. Formerly the fish were packed in snow, or a considerable quantity of snow was placed around the sides of the hold and the fish heaped together in the middle; but for many years this practice has been wholly abandoned, and it is found that the fish will keep equally well without the use of snow.

This trade at present averages about 25,000,000 herring annually, with a valuation of $300,000. About one-third are used for bait by the Gloucester fishermen, and those remaining are sold for food. The market value varied for many years from 75 cents to $3 per 100 fish wholesale, while the retail price was about double those figures. During the past three years the fish have been sold by weight, two scales being on the vessel, one at the main hatch and the other at the aft hatch; the average wholesale price during those years has been from $1.25 to $2.50 per 100 pounds.

Frozen herring form cheap and wholesome food at a season when other fresh fish are obtained with difficulty and only at a high price. They possess a great advantage over ordinary fresh fish in that they can be packed in barrels without ice and shipped to a considerable distance without danger of loss. With the exception of those sold in Gloucester for bait, nearly all the frozen herring are carried directly to Boston and New York, and two or three cargoes are carried to Philadelphia each season. At these places they are sold locally and packed in barrels for distribution, and sent as far south as Washington, and as far west as the Mississippi River.* After reaching their destination, the great bulk are sold fresh, but some are pickled by the dealers, while others are cured as bloaters or hard herring. After being frozen, herring are not especially desirable for either of these purposes, as they become soft and the flesh is rather dark and unattractive in color

ARTIFICIAL FREEZING AND COLD STORAGE.

The artificial freezing of fish and other food products with their subsequent retention in cold storage is one of the most recent methods of preservation, originating about thirty-five years ago; and while it has acquired considerable importance in certain localities, its practical value is scarcely appreciated by the general public. It is applied in the various marketing centers of the United States and to some extent in the countries of Europe and South America. Its greatest development and most extensive application in the fisheries exists along the Great Lakes in freezing whitefish, trout, herring, pike, etc., about 3,500 tons of which are frozen each year. On the Atlantic coast of the United States it is used in preserving bluefish, squeteague, mackerel, smelt, sturgeon, herring, etc., the trade in these frozen fish "tailing on" or immediately following the season for fresh or green fish. On the Pacific coast large quantities of salmon and sturgeon are frozen and held in cold storage until shipped, the trade extending to all parts of America and northern Europe. At various points throughout the interior of the country there are cold-storage houses where fishery products are held awaiting demand from consumers. In Europe there is comparatively little freezing of fish, although the process is applied very extensively to preserving beef, mutton, and other meat products, and the markets of Hamburg and other continental cities receive annually several million pounds of frozen salmon from the Pacific coast. In England large fish freezers were erected several years ago at Grimsby and Hull, but did not prove successful and were finally dismantled.

By the use of ice alone during warm weather the temperature of fish can never be kept below 32° F. While this low temperature retards decomposition, the fish acquire a musty taste and loss of flavor and eventually spoil. To entirely prevent decomposition the fish must be frozen immediately after capture and then kept at a temperature of several degrees below freezing. The belief held by some persons that freezing destroys the flavor of fish is not well founded, the result depending more on its condition when the cold is applied and the manner of such application than upon the effect of the low temperature. Fish decreases less in value from freezing than meat does, but it is especially subject to two difficulties from which frozen meat is free: first, the eye dries up and loses its shining appearance after a very long exposure to cold, and second, the skin, being less elastic than the texture of the fish, gets hard and becomes somewhat loose on the flesh. Frozen fish is not less wholesome than fish not so preserved. The chemical constituents are identical, except that the latter may contain more water, but the water derived from ingested fish has no greater food value than water taken as such. The principal objection to this form of preservation is the tendency to freeze fish in which decomposition has already set in, and the prosperity of

* See Fishery Industries of the United States, sec. V, vol. 1, p. 451-456.

the frozen-fish business requires that any attempt to freeze fish already slightly tainted should be discountenanced. When properly frozen and held for a reasonable period, the natural flavor of fish is not seriously affected and the market value approximates that of fish freshly caught. The process is of very great value to the fishermen supplying the fresh-fish trade, since it prevents a glut on the market, and it is also of benefit to the consumer in enabling him to obtain almost any variety of fish in an approximately fresh condition throughout the year.

DEVELOPMENT OF COLD STORAGE.

The first practical device for the freezing and subsequent cold storage of fish was invented by Enoch Piper, of Camden, Me., to whom a patent* was issued in 1861. His process was based on the well-known fact that a composition of ice and salt produces a much lower temperature than ice alone, this knowledge having been applied for an indefinite period in freezing ice creams, etc.

The following is a description of Piper's apparatus and its application:

The fish were placed on a rack, in a box or room having double sides filled with charcoal or other nonconducting material. Metallic pans containing ice and salt were set over the fish and the whole inclosed. The temperature in the room would soon fall to several degrees below the freezing point of water, and in about 24 hours (the mixture being changed once in 12 hours) the fish would be thoroughly frozen. The fish were then covered with a coating of ice by immersing them a few times in ice-cold water, or by applying the water with a brush, forming a coating about one-eighth of an inch in thickness. After the coating of ice was formed the fish were sometimes wrapped in cloth and a second coating of ice applied. In some instances they were covered with a material somewhat like gutta percha, concerning which much secrecy was exercised. The fish were then packed closely in another room, well insulated against the entrance of warmth, by means of double walls filled with some nonconducting material. Fixed perpendicularly in the second room were a number of metallic tubes, several inches in diameter, filled with a mixture of ice and salt to keep the temperature below the freezing point.

The process was also patented in the Dominion of Canada, and a plant was established near Bathurst, New Brunswick, in 1865, the output consisting almost entirely of salmon, a large proportion of which were imported into the United States. In order to hold the frozen fish in New York while awaiting a market, Piper constructed a storage room in a shop on Beekman street, that being the first cold-storage room for fish in the United States. The walls of the room were well insulated, and around the sides were two rows of zinc cylinders, 10 inches in diameter at the top and decreasing in size toward the bottom, connecting at the lower end with a drainage pipe. The cylinders were filled with a mixture of ice and salt, which was renewed whenever necessary. Whatever may have been the imperfections in his process of freezing, the system of storage was quite satisfactory and differs little from that in use at the present time. Piper refused to sell rights to others for the use of his process, and after maintaining a monopoly of the business for three or four years his exclusive right to it was successfully contested by other fish-dealers in New York, who applied it to storing other fish besides salmon.

The principal objection to Piper's process is that the fish are not in contact with the freezing mixture during the operation of freezing, and, consequently, too much time is required for them to become thoroughly frozen. Several devices have been used for overcoming this objection, among which are covering the fish with thin sheet rubber or other waterproof material, and packing them in the mixture of ice and salt.

---

* No. 31736, dated March 19, 1861.

The greatest improvement, and the one used almost exclusively when ice and salt form the freezing agency, originated in 1868 with Mr. William Davis, of Detroit, Mich.,* the description being as follows:

> Two thin sheet-metal pans, or a box in two parts, are made one to slide over the other, the object being to place the fish in one pan, slide the other pan vertically over it, and the box is then placed in direct contact with the freezing mixture. By having the box constructed in this manner, it is capable of being expanded or contracted to accommodate the size of whatever may be placed therein, and the top and bottom always be in contact with the articles to be frozen. After the fish are inclosed in the pans, the latter are placed in alternate layers with layers of the freezing mixture between and about them. When the fish are thoroughly frozen they are removed from the freezing pans and placed in a cold-storage chamber in which the temperature is kept 12 or 14 degrees below the freezing point.

Another arrangement for bringing the fish in contact with the ice was devised and patented by a fish-dealer of Toledo, Ohio,† as follows:

> A chest is so constructed that cells or compartments of various sizes may be formed in it for the reception of articles to be frozen, this being accomplished by having the walls of each compartment movable and each separated from the other by blocks of sufficient size to closely contain the articles placed therein, usually from 2 to 6 inches, and the freezing being effected by filling the spaces between the cell walls with salt and crushed ice. The sides of the cells or compartments are made of thin metallic plates, so that the upper end of each pair may be brought together, forming a closed compartment, to prevent the entrance of the refrigerating mixture into the fish chambers while filling the salt and ice compartments. To freeze a quantity of fish by this method, they are placed in the open cells or compartments, head downward, as is most convenient. When each cell is full, the top edges of the sides are brought together and hold in that position by clamps. The spaces between the cells, varying from 2 to 4 inches across, are then filled with the freezing mixture and the cover of the chest shut down. The cold produced passes through the wall separating the mixture from the fish and quickly freezes the latter. Provision is made for permitting the water formed by the melting ice to flow in grooves to suitable outlets, the bottom pieces of the cells keeping the fish dry. When the fish are frozen, the cover of the chest is raised, the contents removed, and the fish transferred to the storage freezer.

This method of freezing was used only a few years, the pan process being found much more practicable.

With a view to bringing the fish more closely in contact with the freezing mixture, some dealers at first used thin sheet-rubber bags, or other waterproof sacks, in which the individual fish were placed and then surrounded by the salt and ice, thus exposing the entire surface of the fish, resulting in much more rapid freezing. When the fish were frozen the sacks were removed and were dipped in water, which thawed them sufficiently to permit the fish to be withdrawn. This method was abandoned after a year or two.

To facilitate the shipment of frozen fish in barrels, Messrs D. W. & S. H. Davis, of Detroit, Mich., introduced‡ a process by which the fish are frozen in circular pans of varying sizes suited to the measurements of the barrels. After being frozen, the contents of each pan are removed entire and placed in appropriate position in the barrel, and the barrel headed and placed in cold storage.

In 1877 C. W. Gauthier patented§ a modification of the preceding method, using thin pliable partitions in the circular pans for separating each fish from its neighbor, so that the individual fish in each pan may be packed in the barrel separately. None of the last four processes has ever been used to any considerable extent.

* Letters patent No. 85913, January 19, 1869.
† Letters patent No. 109820, December 6, 1870.
‡ Letters patent No. 165596, dated April 6, 1875.
§ Letters patent No. 187122, dated February 6, 1877.

As the trade developed, the size of the storage rooms was increased and improvements were adopted in the arrangement and form of the ice-and-salt receptacles and in the method of handling the fish. But the freezing with pans immersed in ice and salt, as in the Davis process, and the subsequent storing of them in the manner used by Piper, continued without any great modification until the introduction of ammonia freezers into the fishery trade in 1892. At that time ice-and salt freezers and storage rooms existed at nearly all the fishing ports on the Great Lakes; eight or ten small ones were in New York City, and several were in use on the New England coast. Some of those on the Great Lakes were very large, with storage capacity of 700 or 800 tons or more, and the aggregate storage capacity of all in the country approximated 8,000 tons. Ammonia cold-storage houses had been established at various places along the coast and in the interior during the ten or fifteen years preceding, and in these some frozen fish had been stored. But the first ammonia establishment for freezing fish exclusively was established at Sandusky, Ohio, in 1892. The method of freezing differs from the former process in that the pans of fish are placed on and between tiers of pipes carrying cold brine or ammonia instead of being immersed in ice and salt. In the storage rooms less difference exists, coils of brine pipes taking the place of the ice-and-salt receptacles, the blocks of fish being removed from the pans and stored as in the older process.

DESCRIPTION OF ICE-AND-SALT FREEZERS.

The outfit of an ice-and-salt freezer consists principally of temporary stalls or bins where the fish are frozen, and insulated rooms where the frozen fish are stored at a low temperature. In addition to these there are ice-houses, salt-bins, freezing-pans, and the various implements for the convenient prosecution of the business. The freezing bins are usually temporary structures within the fish-house, and are generally without insulation. The wall of the fish-house may form the back, while loose boards are fitted in to form the sides and front as the bin is filled, in the manner hereafter described. A better way is to build the bins with sides and back 4 or 5 inches thick, filled with some nonconductor, with double or matched floor and with movable front boards.

The storage rooms are commonly arranged in a series side by side and separated from each other by well-insulated partitions, the capacity of the rooms ranging from 25 to 200 tons each. The outer walls of these rooms, as well as the floors and ceilings, are well insulated, made usually of heavy matched boards, with interior packing of some nonconductor of heat. Among the latter may be mentioned planing-mill shavings, sawdust, pulverized charcoal, chopped straw, slagwool, etc. Most of the walls are 16 or 18 inches thick, filled with planing-mill shavings or sawdust, and in some freezers the damaging effect of rats is obviated by placing linings of cement between the shavings and the board walls. Most of these loose materials have their economic drawbacks, chiefly because of their strong hygroscopic tendency, the material losing its insulating power and decaying, this decay also attacking the wood of the walls. Because of this, many of the storage rooms recently constructed are insulated by having the walls made up of a combination of mineral wool, insulating paper, air spaces, and inch boards.

The sides, and in some cases the ends, of the room are lined with the ice-and-salt receivers, consisting of galvanized sheet-iron tanks, 8 or 10 inches wide at the top, narrowing to 3 or 4 inches at the bottom, and placed about 4 inches from the wall in order to expose their entire surface to the air in the room. These tanks open at the top, which extends above the ceiling so that they may be filled without opening the storage room. At the bottom is usually a galvanized-iron slanting gutter, into which the water resulting from the melting ice flows, whence it is conducted through the floor of the room by a short pipe, protected from the entrance of air at its lower end by a small drop cup, into which the brine falls and runs over at the top. In some fish-houses this brine, which is otherwise wasted, runs into receiving tanks, where it is stored and used as required in pickling fish. The ice-and-salt tanks must be cleaned from time to time in order to rid them of dirt and sawdust. Their capacity should be in proportion to the size of the room and the excellence of the insulation secured, and they should be large enough to render it unnecessary to fill them oftener than once a day, even in the warmest weather.

The appliances used in the ice-and-salt freezers are described at length in the account of the processes of freezing and storage (see pp. 377–384).

While crushed ice and common salt are generally employed as a freezing mixture, numerous other compounds are available. The following compilation gives a number of mixtures that may be employed in refrigeration, the initial point in the case of crushed ice or snow being 32°, and in the other mixtures 50° F. Most of these have as yet been employed only in laboratory practice and for certain special purposes, only a few of them having been applied commercially on a large scale. These formulæ are obtained mostly from Leask's Refrigerating Machinery and its Management, published in London in 1895.

Composition.	Parts.	Minimum tempera- ture.	Composition.	Parts.	Minimum tempera- ture.
		°F.			°F.
Crushed ice or snow	2	} − 5	Sodium sulphate	8	} − 0
Sodium chloride	1		Hydrochloric acid	9	
Crushed ice or snow	5		Sodium sulphate	6	
Sodium chloride	2	} −12	Ammonium nitrate	5	} −40
Ammonium chloride	1		Nitric acid (diluted)	4	
Crushed ice or snow	24		Sodium phosphate	9	} −12
Sodium chloride	10		Nitric acid (diluted)	4	
Ammonium chloride	5	} −18	Snow	3	} −23
Potassium nitrate	5		Sulphuric acid (diluted)	2	
Crushed ice or snow	12		Snow	8	} −27
Sodium chloride	5	} −25	Hydrochloric acid	5	
Ammonium nitrate	5		Snow	7	} −30
Ammonium chloride	5		Nitric acid (diluted)	4	
Potassium nitrate	5	} −10	Snow	4	} −50
Water	16		Calcium chloride, crystallized	3	
Sodium sulphate	5	} − 3	Snow	3	} −51
Sulphuric acid (diluted)	4		Potash	1	

DESCRIPTION OF MECHANICAL FREEZERS.

It is scarcely within the scope of the present paper to enter into a comprehensive description of the numerous systems of mechanical freezers. They are all based on the principle that a liquid passing into a gaseous state, or converted into a vapor, carries away a definite amount of heat from the objects by which it is surrounded.

The compression system is in most general use, and consists of three operations following each other in rotation, and which are practically the same in all refrigerating machines. By means of a large compression pump, anhydrous ammonia, which is the gas usually employed, is compressed to a pressure varying from 125 to 175 pounds to the square inch. During this operation heat is developed according to the amount of the pressure exerted upon the gas or to the relative volume to which it is reduced, and this heat is withdrawn from the compressed gas by forcing it through coils of pipe in contact with cold water, the heat being transferred to the water. The gas is now ready to assume a liquid state, and in so doing transfers additional heat to the water surrounding the pipes. The liquid gas thus obtained is allowed to enter coils of circulating pipe at a pressure much lower than that required for retaining the gas in a liquid state, whereupon it reexpands and extracts from the pipes and the substances surrounding them a quantity of heat equal to that which was previously given up by the gas during the period of condensation and liquefaction. The gas is then drawn from the expansion coils by the pumps at a pressure of 10 to 15 pounds above that of the atmosphere, and is again compressed in the condensing coils at a pressure of 125 to 175 pounds to the square inch, and the same cycle of operations is repeated. Various modifications of the above, as well as auxiliary processes, have been introduced, but the principles are the same in all compression machines, the differences being in their application.

The absorption system, which is comparatively little used at present, is based on the fact that many vapors of low boiling point are readily absorbed by water, but can be separated again by the application of heat to the mixed liquid; and the machinery in an absorption system differs from that in a compression plant principally in the substitution of an absorber for the condenser and in applying heat to the ammonia water to drive off the anhydrous ammonia at a high pressure.

Formerly, in order to avoid danger from leakage of gas through the circulating pipes carrying the cold ammonia, those pipes were not passed through the freezing and storage chambers, but were stored in a large tank surrounded by some liquid whose freezing point is very low, such as salt brine, or, when lower temperature is desired, a solution of chloride of calcium, and this cooled liquid is pumped through pipes circulating in the freezing and storage rooms. The improvements in the manufacture of freezing machinery have resulted in the making of much tighter pipes, so that at present in many freezers the ammonia coils pass directly through the freezing rooms, and in some instances they also pass through the storage rooms, but brine or chloride of calcium circulation is yet preferred for the storage rooms.

In the mechanical freezing-houses there is a machinery room containing the boilers, compression pumps or absorption tank, according to the system employed, brine pump, etc. Apart from these and within well-insulated walls are the cold rooms, of which there are two kinds—one for the freezing of fish and the other for their storage after being frozen, the capacity of the latter being usually much greater than that of the former. In the freezing-room the circulating pipes containing the cooling material are ½ inch to 2 inches in diameter and arranged in shelves or nests with horizontal layers 4 or 5 inches, and sometimes 10 inches, apart, ranging from the floor to the ceiling, the entire room being occupied with these nests, except sufficient space for moving about. These pipes are sometimes made in separate coils, so that if desired the brine may be circulated through only a portion of the pipes, and there is generally a vertical row of pipes on each side of the freezing-room. The temperature depends, of course, on the quantity of green fish and the progress of the freezing process, but with direct circulation, or using brine made of chloride of calcium as the circulatory medium, a temperature of −10° F., or even less, is obtainable. In this room the green fish are frozen, and then removed to the storage rooms.

The storage rooms are constructed similarly to the storage rooms in ice-and-salt freezing houses, the only difference being that circulating pipes are substituted for the ice-and-salt receptacles. The pipes in the storage rooms are usually larger, but are not so numerous as in the freezing-room. They are arranged at the ceiling, and sometimes about the upper sides also.

The freezing and storage rooms have well-insulated walls, ceiling, and floors similar to the storage rooms using ice and salt as a freezing agency. The walls are sometimes 16 or 18 inches thick, filled with sawdust or planer shavings; but usually they are made up of successive layers of boards, paper, mineral wool, and air space. In one of the most recently constructed freezing establishments, that of the Cincinnati Oyster and Fish Company, the walls are constructed as follows: Seven-eighth-inch boards, insulating paper, ⅞ inch boards, 2-inch air space, ⅞-inch boards, two sheets of insulating paper, ⅞-inch boards, 4 inches of mineral wool, ⅞-inch boards, insulating paper, and ⅞-inch beaded boards. In the same establishment the ceiling is insulated by ⅞-inch boards nailed against the joists, two sheets of insulating paper, ⅞-inch boards, 2 inches of mineral wool, ⅞-inch boards, insulating paper, and ⅞-inch boards. On top of the ceiling and between the joists there are 3 inches of mineral wool, ⅞-inch boards, insulating paper, and ⅞-inch boards. The floor is insulated by nailing ⅞ by 3-inch strips between the joists and close to the bottom, on top of which are ⅞-inch boards, insulating paper, and ⅞-inch boards, the whole being pitched throughout so as to make it perfectly air-tight. Then come 2 inches of air space, ⅞-inch boards, insulating paper, ⅞-inch boards, 4 inches of mineral wool, ⅞-inch boards, insulating paper, ⅞-inch boards, 3 inches of concrete, and 1½ inches of cement. Resting on the cement floor are ⅞ by 3 inch oak racks, to permit a free circulation of air under the fish stored in the room.

In 1896 there was erected at Goble, Oregon, a freezing and cold-storage plant differing from any other in the United States, in that cold air is used as the freezing medium. There are similar plants at Montreal and Quebec, and there are several used in Great Britain and Australia for refrigerating beef and mutton.

The following is a description of the Goble establishment:

The building is 100 feet long by 52 feet wide, exclusive of the boiler and engine rooms, which are under another roof adjacent to the main structure. The first floor is well insulated and divided into 8 storage compartments insulated from each other, the dimensions of which are 40 feet in length, 10 feet in width, and 10 feet in height, the floor space within each being occupied by two parallel car tracks with an alleyway between. The floors of these rooms consist of 16 inches of sawdust, 3 pieces of ¾-inch hard felt and 5 air spaces, floored over with 1-inch plank; and the walls have two thicknesses of felt and 4 air spaces. The remaining portion of the width of the building, 8 feet on the inside, is occupied by a corridor, in which is a car track with suitable turntables leading to each track in the storage rooms; and the remaining 9 feet net in the length of the building is taken up by an elevator, stairway, and tool room. On the fourth floor there is a tank, 22 feet long and 6 feet wide, filled with brine cooled by ammonia circulation. In this tank there are 5 disks, 4 inches in diameter, revolving on axles running across the tank. The air from the freezing and cold-storage rooms is collected and forced over the revolving disks as the brine drops off, and is then returned by other channels to the freezing and storage rooms.

When the fish, consisting principally of salmon, are received at the dock, they are washed, wiped dry, and placed on cars fitted up with 7 galvanized-iron shelves, 5 feet 4 inches long and 3 feet wide, the capacity of each car approximating 1,000 pounds, and its cost, with the necessary shelves, about $31. When the shelves are filled, the cars are wheeled into the corridor leading to the freezing rooms, and then to the proper compartment, where, still remaining on the cars, they are frozen by the cold air forced over and among them. Each room has capacity for 7 cars on each track, or 14 cars in all, with 7 tons of fish, giving an aggregate freezing capacity in the 8 rooms of 56 tons at one time. The air introduced in these rooms has a temperature of 5° to 10° F., and is changed every 2 or 3 minutes. Freezing can be accomplished in 12 to 14 hours, but ordinarily from 14 to 24 hours are taken for the operation to be completed, depending somewhat on the size of the fish. After the fish are frozen, the cars are removed to an elevator and run up to the second floor, where the fish are glazed, being dipped in water, and thus covered with a coating of ice about one sixteenth of an inch thick. This glazing is effected by transferring the fish to racks and then dipping them in a tank 20 feet long by 4 feet wide, partly filled with water at a temperature several degrees below freezing. The fish are then usually wrapped in paper and neatly packed in paper-lined boxes, usually of 250 pounds capacity, ready for shipment, and transferred to storage rooms on the third floor, having a capacity for about 800 tons.

In addition to the machinery above described, there are two 80-horsepower boilers and two 60-horsepower engines, with a fire pump capable of delivering 18,000 gallons of water per hour, an ice plant with capacity of 20 tons a day, and an auxiliary one with capacity of 3 tons daily. The cost of the plant is said to approximate $45,000. The machinery employees number 4, and usually there are 20 men employed in handling the fish.

PROCESS OF FREEZING AND COLD STORAGE.

In freezing fish, as well as in preserving most food products, the superintendent must give close attention to economy of the process as well as to the excellence of the product, and the costliness of the best process sometimes prevents its use. To secure the very best result, the stock to be frozen should be perfectly fresh and free from bruises and blood marks. It improves the appearance, and therefore increases the value, if the fish are graded according to size, but that is rarely done. All kinds of fish keep and look best when frozen just as they come from the water, with heads on and entrails in, and it is better that the fish be not eviscerated before freezing, except in case of very large fish, such as sturgeon. But since the freezers receive the surplus from the fresh-fish trade, many have been already split and dressed. Generally fish that are frozen with heads off and viscera removed are not strictly fresh; but this rule has many exceptions.

Whether round or eviscerated, the fish are first washed by dumping them into a wash box or trough containing fresh cold water, which is frequently renewed, and stirring them about with an oar-shaped paddle or cloth swab to remove the slime, blood, etc. Some freezers consider it inadvisable to wash flatfish, because of their being too thin. From the wash box the fish are removed by hand and placed in the pans, or, better, they are removed with a dip net and deposited in trays situated on a pan-filling bench. The bench in use in most of the freezing-houses of this country is 12 or 15 feet in length and about 2 feet wide, and at intervals of about 3 feet or so there are square trays 3 or 4 inches deep, with lattice bottoms for drainage, in which the fish are deposited from the wash tank. On either side of each fish-tray is room for a fish-pan, at which stands an operative engaged in filling pans.

The pans are of various dimensions adapted to the size of the fish usually received. In most houses they are about 26 or 28 inches long by 16 or 18 inches wide and 2 or 3 inches deep, with capacity for about 40 pounds of fish, the material used being generally No. 24 Juniata galvanized iron, with the corners turned down, riveted and soldered. In some houses much smaller pans are used, the smallest observed by the writer being 16 inches long by 8 inches wide. Generally at each corner in the bottom of the pans there is a small round hole for drainage purposes, but some pans are made water tight. Each kind has its advantages. In the tight pan the water remaining on the fish from the washing-tank accumulates in the bottom and adds so much to the material to be frozen, but it also serves to hold the fish together in the frozen block. In some freezing-houses using tight-bottom pans the weight of the ice in the bottom of the frozen block amounts to more than 5 per cent of the total weight. This, of course, increases the weight of the frozen fish when they are sold. An erroneous idea prevails, to some extent, that in using ice and salt for freezing, it is necessary to use tight-bottomed pans to exclude the brine.

The fish are generally placed so as to make a neat and compact package entirely filling the pan, so that the cover will come in contact with the upper surface of the fish. It is desirable to have the backs of the fish at the sides of the pan and the heads at the ends, so as to protect the blocks in handling. It is also desirable, when the size of the fish so admits and a cover is used, that the bellies be placed upward, since that portion has greater tendency to decompose, and, as the cold passes down, this arrangement results in freezing the upper portion of the block first, and also in less compression of the soft portion of the fish by removing the weight therefrom. This practice, however, is not by any means uniform. In case the fish have been split and eviscerated it is advisable to place them slanting on the sides, but with backs up, so as to permit the moisture to run from the stomach cavity, but that is not the general practice. Large fish are necessarily placed on their sides, the fish being curved, if necessary, so as to lie in the pan best. Some freezers place herring and other small fish on their sides two layers deep in the pans, while others place a bottom layer of three transverse rows, the end rows with the heads to the edge of the pan, and a top layer of two transverse rows laid in the two depressions formed between the bottom rows. In case of pike and some other dry fish a small quantity of water is sprinkled over them, since they do not ordinarily retain sufficient moisture to hold together when frozen, as is the case with most other species.

Formerly all pans were provided with covers, as described in Davis's letters patent, and this is necessarily so at present, where ice and salt are used for freezing, the cover being required to separate the freezing materials from the fish. These covers are slightly larger than the pans, so as to slip on easily. The cover best adapted to a pan 26 inches by 14 inches by 2 inches is 26½ inches by 14½ inches by 1¾ inches, with the sides slanting toward the base. But in some houses, where circulating brine or ammonia is the freezing medium, the covers are being discarded, resulting in a more rapid freezing of the fish, as the cold does not have to pass through the metallic cover. But in that case the top of the block of fish does not present so smooth an appearance as when the cover is used, for the latter presses the fish down somewhat and unites them more closely, making a firmer package. In order to make a compact block those houses not using covers usually place the small fish bellies down. Only a few freezing houses have discarded the use of the pan covers altogether, and in the more advanced freezers covers are used when the fish can be placed so as to come in contact with the cover, otherwise they are discarded. In many sharp freezers, including the one shown in the illustration opposite page 374, the greater portion of the pans have covers, while the others have none.

As soon as the pans are filled and the covers fitted on they are placed in the sharp freezer. In houses where circulating ammonia or brine is used the sharp freezer consists of a series of coils of small circulating pipes, through which the freezing medium passes, on which the pans of fish are placed, the whole being inclosed in a room of suitable size provided with insulated walls and with doors Where ice and salt are used, as in most of the freezing-houses, the sharp freezer is usually a stall or bin, open in front and sometimes on the sides, the front and sides being built up with loose boards as the bin is filled.

The arrangement of the ice, salt, and fish-pans in the bin is as follows: The ice, after being passed through a grinder, where it is crushed into small particles, is mixed with salt in the proportion of from 8 to 16 pounds of salt to 100 pounds of ice. The mixing is conveniently done by scattering salt over each shovelful of ice as the ice is shoveled from the grinder to the handbarrow. Many varieties of salt are used, most houses preferring a coarse mined salt because of its cheapness. Others use finer salt because of its coming in close contact with the ice and resulting in a lower degree of cold and the more rapid freezing of the fish, although the salt does not last so long.

The amount of ice and salt required in freezing a given quantity of fish depends principally on the fineness of the materials and the proportion in which they are used, and to a less extent on the insulation of the freezing-bin, the amount of moisture in the atmosphere, and the size of the pans and the manner in which the fish are placed therein. The finer the ice and salt the quicker the freezing and the exhaustion of their strength. A larger proportion of salt also results in quicker freezing. The most economical quantities appear to be about 85 pounds of salt and 1,000 pounds of ice to each 1,000 pounds of fish, although some freezers use much more salt and less ice. Much larger quantities of ice and salt are required during warm weather, and also more is necessary when the atmosphere is moist than when it is dry. Some of the ice and salt generally remains unmelted and this may be used over again in connection with fresh materials, additional salt being mixed with it, and as it is weaker than new ice it should be used mainly at or near the bottom of the pile, the top of the pile taking care of the bottom since the cold descends.

In making the freezing pile an even layer of ice and salt, about 3 or 4 inches deep, is placed at the bottom, on which is laid a tier or layer of pans filled with fish, about 3 inches of ice space intervening between the pans and the sides of the bin. This is followed successively by a layer of ice and salt about 2 or 3 inches deep and a layer of pans, the surface of each layer of ice being made even and smooth by means of a straight edge. Sideboards are placed as the height of the pile requires, and a wide board laid on the pile furnishes a walk for the workmen in placing the freezing mixture and the pans. Some freezers place the pans in double tiers between the layers of ice and salt, and in this case the thickness of the layers of freezing material must be increased. In some freezers a light sprinkling of salt is thrown on top of the pans before the freezing mixture is applied. The pile is built up as high as it is convenient for handling the pans of fish and the ice and salt, which usually does not exceed 6 feet. A double quantity of the freezing material is put on top, and the whole should be covered with wood or canvass to exclude the air. The freezing is usually completed in about 15 or 18 hours, but the fish usually remain one day, when they are ready to be placed in cold storage. On one occasion, at a freezing-house in Cleveland, 2,200 pans of herring, each pan containing about 19 pounds of fish, or a total of 41,800 pounds, were filled and placed in the freezing-bin in 14 hours. Twenty-seven men were required, at a cost of 15 cents per hour each, making the total cost of labor $56.70, or nearly 14 cents per 100 pounds.

In the sharp freezer the fish, being moist, are frozen solidly to each other and to the surfaces of the pans. To remove them from the pan the latter is usually passed for a moment through cold water, which draws the frost sufficiently from the iron to allow the fish to be removed in a block without breaking apart. In one or two freezing-houses the thawing of the fish from the sides of the pan is omitted, the cover being loosened and the block of fish removed by hitting the pan at the ends and sides. In several houses each pan of fish is dipped for a moment in cold water, when the top is lifted off. This is usually the case in ammonia freezing houses when the fish are removed from the pans in a cold room where running water would be objectionable.

In most of the houses use is made of a sprinkling trough or tank, $2\frac{1}{2}$ or 3 feet wide and 8 or 10 feet long, with two parallel iron bearings on inclined scantling 6 or 8 inches apart on the inside, on which the pans may slide from one end of the tank to the other. Resting on top and near each end of this trough is a sprinkling box about 36 inches long, 18 inches wide, and 3 inches deep, which usually consists of a box or a block of wood hollowed out from the under surface and with a sheet of metal perforated with many small holes tacked on the bottom. At one end of the box or block is a 1-inch auger-hole, into which the end of a hose may be entered, sending a stream of water into the sprinkling box and through the perforated metal bottom, falling into the trough and overflowing at the lower end.

Some houses substitute a 2-inch pipe with perforated under surface in place of the metal bottom to the box.

The pans of frozen fish are successively placed in the trough under the first sprinkling box, where the water falling through thaws the top sufficiently so that a workman standing at the middle of the trough may remove the cover, and, turning the pan over, he permits it to slide under the second sprinkling box, where the descending stream of water thaws the bottom sufficiently for a workman at the end of the trough to lift it from the block of fish, which remains intact and is removed from the trough and placed on a truck or other conveyance for transfer to the storage room. Three men are required at the trough, one to place the pans under the first sprinkling box, the second to remove the top and turn the pan over and to pass it under the second sprinkling box, and the third to remove the bottom of the pan from the fish and place the block of fish on the carrying truck. In order to avoid thawing the surface of the fish, the water used must be cold and the pans are placed in the trough rapidly, taking but a few moments for the removal of the blocks of fish at the other end. In removing the fish from the pans 8 men are usually required—2 to remove the pans from the sharp freezer and carry them to the sprinkler, 3 at the sprinkler, 1 passing the truck to storage room, 1 handling the blocks of fish in the storage room, and 1 placing them in piles. To sprinkle, unpan, and store 40,000 pounds of frozen fish requires such a force about $3\frac{1}{2}$ hours, and the cost of labor approximates $4.20.

In passing through the trough considerable moisture adheres to the fish, which is frozen by the surplus cold, forming a coating of ice about $\frac{1}{16}$ inch thick, entirely surrounding the irregular block of fish. The process of freezing dries the fish to some extent, the loss in weight amounting to about 2 per cent, but the ice coating placed on them adds about 4 per cent to the weight. Some freezing-houses, in order to make the coating of ice thicker, pass the block of fish, on its removal from the sprinkling trough, through a second trough nearly filled with ice-cold water, which has suspended in it a long box with perforated bottom filled with crushed ice. The blocks of fish pass through a channel in the trough underneath the ice-box, coming out at the other end with a coating of ice $\frac{1}{2}$ inch or more in thickness. In houses where sprinkling boxes are not used the fish are dipped by hand in a tank of water after removal from the pans. In some houses the frozen blocks of fish after removal from the pans are dipped, then cross-piled in the cold-storage rooms, and on the following day, or even the second day after, are again dipped in cold water in order to form a thicker coating.

After the blocks of fish are coated with ice they are passed to the cold-storage room, where they are ranged in neat piles, the blocks being placed vertically in some houses, but more frequently they are ranged horizontally in piles extending from the floor nearly to the ceiling. Strips 2 or 3 inches thick are laid on the floor to keep the fish slightly elevated and allow the cold air to circulate underneath.

Care must be exercised in piling the frozen blocks, lest the piles sag and tumble down. When the room is lofty, to avoid heavy pressure on the lower blocks, a platform or floor is arranged about one-half the height of the ceiling, on which the upper blocks of fish rest. A better way of storing the fish is to pack them in boxes, 3 or 4 blocks to the box, and place these in the storage room. While placing the fish in storage, care must be taken to avoid raising the temperature of the room by the admission of warm air. This is usually accomplished by keeping the door closed as much as possible and in some cases by arranging a woolen flap over the entrance to prevent the admission of a current of air, or by having in the door to the storage room an opening just large enough to permit the passage of the packages of fish.

The quantity of ice and salt required in the establishments which use those materials in the storage rooms is dependent on the outside temperature and the excellence of the wall insulation and is independent of the amount of frozen fish in the room, requiring no more freezing material to keep 50 tons of frozen fish at an even temperature than to keep 2 tons in a room of equal size. With 16-inch or 18-inch walls well insulated, it requires the melting of about 40 pounds of ice per day for each 100 square feet of wall surface when the outside temperature is 60° F. to maintain a temperature of 18° F. inside, this calculation leaving the opening of doors and the cooling of fresh material out of consideration. All calculations as to the quantity of ice used in ice-and-salt freezing are based on the use of natural ice, for artificial ice is rarely used in those freezers. Since artificial ice is usually colder than natural ice, less would be required. The temperature in the storage room should be constant, and about 16° or 18° F. is considered the most economical. Above 20° the fish are likely to turn yellow about the livers, a result generally attributed to the bursting of the "gall."

The storage room should be free from moisture, since the latter offers a favorable place for the settlement and development of micro organisms of all kinds, which tend to mold the fish. To reduce excessive moisture, a pan of unslaked lime, chloride of calcium, or other hygroscopic agency, may be placed in the room, the material being renewed as exhausted. If the storage rooms are very moist, they should be dried out before storing fish in them, this being readily accomplished by using a small gas, coke, or charcoal stove. The storage rooms using ammonia may be dried by passing hot water through the pipes, which of course should, under no circumstances, be done when there are fish in the rooms. In case of mold appearing on the fish it might be well to try spraying them with a solution of formalin, which is a 40 per cent solution of formaldehyde gas in water. The solution, containing 10 parts of formalin and 90 parts of water, should be sprayed over the fish at the first sign of mold.

All fish deteriorate to some extent in cold storage, depreciating both in flavor and firmness. The amount of this decrease is dependent primarily on the condition of the fish before freezing and the care exercised in the process of freezing, and, secondarily, on the length of time they remain in cold storage. The loss in quality during storage is due principally to evaporation, which begins as soon as the fish are placed in storage and increases as the ice coating is sapped from the surface.

Evaporation proceeds at very low temperatures, though not so rapidly as at higher ones; even at a temperature of 0° F. the evaporation during two or three months is considerable. The heavier the ice coating the less the evaporation, but it is almost impracticable to entirely prevent it, and under ordinary conditions it amounts to about 5 per cent in weight in six months, but the loss in quality is greater than the loss in weight.

The method generally adopted of restricting evaporation other than coating with ice is to wrap the fish in waxed or parchment paper and place them in shipping boxes whose length and width are slightly larger than the blocks and deep enough to contain 4 or 5 blocks, or 120 to 150 pounds of fish, the inside of the box being lined with wrapping paper.

A method of largely reducing evaporation was invented and patented in 1880 by Mr. W. B. Davis, of Detroit, Mich., but it is scarcely sufficiently practical for general use, especially with cheap grades of fish. It consists in freezing the fish as above described, except that they are packed in fine pulverized ice in the pans before being frozen, and when taken out of the pans the fish are found solidly imbedded and incased in the block of pulverized ice.

Along the Great Lakes the most popular fish for cold storage are whitefish, lake trout, lake herring, blue pike, saugers, sturgeon, perch, wall-eyed pike, grass pike, black bass, catfish, and eels. In addition to these species the Great Lakes freezers receive considerable bluefish and squeteague from the Atlantic. On the Atlantic coast bluefish, halibut, squeteague, sturgeon, mackerel, flatfish, cod, haddock, Spanish mackerel, striped bass, black bass, perch, eels, carp, and pompano, are frozen. Salmon, sturgeon, and halibut are the principal species frozen on the Pacific coast.

Some varieties of fish are so very delicate that it is not deemed profitable to freeze them, especially shad, but even these are frozen in small quantities. Oysters and clams should never be frozen, the best temperature for cold storage being 35° or 40° F. When stored in good condition they will keep about six weeks. As an experiment they have been kept for ten weeks, but storage for that length of time is not advisable. Caviar also should never be frozen, but held at about 40°. Scallops and frog legs, however, are frozen hard in tin buckets and stored at a temperature of 16° to 18° F. Sturgeon and other fish too large for the pans are frequently hung up in the storage rooms by large meat hooks, and when frozen are dipped in cold water and stored in piles. But when intended for shipment sturgeon are usually cut into pieces of suitable size for packing in the shipping boxes.

In some of the largest freezing houses on the Atlantic seaboard, which freeze and store fish as well as other food products, the fish to be frozen are simply hung up in the sharp freezer, the heads being forced on to the sharp ends of wire nails protruding from cross lathes arranged in series. After the fish are frozen they are removed and piled in storage rooms, where the temperature is about 15° or 18° F. (See plate XII.)

Where the handling of fish is of minor importance compared with other food products, the fish are placed on slat-work shelves in either a special freezing room or in a storage room where the temperature is kept below 20° (see plate XLI, lower half), or they are retained in bulk in baskets, boxes, or barrels in the same room; but these methods are not productive of results even approximating those in the Great Lakes fish-freezers and should not be used where quantities of fish are handled.

The cost of cold storage and the deterioration in quality make it inadvisable to carry frozen fish more than nine or ten months, but sometimes the exigencies of the trade result in carrying them two and even three years. In the latter case they are scarcely suitable for the fresh-fish trade unless the very best of care has been exercised in the freezing and storage, and it is usually better to salt or smoke them.

The rate of charges in those houses which make a business of freezing and storage for the general trade is usually from ⅛ cent to 1 cent for freezing and storage during the first month, and about half of that rate for storage during each subsequent month, dependent on the quantity of fish. However, the cost of running a first-class plant at its full capacity is probably less than one-third or even one-fourth of the minimum above quoted, since it costs no more to run a storage room full of fish than one-fifth or even one-tenth full.

The refrigeration of fish on the Pacific coast, according to Mr. W. A. Wilcox, dates practically from 1890, since when it has steadily increased, the aggregate shipments from Oregon and Washington in 1895 being 236 refrigerator carloads, or 5,872,533 pounds of fresh fish. This consisted chiefly of salmon from Columbia River and Puget Sound, with 1,161,715 pounds of dressed sturgeon and a small amount of halibut and smelt. Mr. Wilcox, on pp. 587–589 of the Fish Commission Report for 1896, describes the process of refrigeration and shipment as follows:

On the reception of the fish at the cold-storage plant they are washed, wiped dry, and then placed on racks attached to trucks; these are run into the freezing rooms where, in a round or undressed condition, the fish are solidly frozen. From the freezing rooms the fish are taken to the packing and storage rooms and packed in cases holding 250 pounds of fish each. In packing, no ice is used. In some cases the fish are "glazed" with ice. This process consists in dipping the frozen fish in tanks of water that are in a room with a temperature of 20° F. On removing the fish from the water they are at once glazed or coated with ice, repeated dippings adding to the thickness of the icy coat. Glazing is an extra precaution to keep the fish from the air. In some cases each fish is wrapped in brown rag paper, in oiled paper, and in brown paper, as an additional protection from the air. The fish having been frozen and packed, the cases are removed to cold-storage rooms and held until needed for shipment. When placed in the refrigerator cars the latter are charged with ice that, except from some unusual delay, lasts the entire trip to the Atlantic coast.

During the past few years the experimental shipment of fresh frozen fish from America to Europe has become of considerable importance. The pioneer shippers had much to learn and their shipments were often under many disadvantages. Sometimes shipments arrived at their destination in prime condition and again were only fair or poor. Frozen fish from America was a new article of food and time was necessary to acquaint the people with them. The markets, as in this country, were often fluctuating and shipments were sometimes sold at a loss. On the whole, results were satisfactory enough to encourage and build up this new branch of the fisheries. The shipments of 1895 included 300 tons of steelhead trout and 200 tons of silver and chinook salmon. Until quite recently the steelhead was but little thought of, but with the increasing demand for fresh fish it has grown to be the most popular of the several species shipped long distances. While not having as much oil as some other species, it is a fine fish, and stands transportation much better than other fish of the salmon family. One case is on record in which steelheads frozen solid and shipped to England, after being received and the frost removed, were placed on the market, and the fish had such a fresh look, as if just from the water, that the dealer was arrested for having on sale fresh fish illegally caught.

Hamburg is as yet the favorite point shipped to, from which the fish are distributed all over the Continent. At New York the cases of frozen fish are transferred from the refrigerator cars or cold-storage rooms on shore to those on board of the steamer, the Hamburg steamers receiving and putting in cold storage any number of cases of fish offered. The distributions from Hamburg are made by packing cases of frozen fish into small truck cars holding from 1 to 2 tons each. The cars are taken upon local steamers that radiate from Hamburg to many far and near ports. During 1895 shipments from Hamburg brought from 30 to 60 pfennig, (or from 7½ to 15 cents) net a pound, freight excepted.

In connection with the present quite large shipments of fish to Europe, notice of a small shipment from New England to Hamburg as far back as 1876 is of interest. During March of that year Mr. J. L. Griffin, then engaged in the fish business at Eastport, Me., made an experimental shipment of fresh frozen salmon that had been taken from the waters of New Brunswick. Mr. Griffin states:

"The salmon having been frozen solid were packed in a box which was inclosed within a second box with an air chamber of 1½ inches between the boxes. These were placed inside of a third packing case with a space of 1 inch between, this space being filled with sawdust. The fish arrived at Hamburg in good condition, but could not find any market as fresh frozen fish, such an article then being unknown. The frost having been removed, the fish were smoked and met a ready sale."

This small shipment not meeting with success, the attempt to introduce fresh frozen salmon from the Atlantic coast to Europe was for the time discontinued. After many years, with new methods of freezing, packing, and shipping, it has been successfully and extensively renewed from the Pacific coast.

For the purpose of economizing in freight charges on ice, the following method is used for shipping frozen salmon and other fish from the Pacific to Atlantic coast points: The frozen fish are first carefully packed in boxes and placed in refrigerator cars previously reduced to a low temperature, the floor of which is covered with several inches of cold sawdust. Between the boxes of fish and the sides, ends, and top of the car is a space of several inches, which is also filled with cold sawdust tightly packed. When filled, the car is at once closed, and no ice is placed in the tanks, as it is found by extensive experience that fish so packed for shipment reach their destination in perfect condition, even after a passage of two weeks or more.

Messrs. Seufert Bros. Co., of The Dalles, Oregon, state:

With frozen salmon, the fish are frozen solid as soon as caught, piled in cold storage like cord wood in a woodhouse, until time gives a chance for packing, which is done at our leisure. Then they are dipped i[n] cold water and taken out at once. This forms a very thin coat of ice on them. They are then wrapped in oiled paper with an extra heavy paper wrapped over that. They are then put in boxes of 300 pounds net fish, lined with paper, and held in cold storage as long as one wishes. When they are shipped in refrigerator cars we put 6 inches of sawdust on the floor of the cars. The boxes are so made that we have about 2 or 3 inches space from the side of cars, which is filled with sawdust, and the top is filled over with 6 inches of sawdust. This sawdust must be fine and dry. Then we put 2 tons of ice in the car boxes with 2 sacks of salt, and the car is ready for a freight run across the continent of from 11 to 14 days. On reaching New York the fish are put in cold storage. If they go foreign, they go directly into the steamer's cold rooms and are not opened until they reach their destination. Frozen salmon will keep perfectly for 10 months; after that they lose ground.

The process of refrigeration applied to sturgeon on the Columbia River when shipped to eastern markets is as follows:

The fish are first beheaded, eviscerated, and skinned. The backbone is then removed and the fish cut into suitable sections for freezing. The sections are packed into galvanized-iron pans 24 inches long, 16 inches wide, and 5 inches deep. The pans are then put into a freezer charged with ice and salt, and their contents frozen into solid blocks of fish weighing about 60 pounds to each pan. The process is precisely similar to that in vogue on the Great Lakes for freezing fish. When frozen, the fish are removed from the pans and packed in boxes, four blocks to each box, and then loaded into refrigerator cars. The cars are charged with ice and salt to keep the temperature below the freezing point. In winter the cars do not usually require to be recharged before they reach their destination, but when the weather is warmer it is sometimes necessary to recharge with ice and salt once or more while in transit. (Report U. S. Fish Commission, 1888, pp. 226, 227.)

A process quite similar is applied to freezing sturgeon on the Delaware River, the frozen fish being then stored for the fall and winter markets.

FREEZING FISH IN EUROPE.

An account of the condition of the frozen-fish trade in Europe has recently been received from Mr. Nicolas Borodine, of the department of agriculture of the Russian Government, of which the following is a partial translation:

In the regions of western Europe, not including Norway and Switzerland, the winters are so mild that fish naturally frozen are not found on the market; hence a strong prejudice has arisen among the Germans and French against fish in that condition. They express the opinion that frozen fish lose their savory qualities, and they esteem them far less than the unfrozen. In consequence of this prejudice frozen fish have not hitherto found much sale in the European markets, although efforts have been made in that direction.

At the end of the eighties an attempt to procure fish in the frozen state throughout the entire year was made at Marseilles. A company was organized there, under the name of the "Trident," which had a sailing vessel furnished with apparatus for freezing fish caught on the west coast of Africa. The selection of Marseilles as a market was unfortunate; in the first place, because the men of the south, never having seen frozen fish and not eating them, utterly refused to buy them; and, in the second place, the inhabitants were entirely unaccustomed to the kinds of fish which were imported. Hence the company was soon compelled to wind up its business. The spread of reports of the worthlessness of frozen fish as food, of which the French were at that time convinced, contributed no little to this failure. In their opinion tainted fish imported in warm weather were better than frozen fish. Hostility was even aroused at first against fish brought in ice from Algiers to Marseilles. These, they said, are not fresh fish, but preserved fish, and therefore it must not be sold in the market as fresh fish.

Another attempt of a similar kind was made by the Norwegians in the Hamburg market in the nineties. The North Cape Joint Stock Company built a special steamer, the *North Cape*, with cold-storage rooms (low temperature being obtained by means of machinery), which were filled with "*Gadus æglifinus*" (haddock). The steamer arrived at Hamburg with a full cargo. In the first year hardly any customers were found for the frozen fish, and a part of the cargo was carried back. According to the reports for 1892-93 (see note by Mr. Heinemann in Pisciculture of the World, 1893, No. 12, p. 385), this company built a large cold-storage warehouse at Vardö, in Norway, at a cost of 200,000 German marks [$47,600]. The fish are caught on the spot, are frozen, and placed in the cold-storage warehouse in a temperature of 5 Réaumur [18° F.]. The shipment on the steamer at that temperature is usually made in the autumn. A cold-storage warehouse for frozen fish, with a capacity of 21,000 poods,* was also built in Hamburg. A railroad goes to the warehouse, and the frozen fish are shipped to any point in Germany by rail.

Fish are sent in a fresh state by rapid transit to Munich, Leipzig, and Vienna. To give an idea of the demand, it is stated in the note already quoted that of 9,000 poods shipped to Hamburg on the steamer *North Cape*, November 11 (26), 1892, 5,200 poods were sold in the first week and the remainder of the cargo in the following week.

Judging by statements made in the German weekly, *Deutsche Fischerei-Zeitung*, the prejudice against frozen fish is rapidly diminishing, owing in great measure to the fact that frozen fish are sold much cheaper than fresh fish shipped in ice; and it may be asserted that the same thing will occur in this case as in the case of the frozen meat from South America and Australia, which is now sold in large quantities in the great commercial centers.

In Paris the cold-storage warehouse is under the Bourse, and is under the management of the Compressed Air Company of Paris. The central station and the office of that company are 56 Rue Etienne Marcel, from which point the compressed air is carried in pipes to many places in the city as a motive power. The employment of compressed air for freezing purposes is based upon its rapid expansion, by which the surrounding atmosphere is rendered extremely cold. For the rarefaction of the compressed air conveyed into the basement of the Bourse a special engine is employed, by which the refrigerated air is carried into the compartments formed by the double wall of the refrigerating rooms. The walls are lined on the inside with tin, painted a whitish color, and on the outside with materials which are nonconductors of heat (layers of moss on wood). Every room is furnished with a ventilator to carry off the damp air. There are 16 rooms, and each of them can be refrigerated separately, and to any temperature near zero or below zero. They are arranged on both sides of the

--- --- ---

* One pood equals 36.112 pounds.

main corridor. They are rented at 2 francs per square meter of surface if the goods are to be subjected to a temperature above zero, and 3 francs per square meter of surface if a temperature below zero is wanted.

In London there are cold-storage warehouses at the Central Market, which belong to a company called "The Central Markets Cold Air Stores, Limited." The immense cellars of the enormous London meat market are used for these warehouses and have a capacity of 1,000 tons. The warehouses are used partly by the company itself, which does business in Australian frozen meat; and are partly rented to butchers at £2 per ton for four days. Besides meat a large quantity of game is stored here, including 100 tons of poultry from Russia annually. The whole underground market consists of a series of large rooms with thick insulated walls. A square wooden pipe, the sides of which are also insulated, passes along the ceiling of all the rooms in the basement; and there are orifices in this main pipe in every room through which cold air can be admitted. Here, too, the refrigerating process is based upon the expansion of compressed air. At first the air is compressed by steam power. It is then driven violently into a vacuum, where, as it expands, it lowers the temperature. From this room the refrigerated air is conveyed into conducting pipes for general distribution. These engines are built on the Bell & Coleman system and are constructed in England by the firm of Holsam & Co., Engineers, Derby. The temperature can be regulated at will. For meat it ranges from 11 to 9° F., but never higher than 14.

The exportation of frozen meat from Australia has lately attained great dimensions. Forty steamers, equipped with apparatus for the transportation of frozen fish, run to London.

Freezing by machinery, for the storage of fish, was first employed in Russia at Astrakhan, by Mr. Supuk, who first built an ice barge for freezing by means of air engines of the Lightfoot system. This occurred in 1888. In this new enterprise Mr. Supuk was subjected to great losses through his failure to induce others to send their fish to his barge. Not obtaining any cooperation on the part of the fish-dealers of Astrakhan, and becoming convinced that he would have to procure fish for himself, Mr. Supuk, in 1891, requested the aid of the Russian Fishery Association, which, however, could not furnish him any material assistance, but which, through its members and by articles in the press, greatly contributed to the establishment of this new business. Recently Mr. Supuk's business has been considerably enlarged and is on a firm footing.

I borrow from Mr. A. K. Heinemann, who inspected Mr. Supuk's ice-barge at the time of its construction, some of the chief details concerning it. Its capacity is estimated at 10,000 poods. It is intended to freeze and to transport in the frozen state not only large but small fish. It is fit for going to sea and for ascending the Volga as far as Nijni-Novgorod. The cold-storage room occupies nearly the whole length of the barge, and is formed by double walls with the space between them filled with sawdust. The walls are at a little distance from the sides of the barge, and the ceiling of the room does not reach the deck, so as to avoid heating. The engines stand, one in the bow, the other in the stern of the barge. The chamber contains five sections, two of which serve for freezing fish and the middle ones for storing fish. In the former the temperature is about 12 R., in the latter it is kept at 2 to 3 R. The capacity of the two freezing chambers is 450 to 500 poods [16,200 to 18,000 pounds]. The largest fish is frozen through and through in 24 to 36 hours. The cold air, in its passage from the engines to the freezing chambers, is conveyed through a separate snow chamber, in which the superfluous moisture of the air is condensed in the form of snow; from there the cold air is conveyed through wooden pipes to the freezing chambers and the cold-storage rooms. The admission of the air is regulated by espagnolettes.

Having procured the capital needed for the whole operation of the purchase and sale of fish on his own account, Mr. Supuk built at Astrakhan, on the bank of the Volga, a stone edifice for freezing fish and for storing the frozen fish, and the barge is used exclusively for the transportation of the frozen fish from Astrakhan to Tsaritsin. On the whole, the undertaking proved very profitable, and Mr. Supuk is extending his business, making improvements, and proceeding to the construction of a second barge, intended for the southern part of the Caspian Sea. The new barge has a capacity of 25,000 poods [450 tons] of fish. The freezing machine is placed in the middle of the barge. It is furnished with electric light.

Freezing by machinery is carried on on a still larger scale by the great fish firm of Vorobieff, at Petrovsk, which has built large cold-storage warehouses, at a cost of 185,000 rubles [$71,865].

A mixture, composed of ice and salt, is employed for freezing fish at Mariupol and Henichesk.

Judging by the printed reports, fish are frozen at Mariupol in tubs made for the purpose [4 by 2½ by 2 arshines] with a capacity of 100 poods [3,611 pounds], in which the fish are kept for days in a mixture of ice and salt [10 per cent of the latter]. When frozen, they are taken out and shipped in baskets by rapid transit to Kharkoff, Moscow, and other great centers.

At Genichesk the freezing of dolphins and sturgeon is carried on in an ordinary cellar constructed on the seashore. In the spring and autumn the cellar is used for salting the fish, and in the summer when the dolphin fishery is mostly carried on in the sea, the chests from which the fish are taken are used to freeze the fish. This is done in the following manner: A layer of ice, 1 arshine * in depth, is laid on the bottom of the chest; this is covered with salt; on this is placed a layer of fish which is covered with a fresh layer of ice and salt of half an arshine, and so on to the top. The fish remain in this condition for 3 weeks, not longer, and are frozen, as in winter, in 24 hours after they are placed in the chest. If it is necessary to keep the fish longer than 3 weeks they are taken out of the ice and placed in boxes containing gratings in order that the cold may penetrate them and to prevent the fish from becoming soft by not being sufficiently salted. If it is intended to take the fish out within a week, the first layer of ice is made only one half arshine deep and the others much less. The capacity is 7,000 poods of frozen fish. The whole shipment from Genichesk amounts to 20,000 poods every summer. The frozen fish is shipped mostly to Kharkoff where it arrives in 24 hours, and may go as far as Moscow. It is sent with great rapidity, in baskets containing 10 poods, with a mat over the top.

At the place of the catch the dolphins are sent direct in fresh condition to the freezing establishment, where they are frozen. On islands at a distance from Genichesk, ice is stored on the spot and the fish are frozen there by being buried in the ice, and are then shipped at night on sailing vessels when the wind is favorable.

The same method has been employed recently at Uralsk for freezing sturgeon, and the frozen sturgeon of the spring catch are shipped to Busuluk and Orenburg. In order to show the fish-dealers the construction of the cold-storage rooms of the American type, for the preservation of frozen fish, a room of that kind was constructed at Uralsk after my plans. The whole operation of freezing the fish and the construction of the room itself have been several times explained in this miniature warehouse; and an opportunity of inspecting it is offered at any time to those wishing to do so.

There is no doubt that freezing fish in the ordinary ice-house appears to be the simplest method, as it requires no special buildings and is done in the common cellar ice-house, part of the ice in which serves for that purpose. The large fish, as I succeeded in ascertaining by personal observation, are frozen very thoroughly by this means and lose nothing of their external appearance. The time during which the fish can be preserved by this method, however, is limited. The fish must be taken out within a week or a week and a half; otherwise the brine acts upon the frozen fish, and it becomes soft and dark. Besides, this method can only be applied where there is a large stock of ice on hand, as is the case in vaults which are also used for other purposes.

In view of what has been said, we must conclude that for Russia, in places where there are fisheries, the most expedient mode of cold storage consists in a combination of the vault and the cold-storage rooms of the American type for the preservation of frozen fish. By means of such combination, in the first place, the space may be used for the ordinary purposes of the spring and autumn salting; in the second place, the largest fish can be easily frozen by direct burial in the ice without any great expense for labor; and in the third place, when the frozen fish are taken out of the ice, they can be stored for the longest period in the cold-storage rooms of the American type.

FREEZING HERRING FOR BAIT.

The demand for fresh herring as bait in the cod fisheries led, in 1890, to the building of a number of freezing houses along the New England coast, where shore herring are frozen during the fall and kept for use during the winter and early spring. Most of these were of the direct anhydrous ammonia absorption system and were designed by M. J. Paulson, of Gloucester. From 1890 to 1893 the following plants, with the designated capacities, were constructed: Gloucester, 4,500 barrels; Rockland, 10,000 barrels; Boothbay Harbor, 4,000 barrels; Provincetown, 3,000 barrels; North Truro, 3,000 barrels. The Rockland freezer did not pay and was dismantled in 1894. The walls of the storage chambers in these plants are thick and well insulated. About the walls on the inside are ranged the ammonia pipes in nests of horizontal rows, the

* One arshine equals 28 inches.

distance between the rows being 4 or 5 inches and between each pipe in the rows 3 or 4 inches. The herring are thrown on hand boards of lattice or small flake platforms, which are placed on the various rows of pipes, and are frozen, the temperature during the process being sometimes 15° F. or lower. They are afterwards stored in heaps on the floor between the nests of pipes and additional fish are placed on the lattice boards.

The plant at North Truro, Mass., is thus described by the engineer, Mr. E. R. Ingraham:

Our building is of wood, 100 feet long, 40 feet wide, and 3½ stories high. Our sharp freezer is on the third story; it is 70 feet long, 30 feet wide, 8 feet high, and contains 10,896 feet of 1½-inch pipe arranged in four coils running the length of the building. The pipes are 12 inches from center to center. Upon these coils are placed wooden flakes, or shelves, 6 feet long and 4 feet wide. Upon these flakes the fish are placed to be frozen. The capacity of the machine is 125 barrels in 24 hours.

As soon as the fish are frozen they are put down through scuttles into the storage rooms, which are four in number and contain 8,400 cubic feet each. Here the fish are held at a temperature of 15 above zero. In our sharp freezer we carry a temperature of from 15 to 15 below. We have two machines of the absorption type—direct expansion. The temperature of our condensing water is 52°. We carry 140 pounds high pressure on the generator, 3 to 10 pounds on absorber, 40 pounds of steam on generator, and 60 pounds of steam on boilers. We burn on an average one ton of coal every 24 hours. The fish are all caught in weirs about one mile from the storage. They are brought in boats to the shore, where they are dressed and washed clean; then they are hoisted to the top of the third story, whence they go down through scuttles into the freezing room, where they are frozen solid.

There are also two ice-and-salt refrigerator plants on the New England coast, one at Gloucester, Mass., with capacity for 3,000 barrels of herring. and the other at Boothbay Harbor, Me., for about 500 barrels. Some food-fish also have been frozen in these refrigerators, but they are used principally for preserving herring for bait.

In the fall of 1898 the schooner *J. K. Manning*, 282 tons, and the barge *Tillid*, 425 tons, were fitted up at Gloucester with direct ammonia absorption freezers and sent to Newfoundland for the purpose of freezing herring and squid for bait. The capacity of the former is 3,000 barrels and of the latter about 2,000 barrels.

A very cheap and ingenious device, known as the Wallems freezer, is used for freezing small quantities of herring for bait in Norway and in Newfoundland. The cost is small, the only materials required being ice, salt, and a stout barrel. Within the barrel four wooden flanges are fastened to the sides and running diagonally or at an angle with the axis. The barrel is one-half filled with a mixture of crushed ice and salt, in a proportion of about three parts of ice to one of salt. In case of ice not being obtainable, snow will suffice. The barrel is loosely filled with fresh herring and headed. It is then placed on its side and rolled on its bilge one or two turns forward and then backward one or two turns, the rolling being continued for about 15 minutes, at the end of which time the herring are generally thoroughly frozen, when they are placed in dry sawdust until used. The flanges inside the barrel are placed at an angle with the axis so as to aid in mingling the fish with the ice and salt when the barrel is rotated.

The rate at which the barrel should be rotated is easily determined by experience, and the salt and ice may be used over again as long as they last, but it is usually desirable to add some additional salt and ice. For convenience in handling, the barrel may be suspended from an axle and rotated by means of a crank, the fish, ice, and salt being admitted and removed through a hinged-lid opening in the side of the barrel.

PRESERVATION OF FISHERY PRODUCTS BY DRYING AND DRY-SALTING.

Next to heat, moisture is the greatest aid in the development of bacteria, and its removal constitutes one of the most important processes of preservation, being applied to a variety of foods, as fruits, cereals, and occasionally to the curing of meats, but its most important application is to the curing of fish. The moisture may be removed in several ways—by exposure to the air, by pressure, by combining with the fish certain substances possessing a strong affinity for water, such as salt, by use of absorption pads, etc.

Drying in the open air is the most ancient method of preserving fish, it being the principal method in vogue among the Phœnicians, and up to the present time it has been employed to a greater or less extent in nearly every fish-producing country, either in its original form or with certain modifications tending to assist in removing the moisture. The use of salt performs a twofold function in curing fish, it acting as an antiseptic as well as diminishing the amount of moisture. Fish can be cured equally hard by resalting with dry salt several times at suitable intervals as by drying in the open air. But the former is impracticable on the score of economy and the latter is not generally employed exclusively, because of the unsuitable climatic conditions, and in America it is most practicable to combine the two methods with the addition of compression, the fish being first salted in butts and afterwards pressed and dried.

The original process of curing fish solely by drying in the sun is very little used in this country, as there are few localities where the air is adapted to it. The Indians of the Northwest dry a number of species, among which are halibut, salmon, cod, eulachon, smelt, etc., and at a few other localities some minor species are dried, but the business is inconsiderable and there is no general traffic in these products.

In Norway, Sweden, and Russia, and other countries of northern Europe, where the air is comparatively free from moisture, cod are yet dried without the use of salt, but Norway is the only country that prepares large quantities in this manner. The fish are beheaded, eviscerated, and cleaned with sea water and suspended in the air on stands about 8 feet high, where they remain for weeks, and even months, until they are hard enough to withstand the strongest pressure of the tip of the thumb in the thick of the flesh along the back without giving way, and it is necessary to soak them in water for several hours before preparing them for the table. This is known as the stockfish cure, and the annual product in Norway exceeds 400,000 quintals, which is marketed in Germany, Italy, Portugal, Spain, Brazil, and other countries, especially those located in the Tropics.

In the United States, the British North American Provinces, France, Iceland, and certain other countries the greater portion of the desiccation is performed by salt, the fish being first salted and then dried; but in each country the methods differ in some particulars, as in the quantity and quality of salt used, the extent of the drying, the length of pressing, etc., depending not so much on the caprice and fancy of the individual curers as upon the market for which the fish are intended, regard being

paid to the appearance and keeping qualities of the product. In curing codfish for
the New England trade about 64 per cent of the moisture is removed—59 per cent by
the salting and 5 per cent by pressing and drying. 100 pounds of split cod contain
about 80 pounds of water; and, in the process of curing, 51 pounds are removed by
the action of the salt and 4½ pounds by pressure and drying. Norwegian stockfish has
been freed from about 96 per cent of moisture by atmospheric drying alone.

The principal marine products to which this process is applied are cod, hake,
haddock, cusk, pollock, mullet, shrimp, channel bass, barracuda, bonito, and salmon,
but as its application to the cod and its related species is the most extensive and
valuable, the methods of treating those species will be first described.

DRIED CODFISH.

In speaking of codfish generally the various species of *Gadidæ* common on the
New England coast are usually referred to, the most important being the cod (*Gadus
callarias*), haddock (*Melanogrammus æglifinus*), pollock (*Pollachius virens*), hake (*Mer-
lucius bilinearis*), and cusk (*Brosmius brosme*). The proportion of these entering into
the dried-fish trade is about as follows: Cod, 83 per cent; hake, 10 per cent; haddock,
3 per cent; pollock, 3 per cent, and cusk, 1 per cent.

The greater portion of the dry-salted codfish in America is obtained from the
bank fisheries, especially from Grand Banks and vicinity. Many are also received
from Georges and other neighboring banks, representing the surplus from the fresh-fish
trade. Gloucester is the principal market for this article, while Portland, Provincetown, Boston, Boothbay Harbor, and other ports handle large quantities.

In large ports, especially Gloucester and Boston, most of the product is prepared as
boneless fish, but in the smaller places the fish are usually dried more thoroughly for
distant markets. During the eighteenth century and the early part of the present century codfish were prepared largely for export, and consequently were made quite dry, so
as to keep for several months in warm, moist climates. At present, however, the great
bulk of the product is intended for domestic consumption and is not so thoroughly dried.

It was formerly customary for United States fishermen on the Grand Banks and in
the Gulf of St. Lawrence to land on the rocky shores of Newfoundland and other British
Provinces to cure their fish. In that case the fish were split, salted, and kenched in
the vessel while on the banks, and on landing they were dried upon the rocky beaches.
After curing the fish many of the vessels carried them directly from the provinces to
Spanish and other ports, where they were sold and a return cargo of merchantable
products was brought home. The privilege of landing to cure fish was considered
quite valuable and occasioned much international controversy. By the treaty of 1818
the fishermen of the United States were allowed to land and cure their fish within
certain prescribed limits, which were increased in extent by the Washington treaty
of 1871. But during recent years this privilege has been of no value, as the fishermen
have brought all their catch directly home to be prepared by regular fish-curers.

The present process of curing cod, haddock, hake, cusk, and pollock, which is
essentially the same as that in vogue a century ago, except in the amount of drying,
is as follows, this description being especially applicable to the fishery on the Grand
Banks and to the preparation of fish for the domestic trade:

The dressing is begun as soon as the dories return to the vessel, or as soon as the
day's work is over when fishing from the vessel's deck or from shore boats. Unless

this be done the cured fish are likely to have a dark color and the flesh be broken and loose, especially near the backbone. This is also noticeable in fish caught on trawl lines when stormy weather prevents the overhauling of the trawls for two or three days. Care should also be taken not to bruise the fish any more than is necessary and to protect them from the sun by means of tarpaulin or otherwise, if they are not to be dressed for several hours. It would improve the appearance of the cured product if the fish were bled as soon as practicable after removal from the water, but this is not a common practice in the New England fisheries, except in the vessel hand-line fishery, when the tongues are cut out, which is more particularly for the purpose of keeping count of each man's catch. The blood is subject to putrefactive action much more readily than the flesh, and if it remains in the pores it causes the color of the flesh to turn dark. The additional trouble of bleeding the fish would be slight, it being done by cutting the throat and the large vein near the neck bone.

A dressing gang on the usual Grand Banks vessel consists of a "throater," a "gutter," and a "splitter." The first named, taking the fish in his left hand by the head and resting it on its back on the edge of the tub, makes a transverse cut across the throat immediately behind the gills, with a strong, sharp pointed knife. Introducing the knife at this opening, he cuts down the belly, laying open the abdominal cavity, and making also one cut on each side downward he separates the head from the sides. Then by pressure simultaneously upon the head and the body, the neck resting on the edge of the tub, he breaks off the head from the body at the first vertebra. The gutter, taking the fish, opens the abdominal cavity with his left hand and with his right hand tears loose and removes all the organs contained therein. The livers are thrown into a separate receptacle, while the stomach and other organs are with the heads thrown into the sea or into the gurry pen on the deck, whence they are discharged into the sea on changing the berth of the vessel.

The fish then passes to the splitter, who is provided with a knife rounded at the end and with the blade slightly curved flatwise. With the back of each fish braced against a cleat or batten nailed on the splitting board, he makes a long incision down the ventral surface, continuing the opening made by the throater, and also makes a straight, clean cut along the left edge of the backbone to the tail, inserting the knife no deeper than is necessary for cutting out the backbone. With a horizontal stroke he cuts through the backbone about two-fifths of the distance from the tail and loosens it so that he can catch the end in his fingers. Grasping this with his left hand he cuts under it toward the head of the fish and separates the upper three-fifths of the backbone from the body, the lower two-fifths remaining in the fish. In this operation the knife should be pressed close to the backbone, so that no flesh adheres thereto, otherwise the fish will be thin through the back. In dressing pollock nearly all of the backbone is removed because of the large quantity of blood along the bone. The French curers leave more of the backbone in the fish than is customary in America and elsewhere, and to remove the blood in the remaining portion they use a small iron spoon. The cut through the backbone should be horizontal toward the head, passing through two or three vertebra, and it should not be deep enough to damage the muscles lying along the backbone and thus weakening the lower part of the fish.

After removing the sounds or air bladders the backbones are discarded. Sounds sometimes sell at such a low price that it does not warrant saving them and they are discarded with the backbones. A slight incision should be made along the remaining part of the backbone to permit the escape of any blood that may remain.

The fish are then washed in tubs of sea water or by sousing water over them, especial care being taken to clean the neck, the remaining portion of the backbone, and the vicinity of the dorsal fins, and to remove the dark membrane that adheres to the napes. Even if the fish are dressed on shore, sea water is preferred for washing them, as fresh water has a tendency to make them slimy. In washing them in tubs the water should be changed frequently to prevent its becoming foul.

After the splitting and washing, with the subsequent draining, comes the salting, which is accomplished in two ways, forming the kench cure and the pickle cure. In the former the fish are placed with dry salt in regular piles or kenches, and the pickle which forms is allowed to run off, leaving the fish dry, and in the pickle cure the fish are salted in butts or barrels which retain the brine. The kench cure makes a drier product and one suitable for export to hot climates, but is rarely used in the United States except in combination with the pickle cure. Generally in this country the fish are salted in kenches on board vessel and in butts on shore.

In the Grand Banks fisheries, the fish, after draining, are passed to the salter in the hold of the vessel. Grasping each fish by the tail, he throws it upon the kench or pile, flesh side or face up, and with a small scoop sprinkles over it a quantity of salt. The kenches are built up in regular order, the fish being laid head to tail, spread out flat, with the back or skin down.

Salting requires considerable skill, the fish spoiling from insufficient salt and deteriorating in flavor from an excess. As a rule, the less amount of salt required for preserving the fish the better, but the salt should not be used sparingly. Some fishermen, in order to make their fish weigh heavy, put on too little salt and at times lose the fish as a result. Thin fish require much less salt than thick ones, and less salt is necessary in cold weather than in warm. Also those fish which are to remain a long time before being used require more than those to be marketed quickly. Experience is the only guide, but as a rule an even layer of salt thoroughly covering the fish and leaving no vacant places or finger marks is sufficient. Coarse salt is preferred to fine, especially for fish that are not to be marketed for a considerable length of time, as it does not go to brine so quickly. Formerly Cadiz salt, by which name all the Spanish salts are called in this country, was used, but during recent years Trapani salt has been employed almost exclusively, the use of the former being abandoned on account of its abundance of lime, which settles on the fish, and also its greater tendency to impart to the fish a reddish color, attributed to some vegetable matter contained in the salt, which develops rapidly during warm, moist weather. On an average, 1 bushel of Trapani salt is used to each hundredweight of fish in the Georges fishery, while the Grand Banks fishery usually requires 1½ bushels to the hundredweight of fish. Vessels engaged in the latter fishery usually carry about 200 hogsheads of salt in pens or compartments, each pen holding 15 or 20 hogsheads, and as the pens are emptied of salt they are used for storing the fish. The Georges Banks vessels, being absent a much shorter length of time, do not carry so much salt.

The fish remain in kenches until the vessel arrives in port, care being taken that no water reaches them. If the vessel is absent three or four months those fish caught during the first month or so are generally rekenched and additional salt sprinkled among them if it appears necessary. As the pickle accumulates and is driven out from the fish in the middle and lower parts of the kenches by the pressure of those above, the vessel must be pumped out frequently. Some fisherman claim that they can determine from the quality of the pickle pumped out whether the fish are keeping in good condition. If the pickle shows signs of being tainted, the fish must at once be overhauled and repacked.

As soon as the cod are landed on the dock they are culled, the principal grades being (1) large cod, which includes all over 22 inches in length salted; (2) medium or small cod, between 16 and 22 inches in length; and (3) snappers, which comprise the lowest grade. If the trip is from other grounds than the Grand Banks, the "scale fish" are separated from the cod; these comprise hake, haddock, pollock, and cusk. Of the cusk, two grades have been made during recent years, namely: (1) large, covering all over 19 inches in length salted; and (2) snappers, comprising all under 19 inches in length. Each grade is weighed separately and the fish are washed with clean salt water in tubs, vats, or old dories. If the fish have been kenched for two months or more they are sometimes rubbed with bristle or palmetto brushes to remove surplus or incrusted salt, etc.

They are next placed in large butts, usually old molasses or sugar hogsheads, each having capacity for about 900 to 1,000 pounds of fish. From 2 to 4 bushels of Trapani salt is sprinkled among the fish in each butt, this quantity depending on the degree and length of salting on the vessel. In the case of Grand Banks cod about 2½ bushels of salt are placed in each butt, whereas for Georges cod, which usually are not so heavily salted in the vessels, about 3½ bushels are required. With the exception of the bottom layer the fish are generally placed with the skin side up, but this is not the universal custom, some curers placing all the fish flesh up except the last two or three layers. Fish in butts take the salt better if placed face or flesh side up, and in case they have been only lightly salted it is best to place them face up; but if they have been thoroughly salted on board the vessel it seems immaterial whether they be placed face up or back up in the butts. The bottom layer is placed back down, to protect the fish from the bottom of the butt, and the top layers have the backs up as protection against dirt, dust, etc. On top of the pile is placed about half a bushel of salt, to strengthen the weak pickle which floats up to the surface. In case the fish have been but slightly salted on board the vessel, some curers hang over each butt a basket containing about a bushel of salt, through which water is allowed to percolate and, thus charged with brine, to flow into the butt, the salt in the basket being renewed as it melts away. The fish remain in the butts under shelter until orders are received, which may be a year or more, in that case more salt being added from time to time; but the sooner they are used after the first few weeks the better, otherwise they have a tendency to turn yellow, or in case of pollock they turn dark.

When orders are to be filled the fish are removed from the butts and placed flesh side down, except the first two layers, in kenches about 3 feet high, for the purpose of pressing out some of the water and giving the fish a smooth surface, this being known as "water-horsing." The following day they are again repiled in a similar manner, but with those fish in the upper half of the first pile placed in the lower half of the second. The water-horse should be made a little higher in the middle than on the sides, in order to facilitate the running of the moisture from the fish, and it should be put on racks about 3 inches high, to protect the fish from the moisture, dirt, etc., on the floor. In order to avoid water-horsing a second time, many of the curers place weights on top of the first water-horse, this being most easily accomplished by placing boards on the pile of fish and rolling empty hogsheads on top; but this is not a desirable substitution for repiling, since the lower layers are thereby compressed more than the upper ones. All handling of fish from the vessel's hold to beginning the water-horse is generally done with pews or long-handled forks with one or two prongs to each handle.

From the water-horse the fish go to the flakes, which are of two kinds, stationary and canting, the former being the more common. They are about 2½ feet high, 8 feet wide, and of convenient length, with passageways wide enough for handbarrows between the stands. The horizontal top, resting upon ordinary wooden horses 8 feet long and about 2 feet high, is in three parts: (1) The long joists 2 by 3¼ inches and 12 or 14 feet long; (2) the cross joists about 1¾ by 2 inches and 8 feet long; (3) the 1-inch triangular strips, upon which rest the fish. Three of the long joists run lengthwise of each stand, and to these are nailed the cross joists, about 12 inches apart, and to the cross joists are nailed the triangular strips, 3 or 4 inches apart. At each end and transversely at suitable intervals over the flakes are placed frames about 15 inches above the flakes, upon which cotton awnings may be stretched when the sun is hot. The canting frames differ from the above in that they are fixed only at the middle and to a horizontal axis, so that they can be turned at an angle with the horizon, in order to expose only the edge of the fish to the sun and to get the benefit of even a slight breeze. With these flakes cotton awnings are dispensed with, but very few of them are now used in this country, and they are practical only in yards running north and south. At Gloucester many of the flakes are on platforms built over shallow water. In Portland, Me., the roofs of the fish-houses are used as resting-places for the flakes, and at Rockport, Mass., the flakes are built on tall posts overhanging the sloping rocks, thus allowing the air to freely circulate beneath the fish. The old style of brush flake is not used at present on the New England coast, nor are codfish dried on the beaches in the United States, as is common in Nova Scotia and Newfoundland. About thirty years ago a form of flake was introduced having screens or slatted frames, like those of window shutters, arranged to protect the fish from the weather or to screen them from the rays of the sun, as circumstances may require, but its use was never extensive.

In carrying the fish from the butts to the flakes they are placed on wheelbarrows or on handbarrows, the latter being usually made of oak and consisting of two sticks about 6 feet long, with 5 crossbars, each 21 inches long, 2½ inches wide, and 1½ inches thick. For carting the fish from wharf to wharf a low platform 4-wheeled truck, called a jigger, is in common use.

The principal troubles in connection with curing fish are flies, sunburning, and softening. Flies are avoided by keeping the vicinity of the flakes clean and airy and free from all putrefying refuse. During some years the flies are so numerous that it is necessary to protect the slack-salted fish by sprinkling lime or salt about the flakes and yard to destroy the maggots. Sunburning is prevented by protecting the fish from the excessive action of the sun, and softening may result from a stinted use of salt on board the vessel, or from wet, cloudy weather during the process of curing. Cod and haddock burn quite readily unless properly protected, but there is little danger with hake and cusk, even on hot sunny days. Because of climatic conditions it is frequently quite difficult to cure codfish during July and August on the New England coast, the air being moist and the sun so hot as to sunburn the fish very quickly; but in October and November little trouble is experienced from this source. On the New England coast the direction of the wind has considerable influence on the drying. Winds from the northwest or southeast are usually dry and good for curing fish, but under the influence of southwest winds the fish are liable to burn, and when northeast winds prevail it is extremely difficult to dry the fish.

Much difference exists in the extent to which the fish are dried. Some are dried for only a few hours and others for a week or more, depending on the market for which they are intended. Some markets desire fish from which 50 per cent of moisture has been eliminated; others 60 per cent, and others 70 per cent, and since a larger per cent of moisture removed represents a greater increase in labor and decrease in weight of product, a curer endeavors to avoid drying them any more than necessary. Those to be used in preparing boneless fish are dried very slightly, 8 or 10 hours of good sunning being sufficient, while the export fish must be dried for a week or 10 days. Every evening the fish are placed, flesh side down, on the flakes, in small heaps of 15 or 20, and a cover of wood, known as the flake box, is placed over each heap to prevent injury from dampness or rain. This cover consists of a rectangular box with a peaked roof and is generally about 38 inches long, 22 inches wide, and 14 inches high, the whole being made of three-fourths inch rough boards. When the air is moist, the fish are not spread out, but if the weather renders it necessary to keep the fish piled up for several days, they are occasionally rearranged.

When preparing fish for export, after they have been on the flakes two or three days they are placed in kenches under cover to "sweat," where they remain for two or three days, when they are again spread on the flakes for a day or two. In some instances the fish are then dry enough for shipping, but usually it is necessary to sweat them once more and again dry them for a day or so. The export fish are usually dried sufficiently hard to withstand the pressure of the thumb in the thick part of the flesh without retaining the impression. During moist weather these fish are likely to sweat and become soft; it is then necessary to "throw them," scattering them over the flakes for a day or so.

Most of the export fish are what are known technically as "kench-cured." This differs from the above only in that the salted fish on removal from the vessel's hold are not placed in butts, but in kenches, skin down, in the warehouse, whence they are removed as required, washed to remove slime, undissolved salt, etc., and dried on the flakes for three or four days in the manner last described. They are next repiled and sweated for two or three days, when they are dried again for a day or two, repiled and sweated for two or three days, and again dried for a day or two, when they are ready for shipment. These fish are slack-salted, but well dried, whereas fish for the domestic trade are generally heavily salted, but only slightly dried. Hake and haddock are rarely kench-cured, but the latter are not often exported from the United States, although there is a steadily increasing exportation of them from Nova Scotia to southern Brazil and to Cuba.

In case the fish are fresh when received at the curing-houses, they are at once beheaded, eviscerated, split, and washed in the manner described for vessels fishing on the Grand Banks. They are immediately placed in butts, with the flesh side up and with about 7 bushels of salt to 1,000 pounds of fish scattered among them. The fish are piled in each butt until they extend a foot or two above the surface. On the second or third day, after they have settled somewhat, a half bushel of salt is placed on top. No pickle is added, as in case of Grand Banks fish, since the green fish will make their own pickle. The fish remain in these butts at least fifteen or twenty days, and as much longer as desirable, when they are removed, water-horsed, and dried on the flakes, as already described. This is the true "pickle-cured" fish, the treatment of the Grand Banks fish combining both the "kench cure" and the "pickle cure."

Pollock which have been salted only a few days on the vessel are sometimes placed in the butts and weak pickle is allowed to percolate through a basket of salt over them for five to seven hours. Or, if fresh, they are split, washed and kenched, skin down, with 1 or 2 bushels of salt to the 1,000 pounds of fish, and on the following morning they are placed in butts, back up, each butt being filled with weak brine, which leaks through a basket suspended over it. On removal from the butts the fish are water-horsed over night and exposed on the flakes, back or skin side up, for three or four days. By exposing them with the face down the danger of sunburning is removed and flies are less apt to injure them. Pollock cured in this manner are always shipped whole for domestic trade and will keep for only a few weeks.

In the vicinity of Jonesport, Me., a cure somewhat similar to the stockfish is applied to haddock, except that the fish are first lightly corned. The method is simple; the heads and viscera are removed, the bellies cut off, and the fish lightly corned for a few hours. They are then tied together by the tails and suspended over a pole or fence to dry, becoming quite hard and solid within a week or two. Small fish are used, the average weight when dried being from 1 to 2 pounds, and the product, which is very palatable, is entirely for local use. About 4,000 pounds of haddock are annually prepared in this manner at Jonesport, yielding 1,300 pounds of "clubbed haddock," worth $125.

Dunfish is prepared in such a manner that the resulting article has a dun or brownish color. It is of superior quality and is designed especially for use on the table uncooked. The manner of curing is somewhat lengthy, and it requires much more care than curers ordinarily are willing to give to the preparation of fish for market. The fish are usually caught in the winter or spring, and immediately after being landed are split and slack-salted, and then laid in piles for two or three months in a dark storeroom, covered for the greater part of the time with salt hay or eel-grass and pressed with weights. At the end of that time they are dried for a few days in the open air and are again compactly piled in a dark room in the same manner as before, for two or three months, when they are dried for two or three days and are ready for market. The process of preparing dunfish made the Isle of Shoals quite noted a century ago, but has fallen into disuse, though some is prepared there each year.

Drake, in Nooks and Corners of New England, says:

The "dun" or winter fish, formerly cured here, were larger and thicker than the summer fish. Great pains were taken in drying them, the fishermen often covering the "fagots" with bedquilts to keep them clean. Being cured in cold weather, they required but little salt, and were almost transparent when held up to the light. These fish sometimes weighed 100 pounds or more. The dunfish were of great esteem in Spain and in the Mediterranean ports, bringing the highest prices during Lent. They found their way to Madrid, where many a platter, smoking hot, has doubtless graced the table of the Escurial. In 1745 a quintal would sell for a guinea.

The foregoing are the principal features in the curing of codfish on the New England coast. The amount of salt required, the time for exposing the fish, the length of the exposure, and so forth, are points which demand practical knowledge obtainable only through long experience. The process of curing hake, haddock, cusk, and pollock, except as above noted, differs in no particular from that applied to cod. They are dressed and split in exactly the same manner and require about the same amount of salting and similar treatment in every particular. Hake are not so likely to sunburn as cod and need not be protected from the sun. Pollock turn somewhat dark, and for that reason are not popular, but among connoisseurs are highly esteemed, especially when slack-salted.

The loss in weight in dressing and curing cod and other ground fish for the domestic market ranges from 50 to 65 per cent, according to the species, the season of the year, and the extent of the salting and drying. The loss is greatest in case of haddock and cod and least in curing cusk and hake. Generally, large fish decrease more than small ones and large Shore decrease more than large Georges. From a number of records made during different seasons, the following summary is obtained, showing the average quantity of each kind of fish required to make a gross quintal (114 pounds) of dried fish suited for the New England markets:

Species	Pounds required to make a quintal cured.		
	Round.	From the knife.	From the butt.
	Pounds.	Pounds.	Pounds.
Haddock	299	206	133
Cod	288	194	131
Pollock	280	184	130
Hake	258	190	131
Cusk	246	178	132

Fresh split cod ready for curing contains about 80 per cent of water and $1\frac{1}{4}$ per cent of salt. A large percentage of this water is withdrawn by salting, some by drying, and a much smaller quantity by compression, the latter process also removing a small quantity of the salt in the form of pickle. The resulting product, when prepared for the domestic trade, contains about 51 per cent of water and $19\frac{1}{2}$ per cent of salt. The stockfish of Norway contains about 17 per cent of water and $1\frac{1}{4}$ per cent of salt. 100 pounds of cod, as they come from the water, will weigh about 66.9 pounds, dressed ready for salting, of which about 53 pounds represent water and 1 pound represents the weight of salt. The process of curing for domestic trade adds about 6.2 pounds of salt and removes about 34.1 pounds of water, of which 31.1 pounds are removed by the salting and 3 pounds by the pressing and drying. This results in 38.8 pounds of dry-salted fish, of which 18.9 pounds represents water and 7.2 pounds salt. By continuing the drying process and removing more water the keeping qualities of the fish are improved, but since it decreases the quality of the flavor as well as the weight of the fish, and adds to the cost of curing, it is not desirable, unless the fish are to be shipped to a warm climate and held there for a long time. To make a quintal of domestic-cured codfish requires 193 pounds of split fish or 288 pounds of round fish, whereas to make an equal quantity of fish suitable for export to Brazil requires about 350 pounds split, and for 114 pounds of Norway stockfish about 474 pounds of split fish or 708 pounds of round fish are required.

The cost of the labor, salt, etc., varies in accordance with the fish being handled, the condition of the weather, the amount of drying required, the facilities for handling the fish, etc., but generally runs from 38 to 50 cents per quintal, of which 15 to

18 cents represents the cost of the salt. It usually costs more to cure fish in July and August than in October, because of the greater difficulty in drying and consequently the increase in number of times that the fish have to be handled. An examination of a number of itemized accounts shows the average cost of handling cod from the round to the cured product to be 43 cents per quintal; haddock, 46 cents; hake, 40 cents; pollock, 43 cents; cusk, 43 cents. If the fish have been salted on the vessel the cost of handling ashore will be reduced by the labor required for dressing, splitting, and salting, and by the decrease in amount of salt used, and ranges between 28 and 38 cents per quintal. When green split cod costs 1¾ cents per pound, a quantity sufficient to make a quintal gross (114 pounds) would cost $3.84, and the cost of handling averaging 43 cents per quintal, it is necessary to sell the cured product at $4.27 per quintal gross to clear expenses. This cost of labor is so small, compared with the original cost of the fish, that it pays to take the utmost care in the process of curing.

The principal grades of dry fish are Georges cod, Shore cod, Grand Banks cod, hake, cusk, haddock, and pollock. Each grade of cod is further divided into large, medium, and small. Georges cod are generally the largest and choicest received, and are taken on Georges Bank, South Channel, Browns Bank, and adjacent fishing-grounds. These fish are usually heavily salted and dried only a day or so. The Grand Banks cod or "Bank cod" are taken on Grand and Western banks and Banquereau and are usually dried longer than the Georges or Shore cod. During recent years cusk have been divided into two grades, large and small, the former comprising all over 19 inches in length as received from the vessel. The prices of codfish vary according to the conditions in which they are sold, and probably the best guide to the comparative values of the different species may be obtained from an examination of the prices received as they are landed from the vessels.

The following shows the prices per quintal of the principal grades of salt fish on the Boston and Gloucester markets in January, April, July, and October, 1898:

Designation.	January.	April.	July.	October.
Georges cod, large	$6.25	$5.50 to $5.75	$5.50	$5.75 to $6.00
Georges cod, medium	4.00	4.00	3.75	3.50 to 3.75
Shore cod, large	5.00 to $5.25	5.00 to 5.25	4.50	4.50
Shore cod, medium	3.50	3.50	3.00	3.00 to 3.25
Bank cod, large	4.00 to 4.25	4.00 to 4.25	4.00 to $4.25	4.00 to 4.25
Bank cod, medium	3.50	3.50	3.25 to 3.50	3.50 to 3.75
Hake	2.25 to 2.37½	2.25 to 2.50	2.00 to 2.25	1.75 to 2.00
Haddock	2.50	2.00 to 2.25	2.00 to 2.25	2.00 to 2.25
Cusk	3.50	3.50	3.25 to 3.50	3.50 to 3.62½
Pollock	2.75 to 3.00	2.75	2.75 to 3.00	2.50 to 3.00

For the local market or nearby trade the whole fish are packed in rough bundles of one quintal, or 112 pounds, each, and tied with cords, or in wooden boxes holding from 100 to 450 pounds each. The 450-pound boxes are 46 inches long, 22 inches wide, and 16 inches deep, inside measurement. At Gloucester, Boston, Vinal Haven, and Portland large quantities are prepared as boneless cod (see pp. 400-405). In packing for export trade the fish are placed principally in drums made of birch staves, with ends of pine and 8 hoops on each drum, and with capacity for 1, 2, 4, or 8 quintals, tightly compressed.

The curing of codfish on the Pacific coast of the United States began in 1864 and has been continued with more or less success up to the present time, the annual yield now amounting to over 5,000,000 pounds, at an average value of about 3¼ cents per pound, most of the catch being made by vessels sailing from San Francisco and operating in Bering Sea or at fishing stations on the islands bordering that sea. The methods pursued in the curing are not dissimilar to those in vogue on the New England coast. A small percentage are marketed hard dried with the skin on and the bones left in, being tied up in bundles of 75 to 100 pounds; but most of them are prepared as boneless fish after their receipt at San Francisco. The product is marketed throughout the Pacific coast and exported to Australia, the Hawaiian Islands, and the Orient. According to Mr. Wilcox, Pacific coast boneless codfish has been most favorably received in Australia, where it has nearly driven the hard-cured stockfish of northern Europe from the market. Large quantities of dried codfish from New England are shipped to the Pacific coast, especially to San Francisco.

REDDENING OF SALTED CODFISH.

Considerable trouble and loss have resulted to the dried codfish trade from the tendency of the prepared fish to turn red some time after it has been dried. It is especially noticeable with fish that have remained in the hold of the vessel for a long time, and occurs but to a limited extent with fish brought in on ice and then cured. It is attributed to various causes, among which are the removal of the gluten by pressure and the oil becoming partly rancid through age, but the most generally accepted theory is that it is due to vegetable organisms in the salt, especially in that produced by the evaporation of sea water by solar heat.

At the instance of the United States Fish Commission, in 1878, Prof. W. G. Farlow, of Harvard University, made an investigation of the cause of this color and the means of remedying it. He attributed the trouble to a minute plant (*Clathrocystis roseopersicina*), consisting of minute cells filled with red coloring matter. This plant was found on the floors and walls of the packing houses, and also in the holds of some of the vessels. It exists to a considerable extent in Cadiz salt, but not in Trapani salt, and when the latter is used the discoloration is not so likely to result. Consequently Trapani salt has almost entirely superseded the use of Cadiz salt in curing codfish. But even when Trapani salt is used the fish is likely to turn red, and in order to destroy the organisms the buildings are usually whitewashed inside as well as outside at least once a year. To overcome this difficulty it has been recommended that boracic acid be added to the pickle in the proportion of not less than 3 per cent of the water used.

With the view to counteract this reddish tendency in cured fish, a method was patented * in 1883 by R. S. Jennings, of Baltimore, Md., by which the salted fish is subjected to the action of superheated steam or hot air to destroy the organic life in the salt with which the fish have been cured. He employed an endless woven-wire apron hung on rollers and having within it a narrow box or pipe with a perforated top. Into this box or pipe air or steam heated to a temperature of 400° or 450° F. is forced and discharged from it against the fish placed on the endless apron, the apron being revolved at such speed as will expose each fish to the action of the heat for about two seconds. It is claimed that by this method the exterior of the fish is heated sufficiently to destroy the germs without injuring the appearance or qualities of the fish, but the process has never been adopted by the trade.

---

* See Letters Patent No. 273074, dated February 27, 1883.

Many curers endeavor to prevent this redness by sprinkling a small quantity of boracic acid and common salt over the fish. About 1 pound of the acid is used to 40 pounds of fish. This article was introduced in the codfish business about 1881 and is now quite generally used, particularly during the warm months, when it is found almost essential in order to keep dry fish in good condition for a few weeks or even days. It is generally employed in the form of a proprietary compound.

In discussing the cause of the discoloration of Pacific coast codfish a prominent fish-curer of New England states:

> The Alaska codfish turn off-color for precisely the same reason that our Grand Banks codfish do, when caught on long trips. The fish are piled very high in kenches, and the pressure of the upper tiers of fish crushes the fibers and sacs between the fibers and all the white gluten is pressed out. In a nutshell, the cause of codfish turning yellow is pressure. Our Georges codfish are caught by vessels which are very seldom out over a month, and a fare of 30,000 pounds in that time is considered a good catch. These fish are all caught with hand lines. They are never piled high in the kenches, as there is no need of it, for there is plenty of room in the vessel for 30,000 pounds without piling them high, and so they are put into very shallow kenches. They hold their weight, which is a saving to the fishermen, and they hold their color simply because the fibers and sacs which hold the white gluten are never crushed. Our Grand Banks fleet are gone from Gloucester all the way from two to four months, and if they have hard luck are sometimes gone six months. If one of them comes in with a full fare in two months, her fish will be very much whiter on the average than one that has been out twice or three times as long. The fish may be piled up to the deck of the vessel and the fish in the lower part of the kenches may be pressed very hard and the fibers and sacs crushed and then fish will turn yellow; but if she stays out twice or three times as long they will be yellower still, because they have been pressed so much longer. But the greater part of her fish, viz, the latest ones that she caught and the tops of the kenches, will be white. The nearer the top the whiter the fish; the nearer the bottom the yellower the fish, and all due entirely to the amount of pressure they have received. I understand that the vessels that fish on the Alaska fishing-grounds are very large vessels, brigs, or barks, and they fish six months and bring in fares of half a million pounds or more and they are pressed too much. The only way that Alaska codfish can be made to hold their color is to send smaller vessels and bring smaller fares. The smaller the fare, if caught quickly, the whiter the fish. The longer the trip and the larger the fare, the yellower the fish. If you will examine transverse sections of Georges codfish and Grand Banks codfish from the top to the bottom of the kenches with a microscope you will notice that the fibers of the Georges are full of white gluten while those of the Bank codfish are crushed and flattened down, and there will be a variation in the fish according to the part of the kench they come out of. Grand Banks codfish and Alaskan, I think, are caught on trawls and they struggle on the bottom; while they are worrying on the hook and struggling to escape, the blood settles in against the skin. They may be very white on the face for all that, if they are well washed and soaked before salting.

PREPARATION OF BONELESS CODFISH.

The preparation of boneless codfish is doubtless the most important development in the handling of dried fish during the present century. For several years prior to 1870 the need was felt of some method of packing dried fish in neat packages of small but definite weight. A number of processes were devised and patented, but very few of them were found of practical value.

In 1868 William D. Cutler, of Philadelphia, Pa., patented* a process by which the fish were divested of skins and bones and run through a machine adapted to grinding, so as to thoroughly disintegrate the fiber of the fish, and if very fat and oily the disintegrated mass was then subjected to pressure to remove the superabundance of oil. It was then spread upon metal, stone, or other suitable surface, heated by means

---

* Letters Patent No. 81987, dated September 8, 1868.

of steam pipes passing beneath the slabs, where it remained until thoroughly dry, probably from ½ to 3 hours, when it was placed in close paper or wooden boxes, each containing 1 pound or other suitable quantity. The product was somewhat similar to the article prepared in Norway from stockfish and sold as "fish meal." Several thousand pounds of codfish were prepared in this manner and sold under the name "desiccated fish." This method was expensive, and the article lacked preservative qualities, being affected by atmospheric conditions to such an extent as to impair its food qualities; yet while the process extended little beyond the experimental stages, it was sufficient to attract the general notice of the trade and encourage the invention of methods of preparing a similar article.

A few months later Elisha Crowell, of New York City, invented * a process that differs little from the present method of preparing boneless codfish.

The following description is given by Mr. Crowell:

The object of this invention is to so prepare cod and other fish that it shall be divested of everything not edible which unnecessarily adds to its weight and bulk, and shall be reduced to the most convenient form for handling and transportation, while at the same time it is sufficiently protected from the action of the air. The usual method of preparing such fish heretofore employed consists simply in salting and drying the fish in large pieces, each piece being generally one-half or the whole of a fish. In this condition it can not be conveniently packed in small boxes, and is therefore exposed to atmospheric influences which injure its quality and taste. From the same cause it is not in a convenient condition for transportation or handling, and the refuse portions add unnecessarily to its weight, while also deteriorating the quality of the article as an article of sale and common use. To obviate these disadvantages one manufacturer has ground up the fish, but when treated thus the air reaches directly every fiber of the fish and soon destroys its taste, besides drying it up to such a degree that it becomes hard and "stringy" and after a time almost unfit for use. It can also be easily adulterated, either with foreign substances or with the ground skin and bones or fish improperly cured. In order to overcome all these disadvantages and produce an article which shall possess and retain all the delicate flavor of the codfish, while entirely clear of useless matter, and in the most convenient possible condition for transportation, I remove the bones and skin, either before or after salting, and then cut up the fish into long, narrow strips. These strips I expose to the drying action of a current of air either naturally or artificially induced, so as to remove the moisture from the fish sufficiently for its preservation. The fish may be cut up or stripped still more between one drying operation and the next. Salt is not usually applied during the operation. The strips thus produced are then cut into suitable lengths and packed in boxes, kegs, or barrels to exclude the atmospheric influences as far as possible. The retailer can pack the article in small boxes containing half a pound or a pound, etc., for the convenience of himself and his customers.

In 1869 Benjamin F. Stephens, of Brooklyn, N. Y., patented† a modification of the process invented by William D. Cutler during the preceding year, this being a saturation of the compressed and granulated fish with glycerin to keep it moist and prevent extreme dryness. During the same year Joseph Nickerson, of Boothbay, Me., introduced a somewhat similar process,‡ consisting in removing the skins, bones, etc., from the salted fish, reducing the flesh to a granulated state, and then steeping it in brine until every particle of the mass was completely penetrated by the brine, after which it was pressed into molds, the pressure serving the double purpose of forcing out all surplus moisture and reducing the fish to hardened cakes of convenient size. Neither of these two processes has ever been used to any noticeable extent.

---

* Letters Patent No. 84801, dated December 8, 1868.
† Letters Patent No. 87986, dated March 16, 1869.
‡ Letters Patent No. 88061, dated March 23, 1869.

A somewhat novel idea is set forth in Letters Patent No. 90334, granted on May 25, 1869, to John Atwood, of Provincetown, Mass. He states:

The object of my invention is to produce a wholesome article of food in the nature of prepared fish that shall be reasonable in price, convenient for cooking, and free from offensive odor. The old method of drying fish is well known, and is open to many apparent objections. The new method of preparing fish by desiccation is so expensive as to make the price of the article to consumers a serious objection to the method. The cause of ordinarily cured fish having an offensive odor at all times, but more especially when the atmosphere is moist, is the mucous membrane between the skin and the flesh, which, when dried and afterwards moistened, becomes slimy and offensive. This is peculiar to the old method of curing and bundling fish. My method of preparing fish, which is particularly applicable to cod and haddock, is as follows: When the fish is fresh I take out the principal bones and fins, the fish remaining whole or split in halves. When partially dried or cured with salt I remove the skin, and with it the entire mucous membrane, the cause of the offensive odor of salt fish. I then pack in light wooden boxes of convenient size—for instance, from 10 to 100 pound boxes. Fish prepared after this method is white, clean, and sweet, and will keep for any length of time. It will not dry up and lose its flavor like the desiccated article, but remains moist and keeps the palatable flavor of freshly cured fish. It can be freshened for the table in a few minutes, and can be brought on whole, as it is often desired to do so. This can not be done by fish cured by the old method, nor by any other modern process.

It appears that these numerous patents were obtained all within a period of nine months, and it is claimed that about the same time other persons prepared dried fish, stripped of skins and bones and packed in small boxes, without applying for patents. At first only the inferior grades of fish were used, but as the new article met with a ready sale greater attention was given to the quality of the preparation. In 1870 three fish-dealers in Gloucester prepared this article, besides several concerns in other parts of New England. The trade increased considerably, and in 1875 over 500,000 pounds of boneless fish were prepared in Gloucester alone. Nearly all of these fish were prepared in a manner somewhat similar to that of Elisha Crowell, and at first the preparers paid a royalty to him. Becoming dissatisfied with certain discriminations made by Crowell in favor of particular firms, several dealers in Boston successfully contested his right to the royalty. Almost immediately the business assumed large proportions, and in 1879 about 12,000,000 pounds of boneless fish were prepared in Gloucester, and over 6,000,000 pounds in other New England ports, giving employment to nearly 400 persons. At present the output of boneless cod amounts to about 25,000,000 pounds annually, most of which is prepared in Gloucester, the remainder being put up at Boston, Provincetown, Portland, Vinalhaven, etc.

The general process of preparing boneless fish is as follows: From the flake yard the dried fish go to the "skinning loft." In skinning, each fish is placed flesh side down on the skinning board, consisting of a soft pine block about 30 inches long, 20 inches wide, and 2 or 3 inches thick, or of an inch pine board of similar length and width resting on two end supports. The dorsal, anal, and ventral fins are first cut away with a knife much like a splitting knife; then grasping the skin at the napes the workman strips it off, usually in two pieces. The nape bone is sometimes torn out in the operation of removing the skin, but generally it is removed with a small iron gaff called a "bone hooker," which is about 8 inches long, with a curved shank and sharp point. The workman then turns the fish flesh up and cutting under the lower end of backbone or tail bone removes it. The dark membrane is then torn from the napes and any dark portion of the flesh cut away. Sometimes in dressing very choice fish a workman removes all the ribs and other small bones, making what is known as "absolutely boneless" fish. And in boning hake and other small and cheap fish the tail bone is frequently left in, and sometimes the nape bones also, this being known as "dressed fish." Hake and haddock are the easiest fish to prepare, and cusk are the most difficult. The cost of skinning and boning ranges from 25 to 40 cents per 100 pounds of prepared fish, depending on the class of fish handled.

The ordinary knives for cutting boneless fish have hook-tipped blades from 6 to 7 inches in length, with white pine handles. For cutting cusk, which have tough bones and skins, a special knife is required, called a "cusk-bone knife," the blade of which is of finely tempered steel, about 1¾ inches long, ¾ inch wide, and ¹⁄₁₆ inch thick at the back, with a square end. The handle is about 4½ inches long by 1½ inches thick at the butt, tapering to a point at the blade end.

The following summary shows the result (in pounds) from skinning and boning a quintal of the various grades of fish:

Condition.	Large Georges Cod.	Large Bank or Shore Cod.	Small Georges Cod.	Small Bank or Shore Cod.	Haddock.	Hake.	Cusk.	Pollock.
	Pounds.	Pounds.	Pounds.	Pounds.	Pounds.	Pounds.	Pounds.	Pounds.
Tail and nape bones in.......	96	95	91	90	90	91	....	97
Tail bone in, nape bones out ..	93	92	85	85	82	85	....	94
Tail and nape bones out.....	88	87	80	82	78	82	82	....

After being skinned and boned the fish are sprinkled with an antiseptic powder composed principally of boracic acid and chloride of sodium or common salt, and placed in many sizes and styles of packages, containing from 500 pounds down to 2 pounds. The boxes are made of spruce or pine, and the small ones, 5 pounds and under, usually have a sliding cover. The most popular sizes are 40 and 60 pound boxes, the dimensions of the former being usually 20 inches by 12 inches by 5 inches, inside measurement, and of the latter 20 inches by 12 inches by 8 inches, inside. In the larger-size boxes the fish are usually placed without being cut. A neat way is to place two halves together, as in the round fish. Others are loosely rolled and placed with the shoulders at the ends of the box and the tails overlapping, and choice Georges fish look very nice in that manner. Frequently when packed in the 40 or 60 pound boxes each individual fish is cut transversely the width of the box and folded over itself. Thick fish are sometimes cut transversely and each piece split and folded over in such a manner that the clean cut appears outside. The fish are also sometimes cut transversely across the fiber, and tightly packed in boxes with the fiber running perpendicularly. In the small boxes the fish must of course be cut in much smaller pieces. The 5-pound boxes usually measure 10 inches by 8 inches by 4 inches.

During the past fifteen years the packing of boneless codfish in 1-pound and 2-pound "bricks" has become very popular. The skinned and boned fish are cut into small pieces 6 inches long and 3 inches wide, as nearly as practicable. the cutting being done either by hand, by treadknives, or by special machinery. Two pounds weight of these are carefully placed in press compartments 6 inches long by 3 inches wide and 3½ inches deep, care being taken to have choice square pieces at the bottom and at the top, and either two or four strings of cotton twine are run through slits in the compartment, so as to pass under and around the brick of fish. The fish are then tightly compressed for a few moments, and on removing the compression the strings are tied and the brick is removed. Many forms of presses are employed, the most usual consisting of a sliding box having two or three compartments, each of the size noted, and so arranged that a hand or foot lever forces a block down in one compartment at a time. The pressure remains while the fish are being placed in the second compartment, and when it is released the box is slid along until the second compartment comes under the press, when the brick is removed. When 1-pound bricks are desired, the 2-pound packages are cut in half. The bricks are then sprinkled with antiseptic powder, wrapped in parchment or waxed paper, and placed in the packing boxes.

A pound of parchment, costing 14 cents, contains about 172 sheets of the size necessary for 1-pound bricks, and 113 sheets of the size necessary for 2-pound bricks, thus making it cost 8.14 cents to wrap 100 pounds of the former and 6.2 cents for 100 pounds of the latter size. One ream, or 4 pounds, of waxed paper for 1-pound bricks costs 30 cents, and 1 ream for 2-pound bricks costs 40 cents, making the cost of using wax paper 6.2 and 4.2 cents, respectively, for 100 pounds of 1 and 2 pound bricks.

Cutting into bricks was greatly facilitated in 1885 by providing a cutting board with pins at stated intervals to hold the fish when pressed down by hand, and with two sets of parallel grooves at right angles to each other cut into the board sufficiently deep to give direction to a knife which is drawn through the fish, these longitudinal and transverse grooves being separated by uniform distances conforming to the size of the bricks. In 1886 a somewhat intricate machine* was introduced at Gloucester for this purpose, and is now used in one or two of the establishments. It consists of a large rotating drum, the surface of which is provided with pins which enter the fish placed thereon and thereby hold them in position. On this the fish are fed, and the drum revolves intermittently, and at regular intervals a knife located above and parallel with its axis descends to cut the fish transversely, the drum rotating intermittently to permit the knife to cut the fish without being crowded by them. These strips of fish are then carried forward by the drum beneath a series of rotary knives mounted upon a shaft, the axis of rotation of which is parallel with the axis of rotation of the drum, the strips being thereby severed into blocks. The length of the bricks is determined by the distance covered by the fish between each descent of the vertical knife, and the distance between the rotary knives determines their width. The pieces of fish are then carried forward beyond the rotary knives and are removed from the pegs by suitable strips or rods entering grooves in the surface of the drum and thus coming between it and the blocks of fish.

A few months thereafter another machine† was introduced for the same purpose, but was never extensively used. This consisted of two revolving drums carrying a platform made in sections with longitudinal grooves, having small pegs in its surface to hold the fish placed upon it, skin side down, and so revolving as to press the fish against circular knives placed at suitable distances apart. The knives were rotated by the motion of the fish, and the latter were cut into longitudinal strips equal in width to the distance between the knives. As these strips passed beyond the knives they were raised off the pins and the platform by rods entering between the strips and the surface of the endless platform. A quantity of fish having been thus cut into longitudinal strips, they were returned to the front end of the machine and by means of a guide were again placed on the movable platform so as to approach the rotating knives at right angles, and as they passed under were cut transversely, forming rectangular blocks. The circular knives were so arranged that alternate ones might

* See Letters Patent No. 346871, dated August 3, 1886, in favor of J. L. Shute and W. O. Taylor.
† See Letters Patent No. 356725, dated January 25, 1887, in favor of Walter S. Moses.

be easily raised for the second cutting, making the length of the blocks of fish double their width.

During 1885 a machine* was devised for splitting or cutting these blocks horizontally, so as to provide suitable layers for the tops and bottoms of the bricks and to give a smooth, regular appearance more acceptable to the trade. This consisted of an endless belt adapted to hold and carry the blocks of fish to an endless cutting ribbon traveling parallel to and an inch or so over the belt, but it was not found sufficiently practical for general use.

The cost of preparing and packing boneless cod in 1-pound bricks is about $2 per 100 pounds, aside from the cost of the cured fish, apportioned as follows: Skinning and cutting, 30 cents; labor at press, 32 cents; wrapping paper, 6 cents; antiseptic powder, 10 cents; boxes, 65 cents; labor for cutting, powdering, and wrapping, 10 cents; miscellaneous labor, 10 cents, and plant and superintendence, 37 cents. The refuse skins and bones are used in the preparation of fish glue and fertilizer, and their sale constitutes an item of considerable importance.

It is stated that of the total quantity of boneless fish, an average of 60 per cent is prepared from cod, 28 per cent from hake, 8 per cent from haddock, and 4 per cent from cusk. Pollock are sometimes prepared as boneless fish, but the flesh is rather dark for this purpose.

During the past six or eight years dried fish have been disintegrated and placed on the market under a number of trade names, such as "desiccated codfish," "fibered codfish," "flaked codfish," and "skriggled codfish." In preparing these specialties the fish are dried somewhat more than in case of boneless fish and all the bones are removed, a quintal of fish as it leaves the butts making about 60 pounds of dried fish for this purpose. By means of a disintegrating or shredding machine the fibers of the flesh are thoroughly separated, giving it the appearance of fine wool. This is spread out under cover an inch or two deep on a platform table for further drying, all dark portions being picked out in the meantime. It is then placed in small pasteboard boxes, usually coated with a varnish of rosin or paraffin and sometimes lined with waxed paper or parchment, each box holding usually half a pound. This product is especially desirable for fish balls and creamed codfish.

In 1885 a process† was introduced by which the disintegrated codfish was subjected to the action of hot water and then formed into cakes or blocks under pressure, in the following manner:

Take cured or salted fish, remove the skin and the bones thoroughly, and then disintegrate the flesh by shredding, grinding, or some other convenient way of reducing it to small pieces. When in this condition, apply heated water to it, and immediately thereafter submit it to sufficient pressure in molds to expel the water and compact the fish and press the small pieces closely together, thereby forming the mass into cakes or blocks, the size of which can be regulated as desired, from 1 pound upward, by the size of the molds employed. Subjecting the disintegrated fish to the action of hot water sufficiently dissolves the gelatin in the fibers to cause the small pieces of flesh to adhere to one another when they are firmly pressed together. Water heated to any temperature above 100 will produce the result, and even steam may be used; but it has been found that the most successful and satisfactory results are obtained by the use of water heated to a temperature between 120 and 200°. If steam is used, the fish will be partially cooked thereby, which should be avoided.

It is claimed that when fish has been treated and prepared in this manner, the salt will not collect upon the outside of the cakes or blocks, as it does upon salt fish prepared by the methods in general use, and that the fish will not become discolored.

---

* See Letters Patent No. 317469, dated May 5, 1885, in favor of Walter S. Moses and Oscar Audrews.
† See Letters Patent No. 326099, dated September 15, 1885.

FOREIGN CODFISH MARKETS.

The world's annual product of dried codfish now amounts to about 600,000,000 pounds, cured weight, the equivalent of 2,500,000,000 pounds of round fish, obtained principally by the fishermen of Norway, Newfoundland, Canada, United States, and France. The chief markets are France, Spain, Portugal, Italy, and Brazil. In 1893 France imported 94,218,948 pounds of codfish, valued at $4,949,037, without counting the large quantity cured by fishermen of that country. During the same year Spain imported 97,811,488 pounds, worth $4.795,278, and Portugal, 43,126,385 pounds, worth $1,789,560. The imports into southern Europe are principally from Norway and Newfoundland; those into Brazil and other South American countries are largely from Newfoundland, and the West India trade is almost monopolized by shipments from Canada. While a steadily increasing export trade has been conducted by Norway, Newfoundland, and Canada, especially with the West Indies and Central and South America, the exports from the United States have very greatly decreased.

A hundred years ago our exports of cod approximated 500,000 quintals annually, at an average value of $4.50 per quintal. In 1804 the exports were 567,825 quintals, worth $2,400,000, the largest quantity ever exported from this country in any one year. The annual exports decreased to about 300,000 quintals during the ten years following the war of 1812, and since that time up to the present they have approximated about 130,000 quintals annually. The exports during the ten years ending June 30, 1894, averaged 16,260,008 pounds, worth $737,084, annually, or 20 per cent of the total quantity cured. Over half of these were sent to Haiti, and much smaller quantities went to Cuba, Dutch and French Guianas, Colombia, Santo Domingo, Jamaica, and various other countries, and especially to ports in the West Indies and South America.

It thus appears that at present this country has only a small share of the trade in the principal codfish markets. Brazil, for instance, consumes about 500,000 quintals of fish annually, of which the United States supplies less than 2 per cent; and none whatever are sent from this country to the Catholic countries of southern Europe, the great fish-markets of the world. A century ago a large part of that trade was controlled by the United States, but since the domestic market will receive fish containing 50 per cent of moisture, while the Brazilian trade requires fish containing less than 25 per cent, greater profit has been found in supplying the home market, and nearly all the curers have contributed to that trade, resulting in a decrease in exports.

CODFISH CURING IN FOREIGN COUNTRIES.

In the British North American Provinces the codfish are cured in nearly the same manner as on the New England coast, except that they are dried much more thoroughly, and in many instances they are not salted in butts, but are spread on the flakes immediately after removal from the kenches. Each morning they are spread out, flesh side up, and at night they are gathered in piles of 15 or 20, with skin side up, and with the largest on top as a cover to the rest. If the sun becomes too hot during the middle of the day, they are turned with the flesh side down to prevent their being burned, but as soon as the great heat is over they are reexposed as before. When the fish are sufficiently dry, large piles are made, containing a ton or more, the whole being covered with birch bark and heavy stones, which serve to express much of the moisture then remaining. After compression for two or three weeks the fish are placed in a dry warehouse awaiting a market. Before being shipped they are spread out on the gravel during one midday to extract any dampness they may have contracted in the warehouse.

For the purpose of comparison with our own methods the following notes on the methods of curing codfish in the principal European countries are presented, the notes being abridged by Adolph Nielsen from volume III of Norsk Fiskeritidende, Bergen, January and April, 1884:

NORWEGIAN METHOD.

As a rule the greater part of the codfish caught at Lofoden is left in salt from three weeks to two months, all according to how the fishery turns out, and how quick the vessels purchasing fish can succeed in getting a full cargo. After the fish are taken from the salt they are generally washed out at the beaches, close to the drying places, which, as a rule, consist of smooth and low rock, in the vicinity of the seaboard. In washing the fish woolen mittens are worn on the hands. After being carefully washed and the black membrane removed from the napes, the fish is put in small sloping heaps on the rocks for twenty-four hours, in order to allow the water to run off. In each heap are put from 6 to 8 fish, the undermost with the skin side turned down, the rest with the skin side up. As soon as the weather allows the fish to be spread, after being in the heaps twenty-four hours, it is carried up to the drying place and spread out, face up. If the weather is fair and safe, the fish is left out the first night, but the skin side is in that case turned up toward evening. Next morning the fish is again turned, face up. After being left out the second day it is gathered together toward evening and put in heaps, 30 or 50 fish in each. The next morning it is spread again, and in the evening is put in a little larger heaps. When the fish has been spread two or three times it is stretched well, especially in the abdomen, before being put in heaps, in order to remove all the wrinkles and give the fish a smooth appearance. This work is considered to be of much importance in regard not only to appearance but also to the durability of the fish, because the dampness always gathers in those wrinkles and is very difficult to get removed entirely if not done away with in time. It is slow work, but they consider it better to devote one day to this than to go through it in a hurry in the evening when the fish are gathered in heaps for the night. Every time the fish are gathered in the evening in heaps these are made larger. After the fish has been spread three or four times, or when it is dry enough to stand pressing (which is noticed on the abdomen of the fish that crack when the fish is bent), it is put into the first pile for pressing. These piles are built round, and a small round peaked roof or cover of wood, about a foot larger in diameter than the pile of fish, is made to cover the piles with. On these roofs weights of stones are applied. These piles or pressing piles, as they generally are named, are built from 3 to 3½ feet high, the first time.

After the fish has remained in those piles from five to eight days, according as the fish was more or less dry when it was put in piles, it is piled over into another and larger pile in this way, that the undermost fish in this first pile is placed uppermost in the second, in which again the fish is left the same length of time as in the first one. If the weather after that time is suitable, the fish in these piles is spread every second day to dry, and for every time it is spread it is set in larger piles. If the weather is not suitable for spreading the fish, it is as often as possible piled over into new piles, in order to accelerate the cure and prevent the fish from afterwards turning slimy. The fish is not reckoned to be properly dry until it keeps itself dry underneath the dorsal fins, or is capable of withstanding the pressure of the thumb without leaving marks in the thick of the flesh. After the fish has been put in pressing piles and has been spread out for drying four to five times, it will, under fair circumstances, be reckoned to be properly cured. The usual time, under favorable conditions, taken to cure fish in Norway is about six weeks.

FRENCH METHOD.

The fish which is brought to France is for the greater part bank fish, caught on the banks of Newfoundland and on the coast of Ireland. With the exception of the fish cured in St. Pierre and Miquelon, all this fish is cured in France, and the greater part of it in Bordeaux. To this port it is brought salted in bulk, in compartments in the vessel's hold, and cured as the orders arrive for certain quantities of fish. In splitting the fish the French cut the backbone a little farther from the tail than most nations do, and for this reason an iron spoon made for the purpose is used for removing the blood in the remaining part of the backbone. The fish is always washed well before it is put in salt. They generally reckon on using 100 tons of salt (Mediterranean) to 2,000 hundredweight of fish, including the salt which is used in preserving their bait. While all other nations either use rock benches, or different kinds of flakes to cure their fish on, the Frenchmen in Bordeaux use scaffolds, on which the fish is hung by the tail. This is held to be the most practical in France, because in this way the fish is cured exceedingly quick (from two to six days) with a minimum of labor expenses, and gives a real good article, although, as before mentioned, not so durable. These scaffolds are made in the following way: A number of sticks are driven down perpendicular in the ground about 1½ yards apart in a straight line from west to east, across which are fastened a row of two laths, far enough apart to admit the tail of the fish to be pushed through. About seven-eighths of a yard above this first row of laths another row is fixed in the same manner. These laths are from three-eighths to five-eighths of an inch thick and from 1½ to 2 inches wide. The arrangements of these scaffolds vary a little; some are fixed in square compartments with laths fastened alternately on the north and south sides of the posts, and with a gangway about 2 feet wide between each row of compartments. On others, again, the laths are all fastened on the north side of the posts and each row of scaffolds about 3 yards apart, in order to prevent the shade from the row in front reaching the one behind. Small cleats of wood are fastened across each pair of laths, for the purpose of keeping them together. The tail of the fish is pushed in between the laths from the north side, with the back of the fish turned upward; by its own weight it will bend down, and the face of the fish show toward the sun, while the tail is jammed between the laths. When the fish is getting a little dry it will hang this way even in a strong breeze of wind. Some have a roof covered with straw over their scaffolds when the sun gets rather hot; others again use no covering, but when the sun threatens to burn their fish they only twist it a little, so that the edge of the fish shows toward the sun instead of the face. In heavy rain, or when the sun is too hot, the fish have to be taken down and put in the stores. In France no pressing of the fish is used; as a rule it is taken down from the scaffolds and shipped to markets after being hung there from two to six days.

SCOTCH METHOD.

In Scotland nearly all of the fish is pickled, very little of it is kench-cured fish. As soon as the fish is caught and unhooked it is bled and gutted. Some of the fishermen bring with them boxes in which to keep the fish; but if boxes are not used the fish is covered up in order to prevent the sun and air from affecting it. When brought to the shore it is headed and cleansed with brushes in fresh water, and split. The backbone is cut slantwise, 20 to 22 joints from the tail, so that the cut extends over two joints, in order to give the fish a better look and strengthen it. A cut is made along the bone which is left, thus allowing the blood that remains in the veins about that part of the fish to escape or be extracted. When split, the fish is again washed in sea water and the black membrane removed. After the fish is cleansed it is pickle salted in tight-covered vessels. To 100 pounds of dry-cured fish is used 45 to 50 pounds of Liverpool salt. If less salt is used the fish is left in the pickle a couple of days more.

In regard to the amount of salt used, the board of fisheries remark that many of the curers salt their fish very heavily in order to increase the weight; but this is a great mistake, because not only is the juice of the fish extracted thereby and the weight reduced, but also, as the drying advances, incrustation of salt forms on the face of the fish, or, in other words, the fish gets salt-burned, and this debases the value of the fish very much. Lately, however, this wrong method of salting fish too heavily is abandoned in Scotland, because the curers find it to be in their own interest not to use too much. The fish is as a rule left in the salt for three days. The salting of the fish is a difficult work when the climate is damp. If the fish under such circumstances gets too little salt it will soon become dun. Whether the fish has taken sufficient salt or not, and whether it has got the required stiffness before being taken from the brine, the curer must be able to judge himself. Quite fresh fish never take more than just the proper amount of salt they claim, no matter how much salt is put on it; whereas old fish very soon get salt-burnt, if too much salt is used.

When the fish has taken a sufficient quantity of salt it is taken up and washed out again in sea water and placed in piles that slope a little for a day or two, in order to give the water a chance to run off before the drying commences. The fish, as a rule, is dried on flakes 3 feet high and 4 feet wide, the top of which is formed of wooden laths 6 inches apart, something like the American flakes. On these flakes the fish is put out the first time with the back or skin side turned down. Toward evening it is turned over, skin side up, and before sunset it is gathered in small heaps, always bearing in mind to leave the back side of the fish turned up. The fish is turned more frequently as the drying advances. The heaps are also made larger every time the fish is spread, and weights are put on top of each heap in order to give the fish a little pressing and a smooth face. The heaps are always covered with mats or canvas. The Scottish curers are also of the opinion that the fish gets frangible or brittle and presents a bad appearance if it is dried too rapidly in the beginning of the cure. When the fish is half-dry, one is able to tell whether it is salt-burnt or not. If the fish is salt-burnt they turn it back up in the middle of the day when the sun has the most power, by means of which the salt gets extracted from the face of the fish. Mr. Ross, inspector of fisheries, recommends always to dry the fish with the skin side turned up from the time it is half dry. After it has been dried a fortnight it is put in large piles for ten days in order to allow the fish to sweat. The piles are covered well. After being taken from these piles it is spread out to dry for one week, after which it is again put in large piles for another four to six days. When after this it gets two or three days' drying it is considered to be properly cured and ready for shipment. The average time for making fish is reckoned to be from six to eight weeks.

ICELAND METHOD.

The cod fishery in Iceland is conducted by the natives in small open boats in the bays, and in some places a short distance from the shore. The gear employed by them consists of hand-line and bultows. As soon as the fish is caught it is bled; brought inshore, it is split in this way, that the remainder of the backbone is left on the opposite side to what is usual in Newfoundland and many other countries. The Icelanders split their fish very deep. After being split, the fish is washed with brushes in clean sea water, the black membrane and all blood being carefully removed. A few also used to wash their fish in fresh water. The backbone is cut slantwise, over two joints, and 18 to 22 joints from the tail, according to the size of the fish. The salting of the fish takes place in sheds as soon as the water has run off it, and it is salted in kenches with one barrel of Liverpool salt to about 350 pounds of large dry fish; if the fish is small less salt is used. After the fish has remained two or three days in this salt it is resalted in new kenches; very little salt (about one-eighth of a barrel of salt to 350 pounds of fish) is used. In this salt it remains for five or six days, and is then ready to be washed out and made, if the weather and the season of the year are suitable. The fish that is caught so late in the fall that it can not be made before the next year, is salted in kenches so heavily that one fish does not touch the other. This fish, they claim, will then, in the spring, be of about the same quality as if it was caught the same year, provided it is washed or cleaned properly and all blood carefully removed.

After the fish has remained a sufficient time in salt it is washed out and laid in small heaps, until the water has run off and a little stiffness is felt in the fish, which generally is so the next day, and if the weather then is fair the fish is spread out to dry; if not, it is relaid in square piles, from 100 to 150 fish in each. If the weather should continue to be wet the fish is piled over in new piles every day, as long as the bad weather lasts, or until it can be spread. When the fish has been spread and got two good days' sun, it is put in pressing piles and the pressing is increased according as the making of the fish proceeds. When the cure is so far advanced that the fish is what they call three-parts dry, it is put in large piles, about 7,000 pounds of fish in each. These piles are covered with mats or boards in shape of a roof, and a weight of stones, which corresponds with the weight of the fish in the pile, is placed on top of the mats or boards. In this state the fish is allowed to remain five to six days, after which time it is spread again, if the weather permits, and the same weight applied every time it is gathered and put back into the piles. In case the weather does not allow the fish to be spread, after it has been put in the first large pressing pile, it is repiled every day and the same weight applied to each pile every time until it is considered cured.

The fish is cured on beaches, which in most places are made of round rocks. The reasons why the Icelanders use such a heavy pressing in their cure of fish are: (1) That the climate is damp and not very warm (as a rule the sun is seldom hot enough to burn the fish, although this may happen occasionally) and that the weather is mostly cloudy or foggy; (2) that their fish is heavily salted; (3) that their fish is rich and thick, and stands a good deal of pressing. On account of the climate being chilly and damp, the pressing is therefore the principal part in their cure, and by frequently pressing and repiling the fish the cure is also accelerated. The Iceland fish is a fine, white-looking, good-eating and durable article which commands good prices in the markets of the Mediterranean. Although it always is a little pliable, or not cured as hard as the Newfoundland and Norway fish, still it keeps well in hot climates, and is preferred to the hard-cured fish, because it is not so apt to get brittle and break. Fish cured in the early spring or in the fall of the year when the climate is chilly, are, if sufficiently pressed and salted, superior to the hard-cured fish in summer time, even if it is a little pliable, and will keep well in hot climates. Complaints of the hard-cured fish being brittle and difficult to handle are often made in the Mediterranean markets.

When the fish is stored in Iceland it is kept well covered with mats or canvas in order to prevent the moist air from affecting the fish. The principal market for Iceland fish is the southern part of Spain, while a part is also exported to Copenhagen and Great Britain.

STOCKFISH.

In preparing stockfish in Norway each fish is bled as soon as taken from the water, care being taken that it be not bruised or mutilated. In dressing, the fish is split from the pectoral fin to the vent, thus leaving the sides connected about the napes and near the tail. The head and entrails are removed, after which each fish is well cleansed outside and inside with sea water. They are next tied together by the tails in pairs and suspended from thick strips about 2 feet apart, on top of stands about 8 feet high. The fish are hung on each side of the strips, care being taken that they are not so close as to prevent the air currents from acting upon them. A two-pronged stick is generally used in hanging them up and taking them down. The grounds about the flakes should be clean, airy, and devoid of all putrefying refuse. During summer the fish are generally split quite through, leaving only enough meat and skin at the tail to support its own weight. The backbone is cut out from about three joints below the vent, and the fish hung up individually by the tail in such a way that one half of the fish hangs on each side of the pole. Large fish, over 28 inches in length, are also split during the winter and spring. By far the greater portion of the stockfish, however, are cured round.

The fish receive no further attention, being left out in all sorts of weather, and are not taken down from the time they are hung up until perfectly dry; that is, hard enough to withstand the strongest pressure of the tip of the thumb in the thick of the flesh along the back without giving away. In taking the fish down from the flakes, dry and fair weather should be selected in order not to leave any moisture about the fish when it is stored. In preparing for export, the stockfish is usually pressed and tied by wire into rectangular bundles, 20 by 24 by 29 inches, containing 100 kilograms or about 220 pounds. Bundles of 50 kilograms are also prepared. Norway is the only country in which stockfish is extensively prepared, about 400,000 quintals being cured annually, the species used being cod principally, but also cusk, pollock, haddock, and ling in smaller quantities. The chief markets are in Italy, Spain, Germany, Holland, and the tropical sections of the continents of America.

The Russians prepare stockfish in a somewhat different manner from that of the Norwegians. The fish is split through the back and left solid in the abdomen. A cut about an inch long is made through the uppermost part of the fish, and through this the fish is tied up on the flakes. They also prepare them in a manner similar to the Norwegian split fish, except that they do not usually take out the backbone.

Before cooking, the stockfish should be softened or disintegrated by beating with a wooden club and all bones removed. The flesh is then soaked for several hours, washed and drained. In Italy thin fish are preferred to thick ones and the disintegrated flesh is placed in cold water over a fire and removed before the water reaches the boiling point, since boiling makes it tough. Steaming is even a better method of cooking, making the flesh white and soft.

DRIED AND DRY-SALTED SALMON.

On the Alaskan coast the Indians dry many salmon for their home use during the winter, and also at times prepare cod and other fish, the work being performed mostly by the women. As soon as the salmon is caught the backbone is broken just back of the head, so as to kill the fish at once and prevent its thrashing about and bruising the flesh. The fish is dressed by slivering the two sides from the head and backbone, but leaving them connected at the tail, the knife being inserted just below the nape bone and drawn closely along the backbone to within 2 or 3 inches of the tail, when a similar cut is made on the other side and a stroke of the knife severs the backbone close to the tail. Frequently the large fish are marked by a number of transverse cuts in the thick portion of the flesh to facilitate the drying. The fish are then suspended from a pole or frame a few feet from the ground with the flesh outward, where they remain until quite dry. Sometimes when poles are not conveniently obtainable the two sides of the fish are separated and laid face up on the beach.

The process of drying requires from ten days to two weeks in ordinary weather. During rainy or cloudy weather the fish are placed under cover or turned with the skin outward. After being thoroughly dried these fish are stored under cover out of reach of dogs and children and form the principal food supplies of the natives during the winter, especially in the villages somewhat isolated from the trading stations.

Salmon when salted are commonly held in pickle in tight barrels, but a few are dry-salted for especially choice trade, in the following manner: The fresh fish is placed in a cool place until the flesh is firmly set, when it is eviscerated and split down the back, so as to lay out flat, and the head and three-fourths of the backbone are removed. If desired, the flesh may then be smeared with blood of the fish to impart a reddish color in the cured product; otherwise it is wiped clean and placed skin down in a salting tub having a layer of salt in the bottom and a layer of No. 2 salt spread evenly over the fish. Other fish similarly salted may be placed in the salting tub, but not so many as to compress the bottom layers too much. The flavor is improved by adding sugar or saltpeter to the salt, about an ounce of either for each fish, and crushed peppers may also be added if desired. The fish remain in the salt from $1\frac{1}{2}$ to 3 days, when they are removed, trussed in the manner so usual in smoking (see page 494), in order to keep them out flat, and suspended in a shady but windy place until dry. The fish should be kept in a cool, dry place until used, which need not be for 2 or 3 months, according to the extent of the salting and the adaptability of the place where it is stored. Some persons pour a glass of cognac over the dry fish and allow it to soak in to improve the flavor, but this is a matter of taste. The preparation of salmon in the above manner is of very small extent, and none of the product goes upon the general market.

The first volume of the Transactions of the Highland Society of Scotland describes the method of drying salmon in vogue in Scotland a century ago as follows:

Kippered salmon are prepared by cutting them smoothly along the back from the tail to the head. The chine, or backbone, is then cut out and all the blood and garbage cleared away. The fishes are then salted and laid above each other, with the fleshy sides in contact, in a trough, commonly scooped out of a solid piece of beech, placed in a cool situation. A lid which exactly fits the aperture is placed above them and pressed down by heavy weights. After the fishes have imbibed a sufficient quantity of the pickle they are stretched upon small spars of wood and hung up to dry where there is a current of air. Sometimes they are hung in the smoke of a kitchen fire, which preserves, indeed, but conveys a bad flavor. Some, in order to communicate a particular flavor, mix spices with the salt, or they rub the fish with spices before they are hung up to dry.

DRY-SALTED MULLET.

In the extensive mullet fisheries of the southern coast from North Carolina to Florida large quantities of this species are dry-salted or kench-cured, the annual output on the coast of Florida amounting to nearly 2,000,000 pounds. Some are pickled in brine, but the majority on the west coast of Florida are dry-salted. The process employed is as follows:

Dressing begins as soon as the fish are landed at the station, which is generally within a few minutes after they are removed from the water. In splitting, each fish is taken in the left hand with the tail toward the splitter, and by means of a knife it is opened along the left side of the backbone from the head to the tail, in much the same way that mackerel are split. All viscera are then removed and a gash or "score" is cut along the right side of the fish, which contains the backbone, in order that the salt may the more readily penetrate the flesh. In some localities the heads of mullet are removed before the fish are split. When roe-mullet are taken the roe bags are carefully removed while the fish are being eviscerated and are salted separately. The blood and black stomach membrane adhering to the napes are then scraped away and the fish are thrown into a trough of clean salt water, where they remain for a few minutes and are thoroughly washed, all particles of blood being carefully removed. On removal from the washing tank or barrel the fish are rubbed thoroughly with salt, Liverpool salt being most commonly used. They are next piled up under cover in kenches, with a sprinkling of salt between each layer, with the backs placed downward, as is the case with green cod, so as to retain the dissolved salt. These kenches are ranged in regular order, with the heads of the fish outward, and extend 3 or 4 feet in height. In some localities, after the salting and before kenching, the two sides of the fish are brought together again, leaving the fish in natural shape, with the abdominal cavities filled with salt.

When a large haul of mullet has been made the work of dressing and salting must be rushed to prevent the fish from becoming tainted; and in warm weather, especially during August, if the fish are not salted within a few hours after landing they are apt to become discolored or to rust. The fish remain in these kenches until they are to be placed on the market, which may not be for three or four months.

In preparing for shipment the salted mullet are placed in boxes or tied up in bundles. No uniform style or size of package is used, resulting naturally in much confusion and inconvenience to the trade. Some fishermen simply bundle the mullet in such a way that the skin side is outward, while others cover the bundles with a single layer of matting or palmetto leaves. When carefully prepared these fish are of excellent quality, except that those cured during the warm weather of July and August sometimes rust.

The following method of curing mullet is recommended to those who wish a really choice product without regard to the cost of preparation:

The fresh fish are cut along the ventral part and eviscerated. They are next soaked for two hours or so in salt water, beheaded and split down the back, and the backbone removed. Four or five cuts are then made transversely across each half of the fish on the inner surface, and the fish are packed in dry salt, where they remain for about one week. On removal they are washed to remove the slime, undissolved salt, etc., and then suspended in the shade, where they are allowed to dry for five or six weeks. Each fish is then sprinkled with fine table salt and carefully wrapped with waxed or paraffin paper to exclude the air, and suspended in a well-ventilated room, where it may be kept for several months under favorable conditions.

DRY-SALTED CHANNEL BASS.

Channel bass or drumfish are caught in considerable numbers along the south Atlantic coast during the late summer, and the small demand for them in the fresh-fish trade results in many of them being salted for local use during the winter, especially along the North Carolina Banks. The following process is employed:

Each fish is split down the belly and eviscerated, the head is cut off with a hatchet or large knife, the backbone is removed, and the fish split in halves. Each half or fletch is then scored lengthwise from the napes to the tail on the flesh side, the cuts being about 2 inches from each other and penetrating the flesh to the skin. The fletches are washed free from blood, etc., and placed in barrels or vats with dry salt sprinkled in abundance among them and with strong brine poured over them. When sufficiently cured the fish are removed from the pickle and placed in the open air on boards, benches, or any convenient place for drying. Care must be taken to shield them from rains, and they should be placed under cover at night to protect them from heavy dews. When sufficiently dried they are stored in a cool, dry place until marketed.

The reduction in weight through dressing amounts to about 50 per cent, and through curing and drying about 33 per cent additional, making the dried weight about 35 per cent of the round weight. The only market is among the coast people—among the fishermen and their neighbors on the mainland. They sell for 8 to 15 cents per side, or 15 to 30 cents per fish, an equivalent of 1 to 2 cents per pound.

DRY-SALTED KINGFISH.

The great bulk of the kingfish caught in this country is sold in a fresh state; but at Key West and some other points on the coast, when the fresh-fish market is fully supplied, the surplus catch is salted. The method usually employed at Key West in drying and salting the fish is thus described by Mr. W. H. Abbott:

If the fish are not disposed of the first day after being caught, they are lightly salted and dried in the sun, in which condition they will keep for a week or two, and if the weather is favorable they will probably keep a week longer, and if quite thoroughly dried the fish will keep a much longer period. Generally the fisherman is a man of very little means and has no capital to work with; consequently the supply of salt which he is able to buy is very small. The salt used is from the Bahama Islands. The fish are laid on a box or bench and the thick part of the fish cut transversely, nearly through to the skin, at distances of 1 to 1½ inches apart. After the fish have been prepared in this way, the fisherman takes a small amount of salt and carefully sprinkles it over the entire surface of the fish and into the cuts, so as to make sure of their being properly cured. He is very careful not to waste the salt by scattering it about otherwise than on the fish. If the fish are to be sold in a few days, they are not exposed to the sun; but if to be kept for a longer period it is necessary to have some of the moisture taken out by the direct rays of the sun.

In a report on the Gulf fishing-grounds and fisheries, by J. W. Collins, the following description occurs:

As a rule, the great bulk of the kingfish taken by the Key West fleet is sold and eaten in a fresh condition, but occasionally some fish are salted on the boats, and a greater quantity are split and salted after they are landed, the surplus being disposed of in this manner. These salted fish are often dried, and to facilitate this and to insure the more thorough drying of the fiber the thick part of the flesh is cut transversely nearly to the skin, at a distance of about an inch apart. There is no systematic method of drying, as in curing cod, but the fish are hung across rails, spread on wood piles, or disposed of in any other manner where they may have a chance to dry, a favorite method being to suspend them by the tail. Cured in this way they make tolerably good food, but it is altogether probable that a much finer article of food might be obtained by smoking the fish.

The amount of kingfish prepared for market in this way is not known, but it is relatively small.

DRY-SALTED BARRACUDA AND BONITO.

On the southern coast of California, in the vicinity of San Diego, from 150,000 to 300,000 pounds of barracuda (*Sphyræna argentea*), about half that quantity of bonito (*Sarda chilensis*), and some yellow-tail or amber-fish (*Seriola dorsalis*) are annually dry-salted and sold in the markets at 3 or 4 cents per pound. As soon as practicable after they are removed from the water, they are split down the belly and dressed like cod, only the backbone is not usually removed, and heavily salted in kenches similar to those on the New England coast. When the weather is favorable the fish are washed and spread on drying flakes, the cure being completed in two or three days in case of barracuda, while a greater length of time is usually required for bonito. 100 pounds of round fish make about 50 pounds dried. When properly cured barracuda present an inviting appearance, being white and dry, and the flavor is excellent; but most of the California product is said to be dark in color and with a strong flavor, due probably to faulty methods of curing.

CHINESE SHRIMP AND FISH DRYING.

In the Barataria region in Louisiana, along the shores of San Francisco Bay in California, and at other points on the Pacific coast, there are camps of Chinamen whose principal occupation is the drying of shrimp and fish, mainly for Oriental markets. Their output also includes miscellaneous varieties of fish, oysters, squid, etc., the aggregate annual product amounting to about $100,000 in value.

The drying of shrimp was begun in Louisiana in 1873 by Chin Kee, whose plant was located on the western bank of the Mississippi River opposite New Orleans. The following season he moved to Bayou Dupont, near the head of Grand Lake, about 80 miles below New Orleans. In 1880 a second establishment or "platform" was built near Bayou Cabanage. A third platform was built in 1885 at Bayou André. The business continued fairly prosperous until 1893, when the severe storm in October completely destroyed the Bayou André platform and camps and severely damaged the other two platforms. The latter, however, were immediately repaired, and in 1897 another platform was erected about a mile above Cabanage.

These shrimp-drying establishments consist of a large platform, on which shrimp are dried, the necessary furnaces and kettles for boiling, warehouses, living apartments, storehouses, wharfage, apparatus, etc. The platforms range in area from 25,000 to 80,000 square feet, and are substantially built of pine boards 1 inch thick with close joints. The number of employees at each establishment ranges from 6 to 12.

Prior to 1886 the shrimp were boiled in kettles over open fires, but since that date greater neatness and economy of fuel have been secured by the use of a grate with a chimney, somewhat similar to the old form of sugar-boiling. The kettles over the grates are 5 feet long, 4 feet wide, and 18 inches deep, with a division in the center.

The shrimp are received from the fishermen each day, thus insuring their freshness and rendering the use of ice unnecessary. Prior to 1888 the price paid was 80 cents per basket, containing about 84 pounds; but since that date the price has been uniformly 65 cents per basket. When measured and received from the boats, the shrimp are rinsed and placed with water and the necessary quantity of salt in the kettles, each kettle holding about 5 baskets of shrimp. About 4 or 5 pounds of salt are used for boiling 100 pounds of shrimp, Liverpool salt being preferred, but coarse American salt is also used.

The brackish lake water suffices for cooking, and it is used for several successive boilings, additional water and salt being added as necessary. The shrimp are boiled for five or ten minutes, when the cover is removed and the shrimp stirred thoroughly with a paddle or other appliance. The cover is then replaced for 10 or 15 minutes longer, when the shrimp are removed with perforated skimmers or shovels and spread on the platform, where they are exposed to the action of the sun, being turned and separated at intervals during the day and covered, when necessary to protect them from moisture, until the drying is completed, this usually requiring two or three days. When shrimp are coming in plentifully the boiling is frequently kept up all night, the boiled shrimp being placed in heaps on the platform and covered with canvas until morning. When thoroughly dry, the Chinamen, with clean shoes or moccasins, tread them for a time to detach the shells and heads from the main part of the flesh. These shells and light particles are fanned off by throwing the shrimp upward through the air, somewhat in the same style as that practiced with wheat, rice, and other similar grains. The meats of the shrimp are then placed in sacks, beaten and thoroughly shaken to complete the breaking up and removal of the shells, after which they are again winnowed or passed through hand sifters, so as to remove all dust and particles of shells adhering to them. They are next made ready for shipment by placing them in flour barrels, containing about 200 pounds each, and are sent to the various markets. In packing, the shrimp should be graded, the whole bright meats being kept separate from those broken or discolored on account of rains during the process of curing.

Each basket of green shrimp yields about 9½ pounds of dried shrimp, which sells for about 14 cents per pound. The market is among Asiatic races almost exclusively. The great bulk is sent to San Francisco, but some are shipped to New York, Philadelphia, Chicago, and Havana. From San Francisco the shrimp are sent to China, Japan, and throughout the west coast of the United States. The quantity dried during the past four or five years in Louisiana has been much less than previous to 1893, and the profits have greatly decreased, owing to competition with Mexican ports.

The following summary shows the quantity prepared in Louisiana during each year since 1886:

Year.	Dried shrimp.	Year.	Dried shrimp.
	Pounds.		Pounds.
1887	304,200	1893	121,800
1888	319,000	1894	83,200
1889	346,400	1895	116,000
1890	293,600	1896	144,400
1891	280,200	1897	151,400
1892	285,200		

The method of drying shrimp practiced by the Chinese at San Francisco is thus described by Mr. Richard Rathbun:

After the day's fishing is over it is usually customary to carry the fresh shrimp to the Vallejo street market in San Francisco in live-baskets covered with a netting, which has a hole in the center closed by means of a puckering string. At the market the live shrimp sell at the rate of about 10 cents per pound, and those remaining unsold are carried back to the Chinese settlement and put at once into boiling brine. The kettle for boiling the shrimp is a rectangular iron tank 6 feet long by 4 feet wide and 2 feet deep, with a fireplace underneath. After sufficient boiling, care being taken to prevent overcooking, the shrimp are taken out and spread to dry upon level plats of hard ground, which have been previously stripped of grass and rendered perfectly smooth. They are spread out and turned occasionally by means of a hoe-like broom. After four or five days, or when perfectly dry, they are crushed under large wooden pedestals, or trod upon by the Chinese in wooden shoes, for the purpose of loosening the meats from the outer chitinous covering, after which the entire mixture is put through a fanning mill for the actual separation of the meats and shells. The fanning mill, a somewhat crude affair, is constructed of wood by the Chinese on precisely the same principle as the one used for winnowing grain. It measures about 8 feet long by 5 feet high, and consists of a square box, divided on the inside for the passage of the separated shells and meats, with a hopper above, and a large fan wheel worked by a crank at one end. The meats are partly used at home or at the various inland Chinese settlements, but are mostly shipped to China. The shells are also utilized as manure to some extent about San Francisco; but, like the meats, are mostly sent to China, where they serve as a fertilizer for rice, tea plant, etc. In San Francisco they sell at about 25 cents per hundredweight. Both the meats and shells are shipped to China in sacks. The trade is entirely in the hands of Chinese merchants, who ship by way of Hongkong. The meats are eaten by all classes in China, but are cheaper and less esteemed than the native shrimp, which are comparatively scarce.

Dried shrimp form a very popular article of food among Asiatic races and are worthy of more extended use in this country. They are very nice rolled in butter and fried, good for making curry, and for "jumbalayer" they are excellent. In China many of the broken shrimp are made into paste by grinding between stones.

In 1885, in connection with their shrimp drying, the Chinese on the Louisiana coast dried a quantity of oysters for market, but the venture was unsuccessful on account of the high price at which it was necessary to sell the product in order to reimburse them for their outlay, 50 cents per pound for the dried oysters being scarcely sufficient to meet expenses.

The tails of the rock lobsters or salt-water crayfish (*Panulirus interruptus*) are frequently dried in the sun without previous boiling or salting by the Chinese on the Pacific coast, but the aggregate of the business is not large.

Since 1885 the Chinese located at Barataria Bay, Louisiana, have dried a quantity of fish each summer. In doing this they make temporary quarters on Timbalier Island, in Terrebonne Parish, Louisiana, using old material from the permanent camps in Jefferson Parish. Tarpaulin or palmetto supplies sheltering for the workmen, and small slat-work frames, about 8 feet long by 4 feet wide, are used for holding the fish, which are turned every few hours by placing a second frame on the one holding the fish, turning both together and then removing the first frame. The season extends usually from May 1 to June 30, and from one to three weeks is necessary for the drying.

All varieties of salt-water fish are there dried except the small bony ones and those excessively fat. Few of the fish are dressed before drying, the head, scales, fins, and viscera remaining, except that the large redfish or channel bass are eviscerated and have the heads and fins removed. The fish are purchased from the seiners at a cost of $1.10 per basket, containing about 98 pounds. A basket of green fish yields about 47 pounds of dried, which sells for 4 or 5 cents per pound wholesale. In 1897 the two companies operating on Timbalier Island received about 1,300 baskets of fish, which yielded 61,100 pounds of dried fish, valued at $2,936.

On the Pacific coast the Chinese usually remove the heads and viscera from the large fish, and in some localities they salt the fish in brine and then dry them, much like the present method of curing cod. Some of the larger sharks and skates are split through the back and hung on poles. The barracuda, albacore, and bonito are split lengthwise along the back, soaked for 2 days in brine, and then dried in the sun, losing about 50 per cent of their weight in drying. The flesh of the dried barracuda resembles codfish somewhat, being white and firm. They usually sell in San Francisco at 3 or 4 cents per pound wholesale.

The redfish (*Trochocopus pulcher*) are dressed by opening the abdomen and removing the viscera, and Chinamen exhibit much ingenuity in giving a picturesque appearance to the head and teeth of this species. According to Dr. D. S. Jordan:

A "junk" with the deck covered with drying redfish seems at a little distance to be full of frogs about to leap. Sometimes I have noticed that the fatty protuberance on the forehead of the redfish has been cut off. This is valued as a delicacy and used for fish chowder.

Squid are dried in small quantities by Chinamen on the coast of California. They are washed and spread out on small slat-work platforms or flakes. The large squid are first split, but the small ones are dried in the condition in which removed from the water. The largest squid-drying establishment is located at Point Alones. There some of the flakes are placed on the ground, but the majority are elevated on posts 2 or 3 feet high, and resemble somewhat the codfish flakes of New England, the principal difference being that the squid flakes have the slats much closer together than those used for codfish. About 10 days are required for the process of curing, and no salt whatever is used. When thoroughly cured they are packed in bundles, each containing about 135 pounds and upward, and each package is covered with matting. They are sent to San Francisco, where some are sold to the domestic trade, and the remainder exported to the Hawaiian Islands and to China.

DRIED STURGEON PRODUCTS.

In the sturgeon fisheries of Russia and of Asiatic countries, and quite recently, to a small extent, in the Columbia River fisheries the spinal cords of the sturgeon have been utilized. After being cleaned and dried this substance is excellent for fish pies, soups, chowders, etc. The method of its preparation in Russia is as follows:

After the fish has been eviscerated an incision is made in the flesh, and by means of a hook enough of the spinal cord is drawn out to furnish a good hold for the fingers, by means of which the whole is extracted in a band, 4 or 5 feet long, consisting of a round whitish substance, marked or slightly disconnected at intervals like sausage links. It is carefully washed in fresh water to remove the blood and slime, and is then drawn by a workman between the edge of the washtub and his left hand, or similarly compressed, to remove the soft viscous matter or nerve tissues contained within; or sometimes it is cut open and those tissues scraped away and discarded. After this operation the substance is rinsed in another tub of fresh water until it becomes quite clear, and it is then exposed in a free circulation of air until it is thoroughly desiccated. For marketing, it is cut into pieces 4 or 5 inches in length, or it is tied in bundles composed of a number of spinal cords. On the Columbia River it sells for about 40 cents per pound, and in Russia it sells for the equivalent of 40 to 60 cents per pound, 25 of the common sturgeon of Russia (*Acipenser guldenstädtii*) being required to furnish 1 pound of véziga or viaziga, as the product is commonly known in the European markets.[*]

The preparation of the spinal cords of sturgeon on the Columbia River is thus described by Mr. W. A. Wilcox:[†]

One product of the sturgeon is used entirely by the Chinese, namely, the spinal marrow. As soon as the fish are landed at the packing establishment a Chinaman, armed with a hook, pulls out enough of the marrow to furnish a good hold; then, seizing it, draws the remainder of it out, hand over hand. In the average-sized sturgeon the spinal cord is 4 or 5 feet long and consists of long, white connecting links resembling sausages. These are cut open and the jelly-like substance contained within is scraped off and thrown away. This marrow is known by the Chinese and the trade under the name of "bone." It is thoroughly dried, and if not sold to the Chinese in this country it is exported to China, where it is much prized for making soups. The Chinamen pay 4 cents a pound for this "bone" and remove it from the fish themselves.

---

[*] See Rapport sur les Expositions Internationales de Pêche, par J. L. Soubeiran, Paris, p. 151.
[†] Report U. S. Commission of Fish and Fisheries for 1893, p. 252.

The following description of the methods of drying sturgeon meat—or, more properly, the dorsal portion of the fish—in Russia is extracted from Alexandre Schultz's, "Notice sur les pêcheries et la chasse aux phoques dans la Mer Blanche, l'Océan Glacial et la Mer Caspienne," St. Pétersbourg, 1873:

For making good "balyk" a large and tolerably fat fish is selected, whose head, tail, sides, and belly are taken off. That which remains, the dorsal part, has to undergo a special salting, while the other parts are salted in the usual manner. The backs of the common sturgeon (*Acipenser guldenstädtii* and of the "sévrionga" (*Acipenser stellatus*) remain entire, while those of the large sturgeon (*Acipenser huso*) are cut, either lengthwise only or else both lengthwise and crosswise. The pieces are placed in a tub so as not to touch each other nor the sides of the tub; and they are left thus after having been covered with a thick layer of salt from 9 to 12 days, and even 15 days when the pieces are large and the weather is hot. The salt is mixed with a little saltpeter, to give to the balyk a reddish color, 2 pounds of saltpeter to 50 poods [1,800 pounds] of balyk. Allspice, cloves, and bay leaves are frequently put into the brine. When the salting is finished, the balyk is put into water for a day or two, in order to detach all particles of the brine from it. Thereupon it is dried, first in the sun and then in the shade, on roofed scaffoldings, which are erected for the purpose. This last-mentioned operation requires from 4 to 6 weeks, and is considered finished when the balyk begins to cover with a slight mold, the absence of which shows that it has been salted too much.

Good balyk must be as soft and tender as smoked salmon; must have a reddish or orange brown color; and must have an odor something like that of the cucumber; it must also be transparent, show no traces of putrefaction, nor have a bitter taste; and, finally, it must not be too salty. There are very few manufacturers who can prepare balyk that has all these qualities.

A pood (36 pounds) of good balyk costs at the manufactory at least 18 rubles ($12.60), and at retail it can seldom be bought for less than 1 ruble (70 cents gold) a pound. The balyk made in March is considered the best. On the banks of the Koura, and in the trans-Caucasian waters, where the sévrionga (*Acipenser stellatus*) is caught in large numbers, balyk is made of at least 300,000 of these fish every year. This balyk, commonly called "djirim," is not of the first quality. It is dry, very salty, and is much sought after by the inhabitants of Kachetia, because it produces thirst and gives them occasion to quench it with the excellent production of their vineyards. A large sturgeon of 20 poods (720 pounds) yields 5 poods (180 pounds) of balyk; a very large sévrionga, 15 pounds; a common-sized sévrionga, 4 pounds; and the common sturgeon, from 8 to 12 pounds.

DRIED TREPANGS.

The preparation for market of the soft echinoderm variously designated as sea-cucumber, sea-slug, bêche de mer, trepang, etc., was once attempted on the Florida coast. The trepang is a very popular food product in oriental countries, esteemed not only by the natives, but by foreigners residing in those countries. China imports annually about 5,000,000 pounds, at an average valuation of 20 cents per pound, from the South Pacific Islands and Japan, where the holothurian is very abundant. In preparing trepangs for market, they are boiled in water for from 10 to 30 minutes according to varieties and sizes, split down on the side, eviscerated, and then exposed to the sun until perfectly cured. In some countries, as in Malay Islands, they are dried over a wood fire, but this product is less desirable than if dried in the sun. It is important that they be kept dry until they reach the consumers, otherwise they become flaccid and decay.

Mr. Silas Stearns described the fishery attempted on the Florida coast as follows:

In 1871 an Englishman came to Key West, Fla., for the purpose of gathering and preparing trepang for the Chinese market. He erected a shed, under which were built fireplaces, with large kettles and other arrangements, and also frames for drying. He arranged with the fishermen, and fishermen's boys particularly, to bring him all the sea-slugs they could obtain, for which he was to pay a certain price. As the slugs were very abundant on the shoals about Key West, and the prices paid for them were liberal, no trouble was experienced in obtaining large supplies. The sea-slugs, still alive and fresh, were thrown into the kettles and boiled a certain length of time, but as to the composition of the liquid in which they were cooked, my informant could not tell me. Then they were taken out, the outer rough skin rubbed off, and the body split with a knife, after which the intestines were removed and the body spread on canvas in the sun to dry. The next operation after drying, and the final one, was to smoke them. This was done in a smokehouse of the ordinary kind, in which they were suspended on slats. After the final process the trepang were packed in bales, covered with sacking, and shipped to New York, where they were probably reshipped to China. For two seasons (winters) this industry was kept up, and apparently with much success; but at the close of the second season the houses and apparatus were sold, and the operator left Key West. Since then nothing further has been attempted in the trepang industry there.*

Trepangs are abundant on several other portions of the United States coast, and especially so on the northwest coast and among the Alexander Islands, and there seems no reason why these supplies should not be utilized and an industry of considerable importance developed.

The following description of the method of curing trepangs among the East Indies is from The Commercial Products of the Sea, by P. L. Simmonds:

The first thing to be done on arrival at an island where the slug is plentiful is to erect on shore a large curing-house about 90 feet in length, 30 feet in breadth, and about 10 feet high. These houses are generally built of island materials and thatched with mats of cocoanut leaves; this thatch must be well put on, so as to prevent the rain from penetrating. The sides are likewise covered with these mats, and a small door should be left in each end. Platforms for drying the slugs are then erected along one side of the house. They should run the whole length and be about 8 feet in breadth, the lower one about breast-high from the ground, and the upper 3 feet above that. The frames are generally made of cocoanut trees and covered with two or three layers of split bamboo or reeds, sized close so as to form a sort of network for the slugs to lie on. Much care and skill is required in the construction of these platforms, so as to prevent the beche de mer from burning. A trench about 6 feet in breadth and 2 feet in depth is then dug the whole length of the platforms for the fires. Tubs filled with salt water are placed at short distances along the side of the trench and a supply of buckets kept in readiness to prevent the fires from blazing up and burning the fish or platforms, as well as to regulate the degree of heat necessary for drying the slugs.

The process of curing is this: The beche de mer is first gutted, then boiled in large pots, and, after being well washed in fresh water, carried into the curing-house in small tubs or baskets and emptied on the lower platform, where it is spread out (about 5 inches thick) to dry. The trench is then filled with firewood, and when the platform is full of trepang the fires are lighted and the drying process commences. From this time the fires must be kept constantly going day and night, with a regular watch to attend to them. On the afternoon of the following day the fires are extinguished for a short time and the slugs shifted to the upper platform, having been first examined, and splints of wood put into those which may not be drying properly. When this is done, the lower platform is again filled from the pots, the fires immediately lighted, and the drying process continued as before. The slugs on the lower platform must be turned frequently during the first 12 hours. On the second day (the fires having been extinguished as before) the slugs on the upper platform are shifted close over to one end to make room for those on the lower platform again, and so on as before for the two following days, by which time the first day's produce will be properly cured. It is then taken off the platform, and, after having been carefully examined, and those not dry put up again, the quantity cured is sent on board the vessel and stowed away in bags. But should the ship be long in procuring a cargo it will require to be dried over again every three months in the sun, on platforms erected over the deck, as it soon gets damp, unless when packed in air tight casks.

If the beche de mer is plentiful and the natives bring it daily in large quantities, 40 men will be requisite to perform the work of a house of the above size, and the pots will want two hands to attend them. These curing-houses consume a large quantity of firewood daily. When beche de mer is cured and stowed away great care should be taken to prevent it from getting wet, as one damp slug will speedily spoil a whole bag.

*Fishery Industries of the United States, sec. v, vol. 2, pp. 815-816.

It appears that there are two ways of boiling bêche de mer, equally good. The first is to take them out when boiled about a minute, or as soon as they shrink and feel hard; the other method is to boil them for 10 to 15 minutes; but in boiling either way the slugs ought, if properly cooked, to dry like a boiled egg immediately on being taken out of the pot. Bêche de mer dried in the sun fetches a higher price than that dried over a wood fire. But this method would not answer in curing a ship's cargo, as they take fully 20 days to dry, whereas by smoking them they are well cured in 4 days. Much skill is required in drying bêche de mer, as well as in boiling it, as too much heat will cause it to blister and get porous like sponge, whereas too little heat again will make it spoil and get putrid within 24 hours after being boiled. There is, likewise, great care and method requisite in conducting the gutting, for if this be not properly attended to by keeping the fish in warm water and from exposure to the sun it will, when raw, soon subside into a blubbery mass and become putrid in a few hours after being caught.

DRYING FISH BY ARTIFICIAL MEANS.

A number of devices have been invented for drying fish by artificial means, by the use of heat, dry air, absorption pads, etc., but none have come into practical use on a large scale. In 1878 there was introduced a drier,* consisting of one or more horizontally revolving wheel-like tables, having two outer rings with a bottom of network on which to place the fish and a corresponding covering of network to overlie and retain the fish during the rotation of the table. The tables are in sectional form to admit of the fish being removed from any portion without disturbing that in any other portion; the whole is supported by converging arms radiating from a vertical spindle. After the fish are placed within the network frame the structure is rotated at a speed to be regulated by circumstances, thus creating a current of air, causing a rapid drying of the fish.

A system modeled somewhat on that used in fruit-driers was introduced in 1877.† The fish are dressed and placed in a tight vessel on a false perforated bottom a few inches above the real bottom. Steam is admitted and the fish cooked until freed from the bones. The flesh is then spread on hurdles, which are introduced successively into a chamber, into the lowest part of which is admitted a current of air heated to about 200°. After the first has been exposed to this temperature about 10 minutes, it is moved up about 4 inches and a second introduced, and so on successively until there are 10 or 12 hurdles in the chamber, and thereafter as each additional hurdle is placed at the bottom the top one is removed.

During this operation the moisture evaporating from the fish forms a vapor which fills the drying chamber, thus keeping the fish in a humid atmosphere and preventing it from becoming suddenly dry and hard on its surface, and the texture is kept loose to allow the water to evaporate freely.

This process did not prove a success, and in 1880 Mr. Alden, the patentee of the above, introduced an improved method, as follows:

Take fresh fish and remove the heads, tails, fins, entrails, and skins, and also the larger bones, leaving the clear fresh fish meat, which should be cut in pieces of suitable size, thoroughly cleansed in pure cold water, and then placed in an evaporating pan placed upon and surrounded by a heating coil or a steam jacket, and having one or more movable blades revolving around on the inside of the pan, so arranged that when in motion the blades will operate upon the principle of the plow, so as to avoid shoving the mass while throwing a furrow in such manner that the fresh fish meat is prevented from adhering to the bottom or the sides of the pan, and is kept constantly in a revolving motion, so as to admit free access of the drying atmosphere for rapidly removing the vaporized moisture, in aid

---

* See United States Letters Patent, No. 207913.   † See Letters Patent, No. 186893.

of which a fan or vacuum chamber may be used, and the mechanism may be operated by steam or other power. When the evaporated fresh fish meat has been placed in the evaporating pan, steam is applied to the heating coils or steam jacket, and the revolving blades are immediately set in motion, the operation or effect of which is within a few minutes to dissolve the fish meat into a jelly-like mass, in which condition, when kept in motion, it soon loses all its free moisture, rapidly disintegrates, and becomes dry or solidified fresh fish fibril, having much the appearance of fine broken vermicelli. Under this method the fish fibril separates and entirely frees itself from the minute or smaller bones, so that they may readily be removed. The fresh fish fibril should be kept in rapid motion until sufficiently dry to remove from the pan (which is determined by its failure longer to throw off vapor), when it should be spread upon cooling screens or muslin until cold, when it may be packed in tin, wood, or paper boxes for keeping or for transportation. Under this process the time required is from thirty to forty minutes, and the temperature must be kept below the cooking point, so as to prevent coagulation of the fish gelatine, and the product, or fresh fish fibril, will, in proportion by weight, be as 1 pound to 10 pounds of live fish, and 5 pounds of prepared fresh fish meat. The greatest possible celerity should be had and care taken that the prepared fresh fish meat is entirely fresh, pure, and without taint. Fresh fish put up after evaporating its free moisture and being fibrilized in the manner described retains its entire nutriment and flavor, is free from all foreign and injurious substances, contains no salt, has no affinity for moisture, and will keep in any climate for a long time.

A plant was established at Gloucester, Mass., in 1881, for the preparation of Alden's specialty, but it did not compete successfully with the boneless codfish, then being prepared in such large quantities at that port.

By a process designed in 1879 by Mr. J. M. Reid,* of Canada, fish previously brine-salted are placed in a tight receiver and subjected for a time to compressed heated air for the purpose of extracting the moisture.

Another method that promised success was put in operation at Gloucester in the spring of 1883, it being the invention of Halifax parties.† Two apartments were fitted up with flakes, the floors being partly open to allow the air to circulate. By means of piston blowers, or of fans operated by steam power, the external air was drawn in from one side of the building and forced over the fish and out on the other side, when the atmosphere was in suitable condition for drying. But when it was loaded with moisture both the inlet and outlet connecting the room with the outside air were closed and communication opened with a cooling room overhead, the temperature of which was lowered by cakes of ice, and the air contained in the drying and cooling rooms was forced over the fish, thence through the cooling room, and back again over the fish and so on, continually keeping the fish cool. The inventor claimed that the fish were brighter and contained less dirt and dust than those dried on flakes, but the expense of the process was considerable and consequently it was soon abandoned.

In 1890 Mr. Cathcart Thompson, of Halifax, brought to notice a process by which he claimed that codfish can be dried by absorbent pads, thereby obviating the dangers and delays of the present method. This process was originally as follows:

A layer of green-salted fish is spread evenly on an absorbing pad. Common gunny cloth makes a good, cheap, and effective one. Another pad is laid over this, succeeded by another layer of fish, followed again by a pad, and so on successively until the whole quantity of fish is spread, a pad being placed over the last layer. A platform of boards is then laid on this, and weights or other appliances are used to cause a slight, continuous, and uniform pressure. The pile is allowed to remain from 24 to 48 hours, during which time the pads become saturated with moisture, which they have extracted from the fish. Repiling then takes place, dry pads being substituted for the wet ones, the latter being dried for further use. Repiling with the substitution of dry pads is continued till the fish have become sufficiently dry, a week or ten days being long enough to effect this object if intended for the home or West Indian market. For more distant markets a somewhat longer period would be required.

---

* Letters Patent No. 221357. † See Letters Patent No. 250382, dated December 6, 1881.

In 1892 Mr. Thompson introduced the following improvement on his method:

A number of light frames of 2 by 1¼ inch lumber, 6 feet in length and 3 feet in width, are constructed. One of these is laid upon the floor and a layer of dry moss and sawdust is spread thereon. This is covered with a sheet of cotton cloth large enough to envelop the frame, a layer of fish is spread flesh down thereon, and the whole is covered by another sheet of cotton. A second frame is placed over the first one and the same process is continued till a height of 3 or 4 feet is attained, then a thick layer of moss or sawdust is placed over the last tier of fish and a cover of boards sufficiently large to go inside the frame is laid over all. Pressure is then applied, by screw or lever, to thoroughly embed the fish in the absorbent. The spreading of the moss and sawdust over the layer of fish fills up the interstices between them and brings every part in contact with the absorbent, and at the same time prevents the fish from being pressed out of shape.

In the twenty-fifth annual report of the Canadian Department of Marine and Fisheries the following account of the above-described process is given:

A quantity of 200 pounds of cleaned fish, put under Thompson's process, gave the following weights:

Duration of test.	Weight.	Percentage of loss.
	Pounds.	
After 72 hours pressure	170	15
After 120 hours pressure	155	22½
After 192 hours pressure	144	28
After 264 hours pressure	134	33
After 312 hours pressure	128	36

Thus, 200 pounds of cleaned fish, after a pressure of 312 hours, is reduced to 128 pounds, 36 per cent moisture being extracted. This seems sufficient to establish the fact that in this way enough moisture can be extracted by simple and cheap means to secure the fish against damage at times when drying under the ordinary process would be impossible. This method could be employed with great advantage by fishermen at the places of catch, as the moisture could be removed from the fish continuously and quite independent of weather. They could then be placed in piles, and the first fine day taken advantage of for final drying. For fish which have been cured by Mr. Thompson's experiment, six hours in the sun should suffice for the United States market, and from 24 to 48 hours to render them suitable for the Brazil markets, where hard and very dry fish are required. Mr. Thompson intends to continue his experiments on a larger scale, to enable him to make the final test of sales in foreign markets, when a further report will be made, and, if successful, a bulletin will be issued by this department. The different experiments were inspected by experienced fish-merchants, who have certified that in their opinion the extraction of 30 per cent will secure the fish from damage until suitable weather offers for their final drying by exposure to sun and air—for the removal of the remaining 10 per cent to 15 per cent. This they consider would not require more than from 6 to 48 hours of good drying weather, according to the market for which the fish are intended.

Mr. Thomas S. Whitman, of Annapolis, Nova Scotia, obtained letters patent on May 10, 1892, in the Dominion of Canada, and on the 13th of February, 1894, in the United States, for an improved process of curing and drying cod by exposing the fish alternately to artificial heat and to currents of fresh, cool air. The inventor claims that by his process fish can be cured much quicker than by the present system and without any of its injurious effects, and that the exact quantity of moisture desired can also be removed from the fish, so as to suit the taste of consumers in different countries.

The following description applies to Mr. Whitman's process:

The wet-salted fish are taken from the kench and washed, after which the surface water and pickle is pressed out of the fish by steam press or otherwise. After having been in press for a few hours the fish are ready to be spread on the wire flakes or trays that are placed in rows about 9 inches apart, the rows of flakes or trays being contained in compartments that are traversed by pipes in which steam or hot water is permitted to circulate. The maximum temperature which the steam or hot water in the pipes should impart to the compartments is about 95 F.

The fish having been spread upon the trays or flakes in the compartments are allowed to remain in a temperature of 90 to 95 degrees for a few hours, until they are thoroughly warmed, whereupon currents of cool, dry air are forced over and under the fish on these flakes or trays. These currents of dry air come from channels or flues that open into the compartments. By opening and closing these cold dry-air flues at proper intervals of, say, two or three hours, thus alternately cooling and heating the fish, from 1 to 2 per cent of moisture per hour is taken from the fish. The products of evaporation are carried off from the compartments by flues running to a chimney, or suitable ventilators may be placed in the tops of the compartments for carrying off the moisture to the roof of the building, or otherwise. It will be perceived that if the heating process were carried on by itself continuously instead of interruptedly, the atmosphere surrounding the fish would soon be charged with moisture to such an extent as to prevent any further evaporations, and the fish, too, would be injured by being warmed for too long a time or too thoroughly. The currents of fresh air which alternate with the heating process described serve to bring down the temperature of the fish and also to carry off the moisture-laden atmosphere which surrounds the fish, bringing into action fresh air which is ready to be charged with new moisture carried away from the fish by the next heating process.

The following account of the application of the Whitman process to curing codfish is abridged from a report made by the owners of the patent in the United States:

The first apparatus for practical working of the Whitman process was put up by the patentee at Annapolis, Nova Scotia, in a building 10 by 80 feet of 2½ stories. This fish-drying establishment has been in constant active operation for four years, and has turned out from the green and kench-salted fish about 10,000 to 15,000 quintals of dry fish annually for export to West Indies, Central and South America, and long-voyage tropical fish markets with very profitable results, giving employment to large numbers of men and fishermen, causing a large increase in Bay of Fundy hake and haddock fishing, and a steady advance in prices, till now these fish are actually commanding higher prices at the Bay of Fundy ports of Nova Scotia than the hake taken by United States fishermen off the New England coast are selling at in Gloucester or Boston; all this being the result of drying the fish suitably for tropical markets, which it has been found impossible to do in the ordinary way by the sun, owing to the humidity of the atmosphere and prevalence of fogs on the Nova Scotia and New England coasts. The same successful results have followed with all the fish-driers Mr. Whitman has had erected for his own account or for others at Halifax, Nova Scotia.

At St. Johns, Newfoundland, for Messrs. Bowring Bros., and for Messrs. Job Bros. & Co., two of the largest fish-exporting firms in America, these drying establishments have been in active operation almost constantly, night and day, since erected, about three years ago, and Messrs. Bowring Bros. have purchased Mr. Whitman's patent right for the island of Newfoundland and dependencies of Labrador.

At St. Pierre, Miquelon, under Mr. Whitman's French patent in 1897, he erected a drier for Messrs. Bonst & fils, of Granville, France, who have a large fish establishment at St. Pierre and a fleet of vessels employed in fishing on the Grand Banks of Newfoundland. This drier they operate on a royalty, and have made a good success of it, drying fish for export to Madagascar and other French colonies in tropical countries, which it was impossible to do on the Newfoundland coast in open air, owing to humid climate and fog. At Paspibeac, Quebec, Canada, Mr. Whitman erected for Messrs. Charles Robin, Collas & Co., Ltd., an extensive drier, costing $5,000, which has been most successfully operated for two years past by Charles Robin, Collas & Co., who, only two months after they had commenced operations, bought of Mr. Whitman the patent rights for the Province of Quebec and Baie de Chaleur coast and paid £1,000 in cash.

A small drier was erected at Halifax, in 1896, for Messrs. Geo. E. Boake & Co., especially to dry fish for their Jamaica trade, which has been in constant use over two years and proved a great success in saving of time and labor. At Halifax Mr. Whitman erected, in 1895, a large drier building, 60 by 120 feet, which has been in very successful operation over since, and has enabled his company to open up large and profitable fish markets in Central and South America, formerly largely controlled by Norwegian fish-dealers. Mr. Whitman is now erecting a fish-drier at Gloucester for Messrs. John Pow & Son, and is about to organize a joint-stock company to operate a large drier at Boston, not only to dry fish for the cutting and other branches of the domestic trade, but for export fish trade.

A correspondent writes as follows in the Yarmouth *Herald* of July 18, 1893, respecting the success of Mr. Whitman's fish-drying apparatus at Halifax:

Within the last few days I have had the privilege of visiting the extensive new fish-drying apparatus that has been put in operation in this city (Halifax) by the inventor, Mr. Thomas S. Whitman, of Annapolis. The building containing the apparatus and storage rooms has been constructed and completed and operations have commenced within the last month. It is a very large building, 50 by 120 feet, and is situated on Liverpool wharf, where there is ample wharfage, and where a large amount of fish can be taken care of. Entering the building a very busy scene meets the eye; thousands of quintals of fish are seen in the various processes of washing, drying, and packing for the largest fish markets in the world. I was particularly struck with the rapidity of the operation. Mr. Whitman buys all the green-salted fish that offers; by his process they are dried perfectly in 48 hours, and are ready to ship in less than a week from kenching. It is certainly a new departure in the handling and curing of fish. The new system invented and introduced by Mr. Whitman is a perfect drier, and at the same time the fish are so kept apart from each other during the entire process of drying that they are also kept cool, the atmosphere by which they are dried being of about the same temperature required in the natural system of drying. It is astonishing to note the vast quantities of fish that can be cured in a short time; several thousand quintals per week is the capacity of this large concern, and it is certainly a busy hive of industry, one of the busiest in the provinces. To-day your correspondent was shown about 8,000 quintals of fish that were being dried, and most of them were in the sea only a short time ago, and before the week closes they will be shipped in perfect order to the fish markets of the West Indies. Considering the large amount of foggy, wet weather that the people of the western counties generally have to meet during their fish-drying season, it would evidently be to the advantage of our largest fish-packers if they were to adopt the methods now used and invented by Mr. Whitman, for it is evident that a vast amount of time is thus saved in the curing of fish, while the uniformity of the curing is maintained throughout, every fish appearing in perfect order as a result of this process. As I stated before, it only required 48 hours to thoroughly dry the fish, and they are then ready for shipment to any part of the world.

It is estimated that the cost of drying codfish by the Whitman process from the water-horse to the finished product is about 30 or 35 cents per 112 pounds for fish suitable for the West India trade, this covering two dryings of 24 hours each and a sweating of 10 to 12 days. In preparing fish for Central America or northern Brazil, 2½ days' drying is necessary, and the cost approximates 40 or 45 cents, while for southern Brazil the fish must be dried for 3 days by the Whitman process, and the cost is about 45 or 50 cents per quintal of 112 pounds.

A number of other processes of artificial drying have been devised, but none of them have been adopted to any extent by the trade.

PRESERVATION OF FISHERY PRODUCTS BY PICKLING.

Pickling foods consists in their preservation and subsequent retention in some antiseptic flavoring solution, such as brine, vinegar, etc. Brine made of common salt is used almost exclusively in pickling fish, while for mollusks, crustaceans, and a few preparations of fish, vinegar with certain spices added is generally employed; but pickling with vinegar is of small importance compared with brine-salting.

A variety of flavoring solutions used for pickling in foreign countries are comparatively unknown in the United States. In Japan small fish are frequently boiled and placed in *shoyu* or *soja*, a sauce made from fermented wheat, beans, and salt. In the same country salmon, cuttlefish, etc., are frequently slightly salted and then boiled and placed in a tight package with rice partly fermented, the development of the ferment being checked by the removal of moisture. The rice, taking moisture from the fish, begins again to ferment and the fish imbibes products of the fermentation, such as dextrine, sugar, and alcohol, and is thereby very delicately flavored.

A number of other antiseptics have been introduced for the purpose of preserving food products, among which are boracic acid, salicylic acid, etc., but as these do not flavor the product, and as they are not generally employed as a solution, their use is not considered as pickling. A discussion of them is therefore reserved for the last chapter of this report, the present chapter dealing with methods of pickling with brine and vinegar.

DEVELOPMENT AND METHODS OF BRINE-SALTING.

The origin of pickling fish with salt is of somewhat uncertain date. It was known to the Phœnicians on the Spanish coast, and was employed by the Greeks to some extent, and the Romans carried it to a high degree of perfection, especially in preserving swordfish from Sicily, tunny from Byzantium and Cadiz, mackerel from Spain, and mullet from Exone. Brine-salting received its greatest development during the thirteenth and fourteenth centuries at the hands of the Dutch in preserving herring caught in the North Sea, and since that time it has become one of the most important methods of preserving fish. Its principal application in the United States is in the preservation of mackerel, herring, alewives, salmon, mullet, cod, lake trout, whitefish, bluefish, shad, etc. It is also used in preserving certain miscellaneous products, as cod tongues, halibut fins, sturgeon eggs, mullet eggs, etc.

The general method of brine-salting is to dress the fish and place them with salt in tight vats or barrels, the salt uniting with the moisture in the fish forming a pickle, in which they remain for a few days until cured, after which they are usually removed and placed in market packages with new brine. But there are many exceptions to this practice, depending on the species of fish and the markets for which they are intended. Some fish, sea herring and river herring, for instance, are usually not dressed at all, being brine-salted in the natural or round condition. Others are gibbed, or split to the vent and eviscerated. But most pickled fish are split either on the back or the belly from the head to the tail, so as to lay out flat; some have the heads removed, and a few have a large portion of the backbone cut out.

However they may be dressed, it is important that the fish be salted as soon as practicable after removal from the water—in the meantime being protected from the sun, from bruising, etc. In case the fish have been dressed they are usually washed and soaked to remove all the blood. In salting, the fish are placed in the barrels or butts, with dry salt sprinkled among them, the quantity used ranging from 20 to 25 pounds to 100 pounds of fish. On the New England coast Trapani salt is generally used, except in the case of mackerel and one or two other species, for which Liverpool salt is preferred. On the Great Lakes, Syracuse and Warsaw salts are preferred, but the other kinds are used to some extent. Along the Middle and South Atlantic coast Liverpool salt is usually employed. The dry salt unites with the moisture in the fish, making a pickle which soon strikes through the fish. If thin, dry fish are being cured, it is sometimes well to add strong brine to aid in forming the pickle.

After a time, averaging for most species about a week or ten days, the fish are cured, and should then be placed in packages suitable for the market with additional salt sprinkled among the fish, and the package completely filled with strong brine. The principal difficulty encountered is the liability of the fish to rust; but by using strong pickle and tight barrels, so that the fish are covered with pickle all the time, this tendency may be easily overcome except during very warm weather.

The quantity of salt used in pickling fish varies according to the size and condition of the species handled, and experience and knowledge of the particular market for which they are intended are the best guides in every instance. A mild-cured fish is preferred to one heavily salted; but if too little salt is used the pickle is likely to slime or sour and the fish become rusty. It is therefore usually desirable to err on the side of too much salt rather than too little. Occasionally, to insure perfect preservation, it is necessary to use so much that the flavor of delicate species is more or less injured. Sugar is sometimes employed to modify the action of the salt and to improve the flavor of the articles pickled when it is not desired to keep the product for a considerable length of time, as in case of pickling salmon. But the use of sugar is sometimes attended with fermentation unless the pickled products be kept at a low temperature; and glucose is now sometimes substituted. The fish are first struck in salt and then packed in a suitable receptacle with a solution composed of about 3 pounds of glucose, 10 pounds of salt, and 5 gallons of water, the glucose being dissolved in the water before the salt is added.

Pickled fish are placed in a great variety of packages adapted to the trade for which they are intended and ranging in capacity from tierces, each containing 300 pounds, to small kegs containing only a pound. Mackerel, sea herring, salmon, cod, and the like, are mostly put up in whole barrels of 200 pounds net capacity. River herring or alewives are generally placed in 160-pound barrels, while the bulk of mullet, lake herring, whitefish, trout, and other lake species are usually packed in half-barrels of 100 pounds capacity. Most of these species, however, are also placed in packages varying from 50 pounds to 10 pounds, suitable for the various requirements of the retail trade, each package being branded with the weight of the fish therein.

Carefulness in the selection of the packages is of great importance. Those used on the New England coast are manufactured mostly in Maine, Bangor being the center of the industry, and the 100-pound barrels or half-barrels used on the Great Lakes are made principally at Sandusky; but while the products of those two cities are the standards, many fish barrels are made at various other points. Wood which imparts a peculiar flavor to the fish should not be used for making the barrels, unless for preparing fish for those markets which exhibit a preference for fish having such a flavor. The staves and heads may be of white pine, white or red oak, spruce, poplar, or chestnut, and they are sent to the fishing ports either ready for use or, to economize freight, in shooks ready to be put together. The Bangor barrel has staves 28 inches in length and the heads 17 inches between the chimes, and is bound with 3 hoops on each bilge and the same number on each chime. In packing valuable fish, such as mackerel, much stouter barrels are necessary than when packing herring, for instance. The average cost of the Bangor barrel used in the mackerel trade approximates 55 cents, and the Sandusky barrel costs about 50 cents.

REGULATIONS RESPECTING BRINE-SALTING FISH.

With a view to maintaining the reputation of the output, and incidentally to preventing fraud on the consumers, statutes affecting the packing of brine-salted fish have been enacted in several of the States, especially in Maine, New Hampshire, Massachusetts, Rhode Island, Connecticut, New York, New Jersey, Virginia, North Carolina, and Ohio. The pickling of mackerel is regulated in Maine, New Hampshire, Massachusetts, and Rhode Island. The laws of Connecticut attempt to regulate the pickling of shad. In Ohio it is required that all pickled fish be inspected except herring, mackerel, and shad; also in Chicago and some other large cities there are municipal regulations relative to the same matter. Few of the State or municipal regulations are strictly enforced; and since there are no national laws protecting inspected fish after leaving the jurisdiction of the State where packed, it frequently happens that fish inspected and uniformly graded according to the regulations of the State where prepared are repacked in other States and sold with short weights and under wrong grades, low-grade fish being sold for choice ones, short fish for long, and even herring for mackerel, much to the injury of the trade. There is nothing to prevent mackerel, for instance, which has been pickled and inspected in accordance with the regulations of Maine or Massachusetts, from being repacked and sold under false brands.

A barrel of fish signifies 200 pounds of fish exclusive of pickle, but without proper inspection many dealers are disposed to place less than that weight of fish, adding brine to keep the gross weight of the barrel's contents the same. The faulty grading of fish is much more frequently practiced, fish improperly cured or those of small size being branded higher than the quality or size warrants. This is the principal reason why so large a proportion of the pickled herring sold in this country are of foreign importation; many dealers preferring to handle those cured and packed under careful foreign inspection, even though the cost be nearly twice as great, since the brand indicates exactly what they are buying.

Inspection regulations are of very early origin, those in Massachusetts dating from 1651. They generally provide for inspectors, who are appointed by the governor or chosen by the towns in which they are to serve. There was an inspector-general in Maine prior to 1875, but the office was abolished that year, and at present the governor is required to "appoint, in places where pickled fish are cured or packed for exportation, one or more persons, skilled in the quality of the same, to be inspectors of fish, who shall hold their office for a term of five years, unless sooner removed by the governor and council."

The regulation in New Hampshire respecting the appointment of inspectors is almost identical with that of Massachusetts. The inspector-general is appointed by the governor, with the advice and consent of the council, for the term of five years, unless sooner removed. He may appoint deputy inspectors, removable at his pleasure, in every town where fish are packed for exportation.

In Rhode Island "the electors in each town shall, annually, on their town election days, choose and elect *   *   * one or more packers of fish."

In Connecticut "the superior court in the several counties may appoint in each town therein not exceeding 15 inspectors and packers of fish."

In each of these States the inspectors are required to give a bond for the faithful performance of their duties, the amount of the bond ranging from $10,000, in case of the inspector-general of Massachusetts, to $100 for the local inspectors in Connecticut. Their duties consist generally in inspecting and branding the fish salted under their supervision, and the fees are: In Maine, 7 cents per barrel; in New Hampshire and Massachusetts, 9 cents, of which 1 cent per barrel goes to the inspector-general. In Rhode Island "the packers [inspectors] of fish shall be paid for opening, assorting, inspecting, weighing, pickling, packing or repacking, heading up, nailing, and giving a certificate, if pickled codfish or mackerel, 20 cents for every barrel, and 15 cents for every half-barrel *   *   *; and for all other, except codfish and mackerel, *   *   * 25 cents for every cask." The Connecticut inspectors receive " for packing, heading, plugging, pickling, and branding each barrel of fish 20 cents, and for each half-barrel 10 cents."

While the foregoing are the fees fixed by law, yet generally, as there is no limit to the number of inspectors, each packing house has one as a member of the firm or employed in some capacity, so that the local fees are rarely paid.

The inspection in Maine is made under the following provisions:

Every inspector who inspects any kind of fish that are split or pickled for packing, shall see that they are in the first instance free from taint, rust, or damage, and well struck with salt or pickle; and such of said fish as are in good order and of good quality shall be pickled in tierces, barrels, half-barrels, quarter-barrels, and tenths of barrels, or kits: each tierce containing 300 pounds, each barrel 200 pounds, and so on in that proportion; and the same shall be packed in good clean coarse salt, sufficient for their preservation; and then each cask shall be headed up and filled with clear, strong pickle, and shall be branded by the inspector with the name and quality of the fish therein. Mackerel of the best quality, not mutilated, measuring, when split, not less than 13 inches from the extremity of the head to the crotch or fork of the tail, free from taint, rust, or damage, shall be branded "number one"; the next best quality, being not less than 11 inches, measuring as aforesaid, free from taint, rust, or damage, shall be branded "number two"; those that remain after the above selection, free from taint or damage, and not less than 13 inches, measuring as aforesaid, shall be branded "number three, large"; those of the next inferior quality, free from taint or damage, not less than 10 inches, measured as aforesaid, shall be branded "number three"; all other mackerel, free from taint or damage, shall be branded "number three, small." The inspector shall brand, in plain letters, on the head of every such cask, the weight, the initials of his Christian name, the whole of his surname, the name of his town, and the letters "Me." an abridgment of the month and the year, in figures, when packed.

Every inspector who inspects pickled alewives or herring, packed whole or round, shall see that they are struck with salt or pickle, and then put in good casks of the size and material aforesaid, packed closely therein and well salted, and the casks filled with fish and salt, putting no more salt with the fish than is necessary for their preservation; and the inspector shall brand all such casks with the name of the inspected fish as aforesaid, but in no case shall the inspector brand the casks unless the fish contained therein shall have been packed and prepared under his immediate supervision.

All tierces, barrels, and casks which are used for the purpose of packing pickled fish shall be made of sound, well-seasoned white oak, white ash, spruce, pine, chestnut, or poplar staves, with heading of either of such kinds of wood, sound, well planed and seasoned, and when of pine to be free of sap, and the barrels to be hooped with at least three strong hoops on each bilge and three also on each chime; the barrel staves to be 28 inches in length, and the heads to be 17 inches between the chimes, and made in a workmanlike manner to hold pickle.

If any person takes from a cask any fish pickled, cured, lawfully inspected and branded, and substitutes therefor or fraudulently intermixes other fish; or any inspector marks any cask out of his town, or which he has not inspected, packed, and prepared himself according to law; permits other persons unlawfully to use his brands, or willfully and fraudulently uses the same himself after the expiration of his commission, he shall forfeit $20 for each cask or box so dealt with.

If any person lades or receives on board any vessel or other carriage, for any transportation from this State, any pickled fish, or cured or salted whole fish, packed or not packed, not inspected and branded as aforesaid, except such as is described in the exception of section 13, he shall forfeit at the rate of not less than $5 nor more than $10 for every 100 pounds thereof; and any justice of the peace may issue his warrant to the proper officer, directing him to seize and secure any such prohibited fish, and convey it to any inspector within a convenient distance for inspection; and every person refusing to give necessary aid in the service of such warrant, when required by the officer, shall forfeit $5 to the person suing therefor in an action of debt; and such inspector shall open, inspect, pack, and brand such fish according to law and detain the same till all lawful charges of seizure and inspection are paid.

The fish inspection laws of New Hampshire follow very closely those of Maine.

The inspection laws of Massachusetts date from 1651, but have been modified from time to time. The following are among the principal provisions at present:

Under the supervision of the inspector-general and his deputies, respectively, all kinds of split pickled fish and fish for barreling except herrings, and all codfish tongues and sounds, halibut fins and napes, and swordfish, whenever said articles are intended for exportation, shall be struck with salt or pickle in the first instance, and preserved sweet and free from rust, taint, or damage; and when the same are found in good order and of good quality they shall be packed either in tierces containing each 300 pounds, in barrels containing each 200 pounds, in half-barrels containing each 100 pounds, or in packages containing each less than 100 pounds, on which the number of pounds therein shall be plainly and legibly branded. Every cask, kit, or package shall be packed with good, clean salt suitable for the purpose, and, after packing with sufficient salt to preserve its contents, shall be headed or well secured, and filled up with a clean, strong pickle.

Casks used for packing or repacking pickled fish intended for exportation, except casks containing less than 25 pounds weight, shall be made of sound, well-seasoned white oak, ash, red oak, spruce, pine, or chestnut staves of rift timber, sound and well seasoned, with heading of either of said kinds of wood, and when of pine such heading shall be free from sap and knots and be planed. The barrels, half-barrels, and tierces shall be well hooped with at least three good hoops of sufficient substance on each bilge and three hoops of the like quality on each chime. The barrel staves shall be 28 inches in length and the heads shall be 17 inches between the chimes. The barrels shall contain not less than 28 nor more than 29 gallons each; the half-barrels not less than 15 gallons each; and the tierces not less than 45 nor more than 46 gallons each. Each cask shall be made in a workmanlike manner, and branded on its side, near the bung, with the name of the maker.

There shall be five qualities of mackerel, three of salmon and shad, and two of other kinds of pickled fish. Mackerel of the best quality, not mutilated, measuring not less than 13 inches from the extremity of the head to the crotch or fork of the tail, free from rust, taint, or damage, shall be branded "number one." The next best quality, being not less than 11 inches, measuring as aforesaid, free from rust, taint, or damage, shall be branded "number two." Those that remain after the above selections, if free from taint or damage, and not less than 13 inches, measuring as aforesaid, shall be branded "number three, large." Those of the next inferior quality, free from taint or damage, not less than 10 inches in length as aforesaid, shall be branded "number three." All other mackerel free from taint or damage shall be branded "number four." Those salmon and shad which are of the best quality for family use, free from rust or damage, shall be selected for "number one" and "number two," the best of them selected and branded "number one," the residue "number two"; all that remain free from taint, and sound, shall be branded "number three." Of all other pickled fish the best, which are free from taint and damage, shall be branded "number one"; those that remain, free from taint and sound, "number two."

Each cask, kit, or package shall be filled with fish of the same kind or parts of the same kind of fish, and whoever intermixes, takes out, or shifts any inspected fish which are packed or branded as aforesaid, or puts in other fish for sale or exportation, shall forfeit $15 for each package so altered. If any casualty renders it necessary to repack a cask of inspected fish it shall in all cases be done by an inspector of such fish.

The inspector shall brand, in plain, legible letters, on the head of each cask of fish inspected by him, the denomination of the fish packed or repacked therein, the initials of his Christian name, and the whole of his surname; and, if a deputy, the name of the place for which he is appointed, the letters

"Mass.," and the year in which the fish are packed; and shall also, when in his judgment it may be necessary, nail in a suitable manner any cask in which fish are packed. Pickled fish duly inspected in the State or country in which it is packed shall not be subject to reinspection in this State.

Small fish, which are usually packed whole with dry salt or pickle, shall be put in good casks of the size and materials required in this chapter for the packing of split pickled fish, and shall be packed close in the cask and well salted. The casks shall be filled full with the fish and salt, and no more salt shall be put with the fish than is necessary for their preservation, and the casks containing such whole fish shall be branded with the denomination of the fish, and a like designation of the qualities as is before described in this chapter in respect to the qualities of other pickled fish.

In Rhode Island provision is made for the election annually of one or more packers of fish in each town, who shall see that all fish packed in the State are properly pickled and repacked in casks in good shipping order, with good salt, sufficient in each cask to preserve such fish from damage, to any foreign port. Other provisions of the law are as follows:

Pickled fish, whether codfish, mackerel, menhaden, herrings, or other fish, shall be sorted and one kind only be put into one cask.

Every cask shall be well seasoned and bound with 12 hoops; those of menhaden and herrings of the capacity to hold 28 gallons, and those for other fish of the capacity, if a barrel, to hold 200 pounds, and if a half-barrel, 100 pounds weight of fish; each cask to be full, and the fish sound and well cured.

Every cask being first searched, examined, and approved by a packer shall, when packed or repacked for exportation, be branded legibly on one head with the kind of fish it contains and the weight thereof; or the capacity of the cask, with the first letter of the Christian and the whole of the surname of the packer, with the name of the town, and with the words "Rhode Island" in letters not less than three-fourths of an inch long, to denote that the same is merchantable and in good order for exportation.

Every cask of pickled codfish and mackerel offered for sale or for exportation from this State shall also be branded "No. 1," "No. 2," or "No. 3," to denote the quality of such fish.

Nothing in this chapter contained shall hinder any fisherman or owners of fish coming to this State from their fishing trips from selling or reshipping their fish to any other of the United States without being packed into barrels or half-barrels.

Connecticut regulations for the inspection of pickled fish relate especially to the curing of shad, and since none of those fish are now pickled in that State except for home consumption the regulations are inoperative. The pickled-fish inspection laws of other States are either inoperative or they relate to certain species, and will be noted in the account of the methods of preserving those particular products.

BRINE-SALTED MACKEREL.

In the preparation of few marine products in this country are such nice distinctions made as in pickling or brine-salting mackerel. Not only has the work been reduced almost to a science by the fishermen and dealers, but it has been surrounded with a mass of legislation qualifying the manner of preparation almost without a parallel in the preservation of food products. Mackerel salting in the United States is confined almost entirely to Massachusetts and Maine, and four-fifths of the product is prepared in the first-named State, Gloucester and Boston being the principal centers of the trade. A few barrels are prepared also in New Hampshire, Rhode Island, and Connecticut.

The pickling of mackerel was of but little extent prior to the beginning of the present century, the annual product on the entire coast previous to 1816 rarely exceeding 15,000 barrels. The first salt-mackerel trip from Gloucester is said to have been made by the schooner *President* to Cashes Ledge, in the Gulf of Maine, about 1819. From that time to 1831 the industry rapidly increased, the output of Maine, New Hampshire, and Massachusetts during the last-named year reaching 449,950 barrels, the largest product in the history of the fishery. The value of the yield during that year was $1,862,793, while the value of the 324,454 barrels packed in those States in 1864 reached $7,001,098. In 1881 the yield was 391,657 barrels, with a reported valuation of $2,447,556. The increasing demand for fresh fish in this country has affected the trade in salt mackerel, a much smaller proportion of the catch being salted during recent years than formerly. Of the 131,939,255 pounds of mackerel taken in the United States fisheries in 1880, 80 per cent was salted, whereas during very recent years the salted mackerel represents less than half of the total yield. The quantity of these fish caught has also decreased greatly, so that at present the trade in salt mackerel is very much less than it was fifteen years ago. In 1887 the domestic product was 93,582 barrels, valued at $1,064,124; in 1890 it decreased to 20,742 barrels, worth $306,731; while in 1892 it numbered 46,946 barrels, worth $611,486. The yield was 24,939 barrels in 1895, 77,464 barrels in 1896, 13,154 barrels in 1897, and 14,286 barrels in 1898, less than 5 per cent of the annual average during the 40 years preceding 1886.

During the last thirty years quantities of salted mackerel have been prepared in the British North American Provinces, the annual product during the past three or four years averaging about 25,000 barrels. The mackerel taken on the coasts of Europe are generally sold fresh, but in Ireland, Norway, England, and Scotland many barrels are salted each year, especially in Ireland. Of the 399,361 barrels taken in those four countries in 1895, 46,500 barrels were salted, nearly all of which found a market in America. In 1897 the European product of salt mackerel was 57,352 barrels, and in 1898 it approximated 50,000 barrels. The European method of salting mackerel was until recently considered somewhat inferior to that in vogue in the United States, differing from the American method principally in that the fish were split down the belly instead of down the back, they were not soaked to remove the blood, and in packing in the barrel they were placed face up. The packers, however, have rectified these mistakes and the foreign mackerel are at present more carefully prepared than formerly, and those received in this country from Ireland and Norway now compare very favorably with the domestic product.

The domestic mackerel that find their way into the salt-fish trade are taken principally in purse-seines, most of those caught by means of lines, gill nets, pound-nets, etc., being marketed fresh. When salted, however, they are prepared in the same manner as those taken in purse seines, except that, the yield being usually much less in quantity, facilities for handling the fish rapidly are not of so great importance.

Mackerel taken by seines or gill nets do not usually keep so well when salted as those taken by lines, as the latter are taken in smaller quantities and greater care can be used in handling them, and they may be readily salted before deterioration begins and very shortly after being removed from the water.

The methods of salting as here given relate especially to fish taken by purse-seines.

When the fish are removed from the seine by means of a large dip net they are thrown on deck; or, if the catch be large, they are placed in a "pocket" or "spiller," rigged along the side of the vessel, where they can be kept alive until the crew have time to dress and salt them. So many fish are sometimes taken at a single haul that if at once removed to the deck many would spoil before the fishermen could properly care for them, and the purpose of the pocket is to provide a receptacle in which fish may be kept alive for several hours. This pocket was introduced in 1877 in a simple form on the schooner *Alice*, of Swan Island. An improved form was invented by H. E. Willard, of Portland, Me., and patented in April, 1881, but valuable improvements and modifications have been made in its construction since that time. The following is a description:

The mackerel pocket is a large rectangular net bag, usually 36 feet long, 15 feet wide, and 30 feet deep, with 2-inch mesh, hung to 1¼-inch rope. On the portion of the rope next to the vessel wooden floats are strung for the purpose of securing the edge of the pocket to the rail of the vessel, this edge being fastened over the rail and between it and a board held in position by wooden pins. The outer corners of the pocket are supported by ropes running through blocks attached to outriggers 4 inches in diameter, by means of which the outer edge of the pocket may be elevated or depressed. To the outer edge of the pocket is attached a rope bridle, the ends of which are fastened about 9 feet from each outrigger. A thimble is attached to the middle of this bridle, and when the mackerel have been turned into the pocket the fore and after staysail halyards are bent into the thimble and the outer edge of the pocket thus supported and the outriggers relieved from considerable strain. In getting the fish into the pocket the latter is slacked down to the surface of the water and the outer edge is fastened to the cork rope of the seine. By gathering the twine of the seine, beginning at the side farthest from the pocket, the fish are readily turned into the pocket, and the edge of the latter is then raised above the surface of the water.

Unfortunately the fishermen have found little use for these pockets during the last six years, the catch of mackerel being so small that they can be readily cared for before any of them spoil. The fish are removed from the pocket in quantities ranging from 25 to 100 barrels at a time. If the weather be warm and moderate the quantity removed at a time is small, but when the air is cool or the water rough or when dogfish are abundant the quantity is very much larger.

For dressing the fish, the crew is divided into working gangs of three men each, one of whom splits and the other two, known as "gibbers," gill and eviscerate the fish. Each gang of men is provided with a splitting board from 6 to 10 inches wide and with two wooden trays about 3 feet square and 6 inches deep, which are generally supported on the tops of barrels. Some crews, especially in the hand-line fishery, have only two men in each splitting gang, the splitter or some one else getting the barrels, filling them with water, and otherwise aiding the gibber. The splitter with his left hand, which is usually covered with a cotton mitten for protection as well as to prevent the fish from slipping, takes the fish round the center of the body, with the tail toward him, and splits it down the back on the left side of the backbone from the head to the tail, so that it will lie open and flat after the viscera have been removed, the knife being held by the fingers and guided by the thumb sliding along the upper side of the fish. On splitting each fish he tosses it to the tray of the gibber, who, with hands covered with gloves to protect them against the bones, opens the fish with a jerk, causing it to break lengthwise along the lower end of the ribs if it is fat, thus making a crease on each side. He removes the viscera and gills and throws the fish, open and face down, into a barrel partly filled with clean salt water, in which the blood is soaked from the fish, whence they are called "wash barrels." There the fish remain until the splitting is finished, which may be 6 or 8 hours or even longer after the first fish have been split. Then the deck is cleaned up and the men proceed to salting.

A good splitter can handle from 2,000 to 3,500 mackerel per hour, and under favorable circumstances 200 barrels of mackerel can be cared for by a crew of 12 or 15 men without difficulty before any of them spoil. Sometimes, when a large haul has been made, the crew may work steadily for 24 or even 36 hours in succession, losing only the brief time given to meals. By practice they can split and dress the fish as well at night as during the day.

If the men have time they "plow" or ream the fish, making a cut in the abdominal cavity on each side near the backbone, in imitation of the natural cracks or breaks which occur in fat fish, thus giving the fish a fat appearance. "Plowing" was begun about 1830, and although for a number of years there was great opposition to the innovation, it is at the present time recognized as a legitimate feature of the trade. The fatness of mackerel as well as the size determines the quality, and the degree of fatness is most readily ascertained by noting the portions covering the abdominal cavity. When the fish is very fat these portions crack open about halfway from the backbone to the center of the abdominal cavity, and the depth of these cracks indicates the relative fatness of the fish. By making the break or crack nearer to the backbone than where it would ordinarily occur and where the flesh is considerably thicker, the fish is given the appearance of being much fatter than it really is. At first these cracks were made by using the thumb nail, and later by the back of the point of the splitting knife, the cut by degrees being made higher than it naturally belonged. The use of the knife led gradually to the introduction of the plow or reamer, of which there are many styles, some made wholly of wood, others with the end tipped with pewter and with fine teeth on the edge, so as to make the crease rough, as though it were broken naturally. A popular form consists of a small cutting blade about 1½ inches in length, cut square forward and tapering to a point at the heel, attached to a curved iron shank, to which a wooden handle is fixed.

In salting, the mackerel are emptied from the wash barrels upon the deck and rinsed by throwing buckets of water over them. A man places them, a few at a time, on a gib tub containing a half bushel or more of No. 2 Liverpool salt, while another man, taking a fish in each hand, rubs the flesh side of the fish in the salt and, with the back of one fish against the flesh of the other, places them in the sea barrel with the flesh side down, except that the two or three bottom layers or tiers have the flesh side up. Formerly it was customary to place all the fish with flesh side up, but this has been abandoned. The salt is carried in the hold in barrels that are subsequently used for packing the mackerel. Liverpool salt is used almost wholly, Cadiz and other coarse salts having a tendency to tear and give a ragged appearance to the mackerel. It is quite important that every portion of the surface of the fish be in contact with the salt, and care should be taken not to leave finger marks where the fingers or thumb cover portions of the fish during the process of salting and prevent the access of salt.

Formerly on some vessels, especially those from Cape Cod, the mackerel were not rubbed in the salt, but were placed in the barrel with the flesh side up and the salt scattered over them. In salting the fish in that manner, Cadiz salt was used principally. The present method is much more rapid and leaves the fish in much neater condition, because the coarse salt pressing against the fish indents and lacerates it. By either process a barrel of mackerel may be salted in from 8 to 15 minutes, about a bushel of salt being used. After standing for a day or so and settling, the barrels are topped up by adding more struck fish to each barrel. When convenient, the barrels are headed and stowed in the hold or secured on deck until the vessel reaches port.

On arrival at port the barrels of mackerel are removed and placed on the wharf or in a storehouse until opportunity arises for repacking them, which may not be for months. Then the top of the barrel is removed, the brine poured off and discarded, the fish emptied out, several barrels at a time, into a culling crib or box of planed boards with slat bottom and usually 5 feet long, 3 feet wide, and 8 to 10 inches deep, placed on legs about 3 feet high. The fish are there culled into the several grades recognized by the trade and thrown into two weighing tubs, each holding about 100 pounds, which rest on a beam scale. These tubs have wooden staves and have the bottom perforated with inch holes to permit drainage, or, better still, a rope net-work bottom, and are bound with two iron hoops and have an iron handle on each side. The diameter of the tub is 24 inches at the top and the height is about 15 inches. When the proper weight of fish is placed in the tubs the fish are removed to a packing crib, somewhat similar to the culling crib, and usually 38 inches long, 26 inches wide, and 14 inches deep, where they are packed in barrels or smaller packages, the various grades being kept separate from each other and placed in different packages.

In packing, a small quantity of salt is sprinkled in the bottom of the barrel, next two or more layers of fish, with the flesh side up and succeeding layers of fish with the back up. Over each layer of fish a large handful of salt is sprinkled, about 35 pounds being used for each barrel of fish, which is required by law to contain 200 pounds of mackerel, exclusive of the weight of the pickle: while half, quarter, and eighth barrels must contain proportionate quantities. The total shrinkage on salt mackerel from the round to the marketable state is about 33 per cent. After being filled the barrel is headed and moved to some appropriate place on the wharf or in the storehouse, where it is "pickled"; that is, a hole is bored in the side or head of the barrel and as much brine as the barrel will contain is poured in. This brine should be made quite strong, at least of 95° salinometer test, and it is conveniently introduced by means of a water bucket with a copper nozzle in the bottom, forming a funnel, the end of which is placed in the hole made in the side of the barrel, a vent on the side permitting the air to escape. The hole is then plugged up and the barrel turned on end and branded. The branding kettle most commonly used is of stout sheet iron, cylinder shaped, 9 inches in diameter and 12 inches high. A rod with a wooden handle at the top passes through the center of the kettle and furnishes the means for handling it. A charcoal fire is made in the kettle and when the brand, usually made of brass at the bottom, is sufficiently heated, the barrels are stamped with the legal inspection marks. Because of leakage and evaporation it is frequently necessary to add additional pickle to the barrel after it has stood several days, the deficiency being noted by the sound produced by striking the barrel with a stick.

The total cost of repacking mackerel, including barrels, salt, and all labor, from the time the fish are received from the vessel, ranges between $1.25 and $1.60 per barrel, depending on the market price for barrels, labor, etc.

The general average of cost approximates $1.44 per barrel, apportioned as follows:

Labor—weighing and culling	$0.25
Labor in packing	.10
Salt in packing	.10
Cooperage	.06
Repickling	.08
Bangor barrel	.55
Supervision, use of plant, etc	.30

The laws of most of the New England States require that the work of culling, weighing, packing, and pickling be all performed under the personal supervision of a State inspector, who places his brand on the head of each barrel or package, indicating the kind and grade of fish, name of inspector, name of town and State where packed, and date of packing. In Maine and New Hampshire it is necessary that the date include the month as well as the year in which the fish are packed, but in Massachusetts the year is deemed sufficient. If by becoming rusty, or the pickle leaking out, the fish require repacking, they must again be inspected. There is much difference in the quality of mackerel taken during different seasons of the year. The early spring catch is generally very poor and shrinks considerably when salted. The fish increase in fatness as the season advances, and those taken during the fall usually improve in weight in pickle. Full-grown fresh mackerel measure 17 or 18 inches in length, but some over 20 inches and weighing 3 or 4 pounds are caught. The average length is about 12 inches and the weight a trifle less than a pound. Salted mackerel measure considerably less, due to the loss of the head and the slight shrinkage in salting.

The grades of salted mackerel are very carefully defined by the statutes of various New England States, and with little difference in one State from those in another. In Maine salted mackerel of the best quality, not mutilated, measuring, when split, not less than 13 inches from the extremity of the head to the fork of the tail, free from taint, rust, or damage, are branded as "number one"; the next best quality, being not less than 11 inches, measuring as aforesaid, free from taint, rust, or damage, are branded as "number two"; second quality mackerel, but free from taint or damage, and not less than 13 inches, are branded as "number three, large"; those of the same quality, not less than 10 inches in length, are branded as "number three," and all other mackerel, free from taint or damage, are branded as "number three, small." The grades required by the laws of New Hampshire and Massachusetts are almost identical with the aforegoing, except that in the last-named States the fish are branded as "number four" instead of "number three, small." The regulations of Rhode Island are somewhat indefinite in this particular, requiring merely that every cask of mackerel offered for sale or for exportation from the State shall be branded "number one," "number two," or "number three," to denote the quality of such fish.

In addition to the grades designated by law, packers of mackerel prepare special grades known as "extra ones," "extra twos," "bloaters," etc. Extra ones are superior in size and fatness to legal ones, and are sold at a much higher price; and the same difference exists between extra twos and legal twos. Bloaters are the choicest mackerel prepared, and only a few barrels are secured each season.

Mess mackerel are also prepared as an additional form of the other grades. These are principally the best and fattest mackerel that would pass as numbers one and two, with the heads and tails removed, and with the slime, etc., carefully brushed off before being repacked. There is an average loss of about 17 per cent in weight in preparing mess mackerel from the customary condition of pickled mackerel, but this varies considerably, depending on the size and degree of fatness of the fish.

The laws of most of the New England States regulate the character and size of the barrels in which pickled mackerel are packed and the materials of which they are made. The law of Maine requires all barrels and casks to be made of sound, well-seasoned white oak, white ash, spruce, pine, chestnut, or poplar staves, with heading of either of such kinds of wood, sound, well planed, and seasoned, and the barrel or cask to be hooped with at least three strong hoops on each bilge, and three also on each chime; the barrel staves to be 28 inches in length and the heads to be 17 inches between the chimes. In Massachusetts all packages, except those containing less than 25 pounds weight, must be made of white oak, ash, red oak, spruce, pine, or chestnut, and the number of the hoops and the size of the barrel staves and heads are the same as set forth in the Maine laws. In each State the barrel must hold between 28 and 29 gallons, and the half-barrel not less than 15 gallons, and the tierce between 45 and 46 gallons. The regulations in New Hampshire are identical with those of Massachusetts as to the quality of the material and dimensions of the barrel, but require that it shall contain between 29 and 30 gallons. Each cask must be made in a workmanlike manner so as to hold pickle, and be branded on its side near the bung with the name of the maker.

Most of the barrels used for pickling mackerel are manufactured in Bangor, Me., but a few are made in various other parts of New England. The price is generally from $40 to $55 per 100, but when an unexpectedly large demand for them exists they sometimes sell as high as $1 each at the fishing port. Barrels once used are sometimes repaired and used over again, but this practice is not commendable.

For convenience in marketing, brine-salted mackerel are frequently placed in half, quarter, eighth, and sixteenth barrels, after they have been prepared in the regulation-size barrels. In recent years a considerable market has been developed for much smaller packages, and when the fish are sufficiently cheap, they are frequently put up in 5-pound and 3-pound tin cans, for a description of which see page 520.

The following summary, compiled from the files of the Gloucester papers, shows the fishermen's price per barrel of the principal grades of mackerel during the first week of September in each year from 1830 to 1898, representing generally the average for the year:

Year.	No. 1.	No. 2.	No. 3.	Year.	No. 1.	No. 2.	No. 3.	Year.	No. 1.	No. 2.	No. 3.
1830	$5.00	$4.50	$2.62	1853	$11.50	$9.50	$7.50	1876	$15.00	$6.75	$5.50
1831	5.75	4.75	2.62	1854	15.00	12.25	5.00	1877	16.50	12.50	8.00
1832	5.00	4.00	2.75	1855	10.00	11.00	6.25	1878	18.00	8.00	5.00
1833	5.72	4.72	2.85	1856	13.00	8.00	6.00	1879	16.00	5.00	3.00
1834	5.72	4.72	3.35	1857	15.00	12.50	8.50	1880	14.00	7.00	4.00
1835	7.00	6.00	4.00	1858	15.50	12.50	8.50	1881	14.00	6.00	4.00
1836	9.00	8.00	5.00	1859	14.50	12.50	8.50	1882	18.00	11.00	8.00
1837	7.75	6.50	4.12	1860	16.00	8.50	5.00	1883	20.00	14.00	10.50
1838	11.00	9.25	5.50	1861	8.50	4.50	2.75	1884	14.00	10.00	3.50
1839	12.50	10.50	7.00	1862	8.25	6.00	4.50	1885	13.75	5.75	3.75
1840	12.75	10.50	5.50	1863	14.00	20.00	6.50	1886	22.00	12.50	9.50
1841	12.00	10.00	6.00	1864	30.00	20.00		1887	17.50	14.00	11.00
1842	9.00	6.00	4.00	1865	22.00	15.00	9.75	1888	22.00	18.50	14.00
1843	10.12	8.12	6.00	1866	22.75	13.25		1889	28.00	25.00	17.01
1844	9.50	7.50	5.50	1867	17.00	12.25	7.50	1890	21.00	17.50	13.00
1845	13.00	10.50	6.87	1868	17.00	13.00		1891	18.00	13.00	8.00
1846	9.12	6.25	3.87	1869	23.00	11.50		1892	20.00	12.00	10.00
1847	12.75	8.25	4.25	1870	23.00	9.75		1893	16.50	14.00	12.00
1848	9.00	6.00	3.37	1871	11.25	7.25	6.25	1894	18.00	14.50	12.50
1849	12.00	7.00	3.50	1872	14.50	9.50	7.00	1895	20.00		
1850	10.12	8.12	5.00	1873	20.00	12.25	9.00	1896	17.50	14.50	
1851	10.00	6.50	5.12	1874	13.25	9.00	7.00	1897	16.00		
1852	9.00	7.00	5.75	1875	16.25	10.25	7.50	1898	17.00		11.50

BRINE-SALTED HERRING.

Several different species of the Clupeidæ family are known locally in the United States as herring. The principal ones are the sea herring (*Clupea harengus*), so abundant in the Gulf of Maine; two kinds of alewife (*Pomolobus pseudoharengus* and *P. æstivalis*), known on many parts of the coast as river herring, and the herring of the Great Lakes (*Argyrosomus artedi*). The sea herring occurs north of Montauk, while the alewife inhabits the rivers and bays all along the Atlantic seaboard, the fishery being of the greatest importance in the tributaries of Chesapeake Bay and Albemarle Sound. The herring of northern Europe are of the same species as those of the New England coast.

In this report the name "herring" refers to the *Clupea harengus*, the other species being known as alewives or river herring and lake herring.

It is impossible to assign even an approximate date for the first salting or pickling of herring. Francis Day, in his well-known work on the Fishes of Great Britain and Ireland, parts V–IX, p. 222, writes as follows on this subject:

At the beginning of the twelfth century there were herring fisheries in the Baltic, to which many foreign vessels resorted; these herring must, therefore, have been *salted*: in fact, in 1155 Louis VII, of France, prohibited his subjects purchasing anything but mackerel and salted herrings at Estampes.

The manner of curing these fish is considered to have been very crude until the time of William Beuckels, or Beuckelzon, a fish merchant of Biervliet, in Flanders, who, during the fourteenth century, greatly improved the methods in use and laid the foundation of the great wealth acquired later by Holland in this business. Beuckelzon died in 1397, and a monument was erected to his memory by Charles V in his native village, Borgo; while Mary of Hungary, during a visit to the Low Countries, is said to have paid a more characteristic tribute to his memory, namely, that of eating a salt herring at his tomb.

The first mention we have of pickled herring in America is by Josselyn, in the seventeenth century, who, in his Chronological Observations of America, says: "We used to qualify a pickled herring by boiling of him in milk." It is almost self-evident, however, that the pickling of herring was carried on by the earliest settlers of America, and possibly by the fishermen who resorted to these shores from Europe before the country was settled, as it was an old-established business in Europe.

The quantity of herring preserved by the process of pickling is greater than that of all other species combined, aggregating nearly 3,000,000 barrels annually, but the yield in the United States (about 30,000 barrels annually) is small compared with the product of Scotland, Sweden, and the Netherlands. Half a century ago the output in New England was many times greater than at present, and there was a considerable export trade, amounting at times to 60,000 barrels; but, owing to carelessness in the method of pickling and absence of uniformity in the quality of the product and in weight of the packages, the trade has been captured by the foreign curers, who now sell about 150,000 barrels annually in the United States, at double the price received for the domestic article. In view of the large quantities of herring on the New England coast and the extensive markets that already exist in this country as well as in the adjacent countries to the south, it seems extremely desirable that more care should be given to the curing of this fish, and the preparation should be governed by fixed standards applicable to both quality and quantity.

The quality of pickled herring varies greatly, depending almost entirely on the quality of the fresh fish and the promptness and care exercised in curing them. Few fish are more difficult to properly cure than this species: the flesh is very delicate and tender, and not only does it injure readily, but it is much less able to take the salt if the pickling be long delayed after removal from the water. If placed in pickle before they have been much exposed, they take the salt quickly and the natural quality and flavor of the fish are better preserved. Another important rule in preparing this as well as other kinds of fish is to have the greatest possible cleanliness in the salting houses and in the tubs and barrels used for salting.

There is no uniform method of curing herring in this country, but most of them are salted just as removed from the water, without splitting or dressing, and are known as "round herring," to distinguish them from the "gibbed herring," which have the gills, heart, etc., removed, and the "split herring," which have the gills and all viscera removed. The method of preparing each will be described separately.

ROUND HERRING.

The bulk of the herring salted on the New England coast are taken in the vessel gill-net fishery, and the fish are usually salted on board the vessels. Some are taken also by small boats making daily trips to the fishing-grounds, and these are necessarily salted on shore. The methods of salting the fish on the vessels and on shore differ only according to the facilities for handling them.

In the vessel fishery, as the herring are removed from the nets they are placed on deck and water is soused over to remove loose scales, blood, etc., and sometimes salt is sprinkled over them. They are next placed in hogshead tubs with about 3 pecks of salt to the barrel of fish scattered among them, when they are covered with brine and left standing for four or five days, or until they are struck. It is important that the salting be done as soon as practicable after the fish are removed from the water, and in the meantime they should be protected from the sun. After being struck, the fish are repacked in market barrels, where they remain for twenty-four hours for settling, when the barrels are topped up by adding a few struck fish to each barrel. It requires about 330 salted fish to fill a barrel, the number of fresh herring required being about 300. In packing the fish in the barrel, they are usually placed with backs slanting upward. Occasionally, however, those in the lower half of the barrel are placed back down, and sometimes a few upper and lower layers are carefully placed and the middle portions arranged with less care. After the barrel has been topped up it is filled with brine, headed, and stored in the hold. On arrival at port, if it appears from sounding that some of the pickle has leaked out, more is added through a hole bored in the bilge and the hole plugged up, when the barrel is ready for branding and marketing. Sometimes at the port the fish are repacked, so as to insure full weight and good fish, new brine being made, if necessary, but if the old pickle is clear it is used over again.

To determine the proper amount of salt required for curing herring requires considerable skill and experience, and the quantity varies according to the condition of the fish, the season of the year, etc. If too much salt be used the fish will soon become hard and dry, with greatly diminished flavor, but if the quantity of salt be insufficient the fish will become tainted and unfit for food within a short time. During warm weather more salt must be used than when the temperature is low, and thin small herring require less salt than thick or large ones. As a general rule, about 5 pecks of salt are required for curing each barrel of herring.

In pickling on shore, the fish on their arrival are dumped into tanks or wash barrels of sea water, from which they are at once removed with a brail net, the fish being rinsed up and down at the same time, and placed in a pickling butt or vat with about 3 pecks of salt scattered among each 200 pounds of fish and a heap placed on top. In two or three days a workman with high rubber boots passes over the butts, treading on the heaps of fish to separate them if massed together. Or, in some localities, the masses are separated by stirring them with a spudger, consisting of a thick board 10 inches long and 2 or 3 inches wide, nailed in the center to a wooden handle. The fish remain in the butts eight or ten days, being examined occasionally and more salt being added to keep the pickle sweet. After being thoroughly struck, the fish are removed with dip nets and placed on a packing table, whence they are packed in the market barrels, 200 pounds to each barrel, sometimes with backs up and sometimes with bellies up, according to market demands, about half a bushel of salt being scattered among them during the process of packing. The barrels are then headed and a hole bored in the bilge or head and sufficient strong brine added to fill the barrel; the hole is then plugged and the barrel is ready for shipment. It is important that the herring be packed so tight in the barrel that their relative positions are not disturbed in handling the barrel, thus removing the scales and breaking the fish. There is little decrease in weight in pickling herring, 211 pounds of fresh fish making a barrel of 200 pounds of round herring. The market price for round herring during recent years has ranged from $3 to $5 per barrel.

The cost of preparing a barrel of pickled round herring in New England is about $2.37, apportioned as follows:

Cost of 200 pounds of fresh herring	$1.00
Labor of pickling in butts	.07
Salt used in pickling in butts	.20
Barrel	.50
Salt used in packing in barrel	.09
Labor of packing in barrel	.08
Cooperage	.06
Cartage	.07
Wear and tear, loss, etc	.30

GIBBED AND SPLIT HERRING.

Although most of the herring taken on the New England coast are salted round, some of them are either gibbed or split before being salted. This method of curing the fish is much more effective than salting them round, the latter being objectionable because all the blood is thereby retained in the fish and undergoes a slight decomposition before the salt thoroughly strikes through the skin and flesh. It is very infrequent that herring are salted round in Europe, the usual practice being to gib them before salting. Less salt is also required for preserving the fish when they have been gibbed, and thus the flavor of the product is improved. Gibbing consists in removing the gills, heart, and sometimes the viscera from the fish by means of the thumb and forefinger. It requires 228 pounds of round fish to make 200 pounds of gibbed fish. Gibbing is very little practiced now in the New England States, because the fish are usually not very fat, and look thin and poor when gibbed. Splitting is performed by cutting the fish down the belly to the vent, removing the gills and viscera, and usually the roe bags and milt of spawning herring. In some localities, especially at Eastport, Me., it is customary to immerse the fish in brine for a few moments before they are cut. This causes the herring to keep their scales better and brighter, and they can also be handled more readily in splitting. Splitting was originally applied only to the extra large herring in order to permit the salt to more readily strike through them; it is now commonly practiced in Newfoundland and Canada, but herring for pickling are rarely split down the belly in the extensive fisheries of Europe.

After evisceration the fish are immersed in tubs of salt water for a couple of hours to allow the blood to be soaked from them, when they are packed in butts or tight barrels, back down, with the stomach cavities filled with salt and with a layer of salt sprinkled between the layers of fish, about 3 pecks of salt being used for each barrel of fish. When properly struck the fish are repacked in the same manner as the round herring above described. In packing in the barrel some fishermen place them on their backs and slightly inclined to one side. Others place them fully on the back. The latter appears to be the better method, as it gives the herring a more round and thick appearance and the pickle has a better chance at the abdominal cavity. The split herring usually sell at about $1 per barrel more than round herring, but in 1898 the price was approximately the same. A 200-pound barrel of fresh herring will weigh 144 pounds when split and eviscerated, losing 56 pounds in dressing. Sometimes, in order to fill out a shipment of split herring, round salted herring are split and eviscerated and added thereto. In this operation care must be taken not to tear the fish.

The present condition of the pickled-herring industry in this country is far from satisfactory. Great improvements in the methods of cure are desirable, and unless the quality of the product be bettered there seems little probability of an increase in the prosperity of the business. The abundance of these fish on the New England coast during certain seasons of the year, and the large market demand for pickled herring in this country, should encourage our fishermen to put up a product that will compare favorably and compete with the foreign cured herring, the great necessity being a fixed standard, applicable to quantity as well as quality, with proper culling and grading of the different kinds. Usually during September and October there is a run of fine herring on the New England coast, which if prepared with that care exercised in the curing of European herring would probably be nearly if not equally as good; but at present their value is depreciated by too long exposure to the sun and air before salting, by carelessness in the manner of salting, failure to separate the fish into the various grades, and by using weak barrels in the packing.

THE IMPORTATION OF FOREIGN HERRING.

The great bulk of the pickled herring consumed in the United States is prepared in the Netherlands, Norway, and Scotland. Of the 1,321,020 barrels received during the ten years ending June 30, 1898, 377,480 barrels came from the Netherlands, 231,098 barrels from Norway and Sweden, and 111,198 barrels from Scotland.

The Holland herring are the most popular of those received in the United States, and the demand for them is constantly increasing. The first of the Holland herring arrive here in June. These are known as "matties," having small roe or milt. The "vol" herring, which have the roes fully developed, arrive usually about the latter part of July. The trade becomes brisk toward the end of August and continues until near the end of November. There is also considerable demand during February, March, and April, and usually all are sold before the 1st of June. The great bulk of the receipts, probably four-fifths, are in small kegs, containing from 45 to 55 herring, about 15½ pounds of fish. These kegs are made of hard wood and measure 9½ inches in height on the outside, 7¾ inches at the bilge inside, and 7 inches at the ends inside, with staves and bottoms ¼ inch in thickness. At the top of the keg the staves are slanted off at the ends, but at the bottom they are of the same thickness out to the ends. They are bound with 6, 8, 14, or 16 hoops, but more frequently with 6 or 8. When only 6 hoops are used the two end ones are usually of galvanized iron and the others of willow or similar material. On the head of the keg is branded the description of the contents, with some distinctive trade-mark.

In addition to the kegs, many Holland herring are received in barrels, containing from 242 to 254 pounds of fish, exclusive of the pickle. A few half-barrels and quarter barrels are also received, but they do not take so well as the barrels and the sixteenths. The price ranges from 45 to 75 cents per keg and $7 to $11 per barrel. Packages containing "milkers," or milt herring only, usually sell for 10 to 15 cents more per keg and from $1 to $2 more per barrel than those containing mixed herring.

The imports of Scotch herring have increased considerably during recent years, from 186 barrels in 1885 to 32,036 in 1898, most of which are received at the port of New York. A few years ago they usually sold for somewhat less than the Holland herring, but during the past two or three years they have sold for $2 to $4 more per barrel. These are also packed in small kegs, but the keg trade is not so extensive as in case of the Holland herring. The consumption of Norwegian herring is also increasing. Practically all the Norwegian and Scotch herring are sold in barrels, the former containing 400 to 800 fish or 221 pounds, and the latter from 350 to 700 fish or 250 pounds, exclusive of the pickle. The demand for these is greatest from September to November and from February to April.

Nearly all the receipts are on consignment, the consignee forwarding the account and remittance as soon as the herring are sold, receiving 5 per cent commission therefor. The expense of handling foreign herring at New York City approximates $2.15 per barrel, made up as follows: Duty (at ½ cent per pound), $1.25; brokerage, 25 cents; cartage, 20 cents; clearance, cooperage, storage, insurance, etc., 30 cents, and commission, at 5 per cent, 45 cents. The following summary shows the ruling prices during October, 1898, for the various classes of foreign herring in the New York market:

Description.	Number to barrel.	Number to keg.	Half-barrels, price.	Barrels, price.
Holland herring:				
Full white hoops, mixed	52 to 54		$4.00 to $4.50	$7.25 to $7.50
Full white hoops, milkers	62 to 65		4.75 to 5.00	8.25 to 8.50
Scotch herring:				
Large fulls, E. C	550 to 675			5.00 to 5.25	9.00 to 9.50
Fulls, E. C	650 to 750		4.75 to 5.00	9.00
Large fulls, W. C	500 to 550		4.50 to 5.00	9.00 to 9.50
Mediums, W. C	300 to 350			8.25 to 8.50
Norwegian herring:				
K. K. K. K	425 to 475			10.50
K. K. K	550 to 640			8.50
K. K	700 to 800			10.00
K	800 to 900			9.50
Cutheads				7.00
Belly cuts				6.00
Bristling				6.00

FOREIGN METHODS OF CURING HERRING.

The importance of improvement in our methods of pickling herring is sufficient reason for introducing in this connection some notes on the methods of curing herring in foreign countries. Mr. Adolph Nielsen, who has had considerable experience with the various methods in use, furnishes the following accounts of the processes of pickling herring in Scotland, Norway, and the Netherlands:

THE SCOTCH CURE OF HERRING.

As soon as the herring is landed and sprinkled with salt the gibbing and cutting take place. This is performed in the following manner: The herring is held in the left hand, stomach up, in such way that the head reaches beyond the thumb and forefinger (index). With the other hand a small straight-edged and sharp-pointed knife is pushed in just below the gill lid on that side of the herring that shows toward the right hand and forced right through the throat, close to the neck bone, so that the point of the knife comes well out on the other side. The forefinger is then pressed against the head and the thumb across the pectoral fins. A little cut with the knife is first made down in the direction of the tail, after which it is given a twist and a cut made close up under the pectoral fins. The throat is grasped between the index and middle finger (on the right hand), with a smart jerk, the intestinals (stomach, crown gut, liver, and heart), along with the gills and pectoral fins, are torn out, leaving only the milt or roe to remain in the herring. When the herring is intended for export to Continental markets the crown gut is often allowed to remain.

Another way of gutting herring, when these formerly have been clipped, is to put the thumb behind the gills and with a jerk from the top and downwards break the gills loose from the head; when these are then taken out the œsophagus (gullet) and the stomach follow, because all these are cohesive; but this way is seldom in use anywhere else than in some cases in Norway.

According as the herring are gibbed and gutted they are sorted in baskets and put into the sousing tubs, where they are rolled in small Liverpool salt before being packed in barrels, generally made of birch. In these the herring are packed fully on their backs, with a small plateful of salt sprinkled over each layer of fish. The one layer of herring is put across the other the whole barrel through, and each layer furnished with two "head-herrings," put on their sides. The barrels are filled until a couple layers of fish extend above the chime or top, and covered with wooden covers made for that purpose. In this way they remain two or three days, after which time, when the herring has settled, the barrels are filled up again with fish from the same packing, headed up, and put down on their sides. Every second day, as a rule, the barrels are given a little turn around until the last packing (bung packing) takes place. A lookout is during this time kept upon the barrels, that none of them are leaking. In order to obtain the official crown brand the herring must have been in salt at least ten days, exclusive of the first-day packing. When the barrels, after such time or later on, are going to be made ready for shipment, the pickle is drawn off through the bungholes and these plugged up; the barrels are then opened and more herring of the same packing pressed into the barrels, either by means of a common press or by pressing the herrings down either by hand or by trampling them down after a small barrel head has been placed on top of the herring. Care is in the meantime taken that the herring is not pressed so tight that no room is left for the pickle. The object with this last and tight packing is to prevent the herring from being shaken about in the barrels during the time of conveyance and to save the recipient from the trouble of repacking the barrels after they have reached their place of destination. After a sufficient quantity of herring is pressed into the barrels, they are headed up and filled through the bunghole with the same pickle that was drawn off formerly, after being strained. The bungs are then put in tight, the hoops driven home, and the barrels blown; after they are joined tight the uppermost hoop is nailed fast, the blowhole stopped up, and they are in a condition ready for shipment. If the herring is to be exported to countries outside the European Continent or to hot climates, it is generally, when the bung packing takes place, emptied out of the barrels altogether; the crown gut is removed if it has been left, and the herring rinsed in clean water and repacked with coarse Liverpool salt. In place of using the original pickle, the barrels are then filled with new pickle made of clean salt. All these barrels are full-banded and furnished with a 1-inch-wide iron hoop on each end. In order to obtain the official crown brand, such barrels should contain no less than 212 pounds of herring, exclusive of salt and pickle. A good many of the Scotch herring are also packed in half-barrels.

The system of culling, along with the official crown mark on all exported herring barrels, has contributed more to the good reputation the Scotch herrings have gained in the continental markets than many may imagine. The dealers, on reception of Scotch herring with the crown brand, are satisfied that the barrels really contain what they are branded for in regard to quality and weight, and this has given them such confidence in the Scotch herring that these are received and approved of without even being opened, while the Norwegian herring barrels, since the official branding was abolished in 1851, must be opened and repacked before the recipient can sell them, which often causes a good deal of inconvenience. The Scotch herring are sorted and branded according to the treatment or cure and the development of the sexual organs.

Crown F, Full Brand: Barrels obtaining this brand must contain all fine, well-cured, large, full herring, and not mixed with herring of a poorer quality, nor with spent herring, nor matties, which have not got their roe or milt fully developed.

Crown F, Matties Brand: Barrels with this brand must contain fine, rich herring, with small milt or roe, must be well cured, and not mixed with full, spent, broken, or dismembered herring.

Crown F, Spent Brand: Barrels with this brand should contain spent herring (herring with their sexual organs more or less collapsed after spawning), properly gibbed and cured, and all full herring, matties, broken or dismembered herring sorted out.

Crown F, Mixed Brand: This brand is used for mixed herring (such herring as can not be sorted as full, matties or spent). The mixed herring should also be properly gibbed, packed, and cured, and no dismembered herring be packed in the barrels.

Crown FF, Repacked Brand: Barrels with this brand should contain herring which have been in salt at least ten days, exclusive of the day of packing and the day of repacking and branding. Further, this herring should, when they are repacked, be emptied out of the barrels in which they were first packed or cured, the crown gut be removed, and the herring be rinsed and repacked, with sufficient salt, in the same barrels, and supplied with new, strong pickle made of clean salt. The barrels should be full-branded and furnished with a 1-inch wide iron hoop on each end.

Lozenge Brand: This brand is used for herring which formerly have been bung-packed and branded but afterwards repacked in the same way as is required in order to obtain the repacked brand. The lozenge is branded just below the crown brand. In case new barrels are employed, they are branded with the crown brand and the lozenge in the same way.

THE DUTCH CURE OF HERRING.

The most of the Dutch herring are caught at sea in drift-nets and cured on board of the vessels. If there is a chance, the dressing (gibbing and cutting) takes place according as the nets are hauled on board and the herring picked out of them. The Dutch way of gibbing and cutting herring is about the same as the Scotch; the only difference is that the crown gut is allowed to remain in the herring, and it is considered that the fat which is attached to this gives the herring a nice flavor. After the knife is put through the throat the cut is at once made up toward the pectoral fins; thus the opening is made smaller than in the Scotch herring.

According as the herrings are dressed they are sorted in baskets, and from these about 200 at a time are put into large trays filled with fine Lisbon or St. Ybes salt and rolled in this salt. After the herring is carefully and well rolled in the salt it is packed in barrels, back down, the same way as the Scotch, with Lisbon salt sprinkled on the top of each layer of fish. When the whole catch is salted down a bucket of *blood-pickle* (made of sea water and the intestines from the dressing) is put over the herring, and the barrels headed and put down in the vessel's hold.

The object of heading up the barrels so soon is to prevent the herring from being affected by the air. In this state the barrels remain for from six to ten days, when they are taken up and filled with herring of the same packing, after the pickle is first drawn. This filling or sea-packing is pretty compact, and it takes about four barrels to fill three of them. After the barrels are filled, the original pickle, after being strained, is put over the herring, and the barrels headed up and blown, and if found tight, put back into the hold of the vessel. It is considered of importance not to undertake the filling of the barrels too early, as the herring in such cases will be shriveled; but, on the other hand, it should not be performed too late, because if so the herrings, by being tossed about in the pickle while the vessel is rolling in the sea, lose a great deal of their scales. After the vessels arrive home, the barrels are again filled with herring and supplied with the original pickle. In this last filling, it is generally estimated that thirteen barrels of herring in a fit state for shipment are obtained from fourteen sea-packed barrels. A large quantity of herring is also repacked in small kegs—one-sixteenth part of a barrel—and containing from 45 to 50 herrings. This is especially for the American markets. A smaller quantity is also packed in half-barrels, but these do not take very well. A Dutch barrel of full, selected herring, with milt and roe, contains about 800 herrings, which weigh from 110 to 115 kilograms (242.2 to 251.6 pounds) net, exclusive of salt and pickle.

Dutch herrings are sorted, according to the development of the sexual organs, in four qualities: Full herring (vol herring, branded VOL); matties (maatjis, branded M); spent herring (Ijlen, branded IJ or IJLE), and herring which have recently spawned (Ruit, branded KZ). Besides, herrings which have not been packed before the day after they were caught are branded O. Each of the first three brands are again sorted in three or more qualities, and branded No. 1, 2, and 3. All herring which on account of so large a catch could not be cured the same day, but had to be left over a night before they were packed, together with torn bellies, or chafed herring, are assorted according to quality, as No. 3, while all herring in good condition and free from faults, as No. 1 and No. 2, according to quality and treatment. Distinction is also made between herring caught in the open sea, near the coast, and in the Zuider Sea; and the barrels are generally furnished with a mark signifying the place and the year in which the herrings were caught. Before the official system of culling was abolished, in the year 1878, this was branded on the bilge of the barrels (or if the herring were packed in smaller packages on the most convenient place) by the culler in such manner that a royal crown was branded in the middle, and the other directions in letters on either side of the crown.

THE NORWEGIAN CURE OF HERRING.

The largest quantity of herring in Norway is caught in the fjords by seines, and kept barred until what food the herring may contain is worked out in the natural way before they are taken up, dressed, and salted. As a rule the herring are salted in the vicinity of the places where they are caught, so that they can be put in salt almost alive, which is of the utmost importance in order to obtain a good article. In this way they have an advantage over the Scotch, who have to go far off the coast for their herring and can not get them in salt before they reach the shore, which often takes a long time. They have an advantage over the Dutch, because, although they salt their herring on board of their vessels soon after they are caught, still they may have been dead several hours in the nets before they are hauled on board, and at all events none of the herring taken in drift-nets or other nets can be deprived fully of the injurious food they may contain, as they can when barred in a seine. When brought to the shore or salting places from the seine the herrings are gibbed in this way: That a triangular piece of the throat, large enough to admit the heart and the pectoral fins to be removed, is cut out by means of scissors made for that purpose, or by a small knife (some also use the fingers). This cut should be made deep enough to divide the large blood veins, situated close to the neck bone, in order to remove the blood they contain. Sometimes, also, the gills are removed, especially on the full herring caught in the spring.

Generally a large enough crew is employed to admit the gibbing and the salting to take place at the one time. On account of the herring caught in seines being always mixed, every gibber has got placed before him or her so many barrels or tubs as the herring are to be sorted in (from three to five); and, according as they are gibbed, every herring is also at the same time, by the gibber, sorted and placed in the various barrels or tubs to which they belong. The salter then takes the herring and packs them in new barrels, which lately have been soaked in sea water, slantwise on their back, with one-fourth barrel of St. Ybes salt to 1 barrel of herring. The herring are packed loosely, one lies across the other the whole barrel through. The uppermost layers are packed sometimes slantwise back up. Some packers put from 1½ to 2 gallons of pickle (made of one-fourth barrel of salt to 1 barrel of sea water) on the herring soon after it is salted, and head up the barrels immediately. Others, again, let the barrels remain unheaded for one day before they fill them with pickle. Before the barrels are headed up a layer or two of herring are generally put into the barrels in order to fill up the empty room caused by the shrinking of the herrings. By putting the pickle on the herring soon after they are packed the salt dissolves quicker and saturates the herring more speedily, so that the contents of the stomach (provided the food is liberated) hardly has any injurious effects upon the durability of the herring. After the barrels are headed up they are broached in the head and blown into by means of a brass pipe containing a valve, which is put down in the hole, and, if found tight, the hole is plugged up as soon as the air has escaped; if not, they are made tight in the places where they are leaky, and blown over again before they are stowed down on board the vessels.

After reaching the port of shipment and before being exported the herring are repacked and the barrels filled with the original pickle which was formed first; and if this does not hold out, new pickle is made to supply what is wanting. As a rule the herring are repacked in such a way that out of 4 barrels salted in the fishing-places from 3¼ to 3½ barrels of herring are obtained when packed for shipment. They never, as a rule, pack their herring as light as the Scotch or Dutch do, except the herring is specially to be put up in such style.

The following notes on pickling herring are from Nielsen's Report on the Cure of Herring:

Qualities of the good herring. —Concerning the nature of the fresh herring, it is required, in order to obtain a good article, that the herring also possess certain qualifications, such as sufficient size and maturity, fleshiness, and fatness. A lean, dry, dismembered or half-rotten herring can never give a good article, even if it is cured ever so well. A small herring, which has not reached the full state of maturity, fetches only small prices in the markets. Of much importance is also the development of the sexual organs. If these are in a far advanced state, the herring loses in fatness and flavor. These should be firm and the whole flesh penetrated with a certain quantity of fat. Large amounts of fat around the blind-gut is a sign of the herring being fat right through the flesh. As a rule, ocean herring (such herring as pass most of the time in the ocean, and only approach the coast for reproductive purposes) are considered superior to the herring that keep themselves close to the coast or in the bays all the time. Of these herring, again, those which are caught in deep water are better than those caught in shoal water. A first-class herring is known by its small head, short and plump body; is broad across the back, and plump toward the tail, and has got a great depth from the back to the abdomen, which gives this well-rounded shape.

Importance of early salting. —In order to obtain a good article of salt-cured herring it is necessary that the herring be liberated from its food and put in salt as soon as possible after being brought out from the water. Even if the quality is ever so fine, a good article can never be had if it is not properly treated during the whole cure. The Scotch herring can not get the official crown brand except they are salted at least twenty-four hours after being brought out from the sea. As a rule the curers do not care about herring which is more than twenty-four hours out of the sea before they are landed, and only take those at a low figure. In Holland, even, a distinction is made between herring which are taken out first and those which are taken last from the nets. The cure of herring on board the fishing crafts commences, therefore, if circumstances allow, soon after the nets are hauled in. To leave the herring exposed to the hot sun while being conveyed to the salting-place, or to leave the herring in the nets until the shore is reached, if the catch has taken place a long distance off the shore, is objectionable. An old law in Norway, of 1775, even prohibited people from taking such herring from their seines in the summer time before 10 o'clock in the evening and after 5 o'clock in the morning, if it was going to be salted for export.

Best salt for herring.—In regard to what kind of salt is the most suitable for salting herring, it is difficult to give any one sort the preference. The choice of salt depends much upon how the herring is going to be cured, and upon the size and quality. The main thing is that the salt is clean, and that it is used in proper quantities. Fine and watery salt melts quicker, but gives weaker pickle. In cases where it is of importance to form pickle speedily, fine salt is preferable, while coarse salt is better for use in filling and repacking, or when the herring is intended for export to hot climate, or to be kept in stock for any length of time. The Scotch curers use Liverpool salt, the Dutch light Cadiz or Lisbon salt, while the Norwegians use St. Ybes salt. It is of much importance, as formerly stated, to put the herring in salt as quick as possible, if a first-class article shall be obtained. For this reason the Scotch, as the herring is landed, sprinkle it heavily with salt in bins or vessels made for that purpose, before it is gibbed and gutted. Generally they use 1 barrel of Lisbon or coarse Liverpool salt (or sometimes both mixed) to about 10 barrels of herring. By this means the herring keep their scales better and brighter, and can also be handled better and quicker when they are afterwards gibbed and gutted. The Hollanders roll their herrings in trays filled with fine Liverpool or St. Ybes salt as soon as they are gibbed and gutted, before they pack them in barrels; and this work is done very precisely. In Norway no sprinkling with salt, as a rule, is used before the herrings are gibbed or packed in barrels, but instead thereof they have to use more salt in packing than the Scotch and Dutch. The sprinkling of herring with salt as soon as they are landed or brought on board of the vessels is considered also to improve the flavor of them very much.

Packing herring.—In packing the herring in barrels it is recommendable not to pack them too tight before they have shrunk in the salt, and also to pack herring of the same size and quality right through the whole barrel. The packing is performed differently among different nations. In Norway the herrings are packed slantwise on their back, while the Scotch and Hollanders pack them fully on their back. By this last mode (which no doubt is the best) the herring get a more round and thick appearance in the pack; and it has also this advantage, that the pickle has got a better chance to get in and saturate through the abdominal cavity of the gutted herring. After the herring has shrunk in the salt the barrels are filled up again and put away, but care is taken that the herring is not packed too hard. As long as the barrels are left to remain still there is no need of hard packing, but when they are to be shipped it is recommendable to repack the herring so tight that they do not move about, even if the barrels are handled ever so roughly, so that the receiver may be exempted from filling the barrels again after they have reached their place of destination.

Herring barrels.—The quality of barrels used for salting herring in is of much importance in order to obtain a desirable product. If too soft wood is employed, the pickle will work through the staves, the herring become dry, and be damaged within a short time. Among the foliferous wood in Europe the populus (poplar) is considered least answerable, and among the conifers the spruce or fir are less suitable than the red pine wood, because the former is generally knotty and more ready to get saturated with pickle or water. Good hard and clean spruce, which is cut fresh and has not been soaked in water, may compete with the pine when it gets properly seasoned. The Hollanders use mostly barrels made of oak; the Scotch use barrels of birch or beech, and the Norwegians use barrels of spruce and red pine wood. Staves made of birch are brittle and apt to twist. In Scotland the regulations for making herring barrels are that the staves shall be not less than half an inch thick and not wider than 6 inches, except the oak staves, which may be 7 inches wide, and that the bottoms should be at least of the same thickness as the staves, and none of the pieces of which it consists be made wider than 8 inches. The usual thickness of the staves in the Scotch herring barrels are from nine-sixteenths to ten-sixteenths of an inch, and the bottoms are generally made three-fourths of an inch thick. In Holland there was a law passed enacting that a herring barrel should be manufactured of at least 13 staves (which makes every stave on an average 5¼ inch wide), and that no stave should be less than three-eighths of an inch in thickness.   In Norway the staves, as a rule, are made five-eighths inch thick, and the barrels are mostly made by machinery.   Scotch barrels are generally full banded for export to hot climates in the summer time.   They are also furnished with an iron hoop on each end. The Dutch barrels are furnished with 18 to 20 hoops, divided 5 or 6 on each end and 4 on each side of the middle, while the Norwegians have only 12 to 16, divided 3 and 3 or 4 and 4.

In Norway a movement has been made lately to get a law according to which all barrels for shipment of herring should be manufactured of a certain kind of wood and of certain dimensions, but, so far as I am aware, such act has not been passed yet.   If the barrels are made of fat spruce or red pine, and also of oak, the herring will take a flavor from the barrels, which some people like very much, but others, again, do not care for.   In Scotland it was prohibited to salt herring in barrels manufactured of red pine until the year 1874, and for many years back a similar act or law existed in Norway; but this law was repealed again on account of the Russians, who consumed large quantities of salted herring, and valued the Norway cure very much on account of the resinous taste the herring got from the red pine barrels.   Some people, again, value the Dutch herring very much on account of the taste or flavor they receive from the barrels.   To keep the barrels from shrinking, it is recommended to put a little pickle in them while they are kept in stock.

In 1889 the government of the Dominion of Canada appointed delegates to visit Scotland and the Netherlands for the purpose of studying their methods of curing herring.   After making a careful investigation they summed up their conclusions in part as follows:

We consider the Scotch system of treating herring, as an article of commerce, to be as perfect as any system can be when honestly carried out in all its integrity, and that improvements in the herring industry of Canada can safely be made after the Scotch model, so far as our somewhat different circumstances may permit.   In one most important respect the Scotch fishermen have an immense advantage over the Canadian fisherman, and that is in having all the curing and packing of herrings performed by a distinct and independent class of merchants known as fish-curers.   We have seen that the herring industry of Scotland never amounted to anything—never prospered—until an enterprising and energetic body of men came forward and assumed the position of fish-curers, taking that part of the business entirely out of the hands of the fishermen, to the great relief of the latter, and the promotion of this important business.   In order, therefore, to place this industry in Canada upon a satisfactory and permanent basis we are of opinion—

(1) That while some slight degree of improvement in the details of curing and packing herrings may be at once attained by changes in the present methods, yet no real permanent improvement can take place, nor can the herring industry in Canada be placed upon a satisfactory commercial basis until the fisherman ceases to be his own fish-curer, and until the business of curing is taken in hand by a class of merchant fish-curers, as in Scotland—men of energy, business experience, and capital, with all the necessary appliances to carry on the business on fixed principles and in accordance with such regulations as may be promulgated for the benefit of the trade from time to time.   In Scotland nearly all the curers devote their whole time to the supervision of their own curing operations.   In former times if a fish-curer did not so superintend his own curing business he made arrangements with a master cooper to furnish the barrels, and generally to superintend the cure and putting up, guaranteeing that the same would pass the Crown brand; or he hired journeymen coopers and a foreman, laid in materials for the manufacture of barrels, and the foreman attended to the curing for the market.   This was the method to a great extent in Scotland 55 years ago, and to some extent still.

When a fisherman cures his own fish, it is done with the least possible labor and cost, and once they pass into the hands of the merchant or trader, paying up so much of his indebtedness to the former, he feels no more interest in the matter.   But it is different with the merchant or curer; his capital, his credit, his good name are at stake, and unless his fish are properly cured he will be the loser. Thus the strongest of all motives, self-interest, would induce him to cure and put up for market an article calculated to render a return for his outlay.

(3) *The barrel.*—That the present Canadian barrel, being too weak to stand the rough handling to which it is exposed on the railways in transportation, should be greatly improved.   We think the present capacity of the barrel should be retained; that it should contain not less than 200 pounds of

herrings, exclusive of salt; that it should be made of stronger material; that the staves should be of hard wood wherever possible, and that if spruce is permitted to be used, the staves should be thicker and stronger than the present stave. Fir and pine should be prohibited. The rule in Scotland has been that the barrel might be made of any kind of wood, fir excepted. It seems very probable that a fir or pine barrel may "sour" fish or other animal food packed in it for the first time. If a new pine barrel or cask be filled full of pure water and covered up and allowed to stand for some weeks there will be formed in the water a pretty firm gelatinous substance, which, if allowed to remain for some time, becomes very offensive, as has been the case often in regard to pipes made of tamarac and pine used for conveying water underground, where frequently pipes with a 3-inch bore have been nearly clogged up from end to end with this offensive matter. If cleaned out, however, and the pipes relaid, the gelatinous substance will not again form.

It is universally conceded that oak barrels are the best of all, and, where procurable, the Dutch use no other kind. The present barrel we think, after sufficient notice, should be prohibited. We are of opinion that a little more bilge would add to the strength of the barrel. The hoops should be of better quality than they are now and there should be more of them; and all barrels intended for transportation beyond the province where the fish are put up should have at least one iron hoop at the top; and should the barrel be of spruce or Norway larch, then, in addition to a thicker stave, there should be an iron hoop at each end. If, however, hard-wood staves can be procured in sufficient quantity, we think that soft-wood barrels should not be allowed. The knot of the hoop should be longer than it is in the case of the present hoop, to prevent it suddenly springing off. There should be in the side of the barrel, above the bottom hoops, 15 or 16 inches from bottom, a bunghole about 1 inch in diameter, with well-fitting bung, and the usual rule as to crossing the heads should be observed. The new standard barrel, when decided upon, should be made a legal standard by act of Parliament, as is intended to be done in Britain at the next session of the Imperial Parliament. There should also be a legalized half-barrel built of the same material and in proportion to the size of material used in the large barrels.

(4) *Small packages.*—We are of opinion that in addition to the barrel and half-barrel, there should be established a grade of small packages in which to put up repacked herrings, after the manner of the Dutch and Germans; that these packages or kegs should be integral parts of the large barrels, say one-fourth, one-eighth, and one-tenth parts of the full-sized barrels. We believe that kegs in every way suitable for this purpose may be procured in Canada. A gentleman largely interested in the sugar-refining business told us that he gets a very neat, small keg or kit, for holding sirup, made in Ontario, at reasonable prices, and he considers that the small herring keg can be made here as well and nearly as cheaply as in Holland. If so, then a very great step is assured toward the establishment of what we believe to be a most important and profitable branch of the herring industry of the Dominion.

We are of opinion that the very first movement toward improvement in this industry should be in the direction of improving the herring barrel and consequent discouragement to the manufacture of the present barrel.

As already stated, there can be no doubt that a very large business can be done in the small package line, if properly gone into and taken hold of with energy and in a business manner. Canada should be able to largely supply the demand on this Continent for herring put up in this way.

(5) *Curing.*—That next in importance to the catching of the herring is the proper curing of them. This process commences in having the salt brought into contact with the herring as soon as caught; and if it could be done at sea as soon as the herring come out of the water, so much the better and so much more thorough the cure. All the printed evidence, all the experience of the Dutch, the Yarmouth fishermen, and others, and all that we heard on the subject goes to confirm this. If attention to this preliminary salting be necessary, and so salutary in results in Britain, how much more necessary is it in the maritime provinces, where the temperature in summer is so much higher? In order, therefore, to preserve the herring from incipient taint and to retain all the delicate flavor and natural excellence of the fish, it is absolutely necessary that at the earliest possible moment the curing process shall commence, and that the herring be scrupulously shaded and sheltered from the damaging effects of the sun throughout all the stages of gutting, curing, packing, etc. This is most important because injury caused to the herring by exposure to the sun can never be remedied. The injury sustained by herring in this way is beyond the power of man to remedy. Inordinate quantities of salt, soaking and washing in water may cover up the damage done and prevent further taint, but the lost

excellence of the fish can never be restored. There can be no doubt that the great bulk of the injury sustained by the herring of the Maritime Provinces is caused in this way; and if this can be avoided in the future by the adoption of some feasible inexpensive measures to protect the herring from the sun, from the time of catch to the time of shipment, a great point will be gained and much done to redeem the character of the Canadian herrings. The next step in the curing process is the "roosing" of the herring after gutting and the proper salting of the fish when being packed. Should it happen that the preliminary salting can not be effected before delivery, then the herring should be well sprinkled with salt during delivery. The western consumer is about tired of eating herrings, out of which all excellence has been extracted by soaking in water and oversalting, and if these western markets are to be retained the quality of the fish must be improved, and that at once.

(6) *Gutting*.—That we consider the Scotch mode of gutting to be as good as any for all commercial purposes, and all that can be desired when properly carried out and the curing properly attended to. In Scotland the early herrings are very fat, and are not branded. These are the herrings which are in such request by the wealthy families of Russia, and they are hurried over to Stettin for immediate sale and use. A fish-curer told us that some of these herrings sold in June, 1889, for £10 sterling ($50) per barrel. There is no material difference in the mode of curing these herrings. In Canada, however, it appears that in the case of fat herrings caught in July and August special treatment has been found necessary. Mr. Gordon, of Picton, who has had much experience in the herring business in Scotland and in Nova Scotia, says:

"I beg to advert to the only additional detail, which, in my experience, I have discovered as applicable to the perfect cure of herrings in the months of July and August, on the coasts of Nova Scotia. Having engaged on my own account in a sailing vessel trading on the coast of Nova Scotia and Cape Breton, and provided with salt and barrels, I preferred purchasing the herrings in their green state, and cured a few barrels after the Scotch manner. On examination of the fish after being struck, I discovered an incipient taint along the backbone of the fish, which would increase with age, so as to render them unsuitable for a distant market in a tropical climate. I came to the conclusion that the taint was owing to excess of temperature here over that common on the Scottish coasts, and besides the herrings are larger and fatter in the months of July and August on this coast than on the coast of Scotland. Thereafter I ripped with a sharp knife the belly of the fish, and filled the belly with salt, and immediately packed them in tight barrels, with one bushel of Liverpool salt to each barrel, and protected the barrels from sun and rain."

Another gentleman, referring to the same subject, says: "Herrings should be all opened with a knife and filled with salt; otherwise they can not be properly cured." This latter statement, as applying to all herrings, seems rather general. Mr. Gordon only recommends this treatment in the case of herrings caught in July and August on the coast of Nova Scotia, when the fish are very fat. Even then it can hardly be possible that the belly of every herring need be filled with salt. To fill a herring with salt must effectually destroy the flavor of the herring and leave it as innutritious as a piece of basswood. It may be, however, that some of these July and August herrings may be utilized for the manufacture of kippered herrings. In October, 1889, a very fine and well-flavored kipper, said to come from Baltimore, U. S., was for sale in Toronto at high prices. It was very fat. The "ciscoes" of Lake Ontario are very fat, fully one-fourth or one-third oil, yet they make most delicious kippers, are in great demand, and sell at high prices. The "ciscoe" is a herring and is taken in deep water in the fall of the year. Many of them are put up as bloaters.

All of this shows how necessary it is that some one or more competent men, experts, technically and practically, in all pertaining to the classification of herring, should fully investigate all the different kinds of herrings on our Atlantic coasts, and decide upon the different modes of cure adapted to special kinds of herrings in special localities and at special seasons of the year.

(7) *Round or ungutted herrings*.—That between the mode above recommended by Mr. Gordon and the mode of putting up round or ungutted herrings there must exist many degrees of difference. We have already referred to ungutted herring, and to the fact that the Scotch curers strongly disapprove of packing ungutted herring, and expressed their surprise that any people of the present day would waste salt and time for such a purpose. In the case of the best packed herrings, if a stave breaks and the brine runs off, the herrings undergo very serious deterioration, but in the case of ungutted herrings, under similar circumstances, total destruction of the contents of the barrel would take place from the setting free of the elements of decomposition contained in the ungutted herring theretofore held in check by the preservative qualities of the salt and pickle. Round or ungutted herrings put on the market in any quantity can only do harm by damaging the character of the whole catch of any given locality. We therefore consider the putting up of all such herring for commercial purposes should be strictly forbidden.

(17) *Washing of herring.*—As stated elsewhere, the washing of herring before curing is not practiced in Scotland, and so far as we can learn never has been. The Scotch curers with whom we conversed on the subject were surprised to hear of herring in Canada being washed before curing. They could hardly believe such a practice possible. To show what has been the practice in the Maritime Provinces in time past, we quote from the answer of Mr. Gordon, Picton, to question No. 1, in 1869. He says:

"It is the universal practice of Nova Scotia fishermen to steep the fish for hours before salting down, and expose them to the action of the sun during the hottest period of the season until the water becomes warm, under the erroneous impression that they are thus benefited by the extraction of the blood. Under this treatment herring part with their scales and juice, and are deprived of that flavor peculiar to herring properly cured. Besides the body of the fish thus saturated with water is rendered tasteless, brittle, and short, and not calculated to turn out satisfactorily at the end of a long sea voyage. The Scotch curers take every precaution to keep the herring from contact with water before and after salting. Salt the fish in their blood and the salt will extract the blood."

Now, it must be quite clear to every man who realizes the importance of retaining intact all those qualities of substance and flavor which render the herring so valuable as a wholesome and pleasant article of food, and which are so highly prized in Europe as already stated, that there could be no more effectual mode devised or adopted for the total destruction of all those qualities than the mode above described by Mr. Gordon. What would be thought of any man or body of men who should treat any kind of animal flesh, beef or pork for instance, in such a manner, and what would be thought of any one who would subject those articles of food to such treatment? Who would think of selling or buying meat so treated? Then, to cover up the damage done by the water soaking to the herring, inordinate quantities of salt are used, and this extracts any vestige of flavor the water may have left, and destroys all the nutritive qualities of the herring.

Here we may be permitted to refer to another pernicious habit, already adverted to—the practice of putting brine on the newly packed herring, in addition to the salt in which they have been cured. This practice is highly objectionable, and is one also that effectually destroys the good qualities of the herring in flavor and substance. This practice may arise from the custom of putting herring down in large casks or vats, and then, after a time, repacking into the common tight herring barrel. Herring treated in this way can not be much better than those soaked for hours in warm water. The herring should be packed in tight herring barrels in the first instance, and fully salted, when the necessary quantity of pickle will be formed from the dissolution of the salt caused by the moisture in the fish. The addition of newly made pickle not only interferes with the curing process going on in the barrel, but, as already stated, as effectually destroys the natural qualities of the herring as soaking in water, or the action of the sun's rays acting through the medium of water. We therefore consider that the practice of washing and soaking herring in water and of adding newly made pickle to newly packed herring should be strictly prohibited.

Having already described the Scottish method of curing and packing herrings, and having expressed our opinion of that system as being entirely adapted to the curing of herrings on our Atlantic coasts, we feel convinced that its general adoption and its legal enforcement would, in a very short time, give Canadian herrings a very high standard and character in the markets in which they are now held in very low estimation. We have given this subject our best consideration, and so convinced are we of the vital importance of proper curing and protection from sun and rain, that we would again urge that every diligence and care be exercised in these respects, as well as to the quality, selection, and separation of the fish in the first instance. It is most desirable that the fish, especially during the hot season, should be handled and shifted as little as possible, as every time they are turned over they part with a portion of the scales and become softer and softer, more flabby, and less ready to absorb the salt.

In the matter of packing pickled herrings for the purpose of repacking into small packages, very great care should be taken to have the fish of the best quality in every respect. In Holland and Scotland, as we have shown, "fulls" and "crown fulls," the highest brands, are taken for this purpose. The repacking should be done well and neatly and the kegs, as already stated, filled up with the brine from the large barrel out of which the fish has been taken. A slight sprinkling of salt on the bottom of the keg and the top tier of the herring should be given.

This branch of the business is worthy of the special consideration and effort of all who may engage in the curing of herring. We know that in western Ontario there is now a demand for herring put up in small packages, especially in the rural districts. It can readily be seen that in a country where there is so much beef and pork not many families will purchase whole barrels of herrings, while many would gladly purchase herrings in half-barrels, quarter-barrels, and the smaller packages

If the Dutch and Germans can afford to pay freight and shipping charges on herrings from Scotland to Holland and Germany, unpack and repack into small packages, pay freight and shipping charges to New York, and sell these herrings in Quebec and Ontario, with a good margin of profit, surely the herring traders of the Maritime Provinces should be able to supply herrings in this shape as good in quality, at lower prices, and with a better margin of profit. By supplying a proper article this branch of the trade can be increased immeasurably both in the United States—the Western States especially—and in the inland provinces of the Dominion.

BRINE-SALTED ALEWIVES OR RIVER HERRING.

At various points along the Atlantic coast more or less alewives or river herring are brine-salted each year. They are prepared in greatest abundance in the tributaries of Chesapeake Bay and the coastal waters of North Carolina, where they are known only as herring, and also to a less extent in Maine and Massachusetts. At the head of Chesapeake Bay 30,000 barrels of herring are brine-salted annually, the number of fish required for the pack approximating 20,000,000. It is not unusual for 300 or 400 barrels of pickled herring to be prepared as the result of a single haul of a seine, and 900 barrels were salted from one haul in 1893. The Chesapeake product is used mostly in the South, and is distributed principally from Alexandria, Fredericksburg, and Richmond. The alewives salted in New England are sold also through the South to some extent, but many of them are sent to the West Indies and South American countries.

The methods of pickling river herring or alewives do not differ greatly from those applied to the sea herring on the New England coast, except that the market price being lower necessitates that they should be prepared in a cheaper manner. The flavor of the alewife does not equal that of the sea herring, consequently there is little need for the nice discrimination required in case of the latter. Usually more salt is used in preserving them than for sea herring, and as a result they will keep much longer. Mr. Joseph Farris, of Eastport, Me., states:

The chief difference between the alewife and the herring is their capacity to keep for a long time is that the alewife has less flavor than the herring. It is almost without flavor. When the herring loses its flavor it becomes insipid and unpalatable, although it may be sound; but so long as the alewife is sound it is as suitable for food as at any time in its preserved condition. Alewives are sometimes kept on hand three years before being shipped, but if herring are not shipped within one year after being cured they are usually turned out of the barrels and used for fertilizer.

The three principal classes of pickled alewives are, (1) "gross," the entire fish being salted, corresponding to the round herring of the New England coast; (2) "split" or "cut," the head and viscera having been removed before salting; (3) "roes," the head being removed and the main gut drawn, but with the roe left in the fish. Each locality has its particular process of preparing the different grades.

THE CHESAPEAKE PROCESS.

The main object is to get the fish in salt as quickly as possible after they are removed from the water, but first the scales must be removed and the fish washed. In case the seine is hauled on a sandy beach the movements of the dying fish about the sand are sufficient for removing the scales. But when the seine is hauled on a float, sand is sprinkled among the fish, and a few workmen, with high rubber boots, shuffle about among them or they are dredged back and forth by means of a board attached to a long handle. The fish are next washed or rather rinsed to remove the sand, loose scales, etc. This may be accomplished by dipping basketfuls of them in the water or by placing them in a slat-work box and running water through the mass, stirring them about in the meantime. Some curers scale and wash the fish at the same time, the fish being placed with sand in tubs of water and washed with brooms and then placed in half-barrel tubs with holes in the bottom and sides for draining. About thirty years ago a machine was introduced for scaling fish which was used for a while, but is now discontinued. This consisted of a revolving lattice-work cylinder, having projecting metallic blades arranged upon its inner periphery and sides. The cylinder was filled with fish and revolved in a tank of water, the scales and slime falling through the lattice work and being carried away by the water.

The salting is done in large vats or hogsheads, a convenient size for the vats being 16 feet in length, 5 feet in width, and 2 feet deep, having capacity for about 32 barrels of fish. The bottom of these is first covered with 4 or 5 inches of very strong brine; then put in 8 or 10 barrels of fish, stirring them about as they are being dumped in, and sprinkling more salt on top, following this up with fish and salt, with a heavy covering of salt on top, 6 barrels of salt being used for 32 barrels of fish. In case the fish are being cured in hogsheads, the latter should be half-filled with strong brine, then 4 half-barrel tubs of fish are dumped in, and these covered with half a barrel of Liverpool salt. More fish and salt are then added until the hogshead is filled. After remaining thus for twenty-four hours the herring are stirred with a "breaker," a long stick or pole, flattened at the end, which is about 3 inches in width, and twelve hours thereafter the fish are again stirred. In stirring or "breaking" them in the vats the breaker is run under the mass and then elevated to the surface, the object being to bring the lower layer of fish to the surface and break up the masses which have become bunched together, so that the salting may be uniform throughout.

After remaining over night the fish are "muddled," for the purpose of "pumping" or drawing forth the blood from the gills. This consists in pushing them back and forth with a rectangular board, 5 inches long and 3 inches wide, attached at the upper surface to a long handle, and is done twice daily for six or eight days. Each time the fish are "muddled," during the first four or five days, a quantity of salt is sprinkled over them, about 2 bushels being used the first day, and the quantity gradually decreased.

At the end of seven or eight days, when the fish have become thoroughly struck or cured, they are removed with scoop nets and thrown on racks or stands having open-work bottoms, where they drain for one or two days before packing. When the herring are very abundant and the workmen exceedingly busy the fish sometimes remain on the racks for eight or ten days, but in such cases they are liable to rust. The fish are packed in barrels, with layers of salt between the layers of fish, from 2½ to 3 pecks of salt being used for each barrel. Turk's Island salt is preferred, but Liverpool salt is used to a considerable extent. The former is larger-grained and does not dissolve so quickly, and it also makes the fish sweeter. The fish are placed backs down, excepting the top layer, and those in each layer are placed at right angles to those in the preceding layer. When the barrel is filled it is allowed to settle for a day or two, then topped up with another layer, strong brine added, and the barrel coopered and stored ready for market. The usual number of herring to each barrel is 400, and the weight is generally 160 pounds.

The cost of preparing a barrel of river herring in the Chesapeake region approximates $1.10, of which 35 cents represents the cost of salt, 50 cents the barrel, and 25 cents the labor; the transportation to market costs about 10 cents per barrel, and the commission for selling is 10 cents, making a total cost of $1.30 for preparing and placing the fish on the market. In 1897 pickled river herring sold for about $1.40 per barrel, giving the preparer only 10 cents per barrel for the cost of the green fish, the superintendence, use of plant, and outlay of money. The average selling price in 1898 was advanced to $2.19 per barrel, which gave a fair margin of profit to the curer. Select "all roe" herring sell for about $6 per barrel.

The preparation of "cut" or dressed river herring differs from the above mainly in that the heads and viscera are removed before the fish are washed preparatory to salting. In dressing, the fish are held in the left hand on a cutting board, with the back from the workman, and with one stroke of a knife held in the right hand the head is removed, and a continuation of the stroke cuts off the edge of the belly, laying the fish open from the napes to the vent, the viscera being extracted by a single movement of the fingers. The cutters in the Chesapeake fisheries receive usually 20 cents per 1,000, and an experienced workman can dress 12,000 to 20,000 per day. The fish are then washed and soaked, and in every other particular of salting and packing in the barrel the process is the same as in preparing the gross or round herring. Less salt is required for cut herring, and the number of fish placed in a barrel is about 650, the weight being 160 pounds, as in case of gross herring. The price in 1897 approximated $2.15 per barrel, but in 1898 it was advanced to about $3 per barrel.

THE NEW ENGLAND PROCESS.

About $15,000 worth of pickled alewives are prepared annually on the Maine coast, mostly in the vicinity of Eastport. To Mr. Ansley Hall I am indebted for the following account of the methods in use at that point:

For packing in barrels the alewives are salted round as they come from the water. The parties who handle them at Eastport usually have an agent in the locality where the fish are caught who buys them from the fishermen and salts them temporarily in barrels to preserve them until they reach Eastport, where they are taken out of the barrels and properly cured before being finally packed for shipment. Occasionally the agent cures and packs them, but more frequently he does not. In some instances they are placed on board a transporting vessel which has been sent especially for them and are salted in tubs or hogsheads by the crew. In such cases they are afterward cured and repacked in barrels by the dealer at Eastport.

In salting them temporarily a small quantity of water is first put into the barrel or hogshead, about one bucket in a barrel and four or five buckets in a hogshead. The fish are then put in loosely in layers with salt between each layer and well covered with salt at the top. The quantity of salt required is about half a bushel to each barrel of fish. If they are packed by an agent to be sent to Eastport for curing and repacking, barrels are used. They are allowed to stand about four days before being headed up and the salt is renewed at the top of the barrels as fast as it dissolves. This first salting is done with Liverpool salt. The effect upon the fish is not to permanently cure them, but to strike them so they will keep in good condition for a short time.

When they reach Eastport they are taken out of the barrels and put into the large herring tanks, which hold about 4 hogsheads or 20 barrels each. Each layer of fish is covered with a layer of Cadiz salt and a heavy layer of salt at the top of the tank. In all, about one-half bushel of salt is used to each barrel of fish, or approximately 10 bushels to the tank. A strong pickle is then made and turned in. The pickle contains about 1 peck of Cadiz salt to the barrel of water, and 5 barrels of pickle are necessary to cover the fish in the tank. Boards are then laid across the tank, with heavy stones on them for weights to keep the fish down under the pickle. If the salt on top dissolves, more has to be added. The fish remain in the pickle about 8 weeks. They are then taken out and closely packed in fish barrels, 200 pounds of fish to the barrel. A layer of Cadiz salt is placed between each layer of fish and a heavy layer of salt at the top of the barrel. About 3 pecks of salt are used to each barrel of fish. The barrels are filled above the chimes and have to stand two or three days to settle before the head can be put in. They are then headed and are ready for shipment. In some instances buyers desire pickle put in, and if so, the barrels are turned over on their side and a 1½-inch hole bored in the bilge, through which pickle is turned in by means of a funnel until the barrel is completely filled. The hole is then plugged. If the buyer does not request it no pickle is put in. It is generally considered that the fish will keep longer without the pickle than with it. It is claimed that the pickle has a tendency to make the fish soft. It is estimated that from the time the alewives are taken fresh until they are cured and packed for shipment about 3 bushels of salt are used to each barrel of fish.

At Waldoboro they are not kept in pickle longer than two or three weeks at most, and are considered in a suitable condition for packing after being in pickle eight or ten days. The quantity of salt used in curing and packing was estimated not to exceed 2 bushels to the barrel.

PICKLING RIVER HERRING IN RUSSIA.

The following method of pickling the large, fat river herring of Russia prevails on the Caspian Sea, this description being furnished by Mr. Schröder, of Stettin, Germany:

The fish are salted in layers, in large reservoirs dug in the ground, protected by a wooden shed, and holding from 60 to 100 barrels. No pickle is poured on them, as it forms of itself, after a few days. In the course of six or eight days the fish are taken out of the reservoirs and packed in barrels, a little salt being sprinkled over each layer. When about three-fourths of the barrel is filled, a mat or sack is laid over the fish, the packer gets into it and tramps them together, the vacant space is packed with tang [a seaweed], the end pressed in by means of a screw and closed. In the uppermost end a bunghole is made, through which is poured a quantity of new boiled pickle, containing from 20 to 25 per cent of salt, the whole being then ready for the market. When the fish are to lie for an indefinite time, ice cellars are made, very conveniently fitted up, and the reservoirs dug beneath them. The Norwegian method, which is much superior, has been tried during the last year and is found to give good results. The salt used in the curing is obtained from salt lakes on the banks of the Volga, and costs from 10 to 15 kopecks per pood. It is found in great abundance, especially on the banks of Lake Baskuntschak. The barrels are made of lime wood, about three-quarters of an inch thick, and are fastened with 14 to 16 wooden or 4 iron hoops. When used for sending a long distance they are made of oak, and it is purposed to try birch-wood barrels, as it is thought that birch keeps the pickle better than lime wood. In size they are a little larger than the Norwegian barrels, and contain about 400 fish of average size. A barrel with wooden hoops costs a ruble, iron hoops costing 10 to 15 kopecks more.

BRINE-SALTED COD AND HADDOCK.

The trade in brine-salted cod on the New England coast is small and is confined exclusively to the small fish, under 16 inches split, measured on the back of the fish from the hollow of the nape to the hollow of the tail. A few haddock are also pickled, but hake, pollock, and cusk are rarely placed on the market in this condition, except possibly a few barrels representing a surplus from the Fulton and other fresh-fish markets situated in places where it is not convenient to dry-salt the fish. In pickling cod or haddock the fish are dressed, split, washed, and salted in butts with about 2 pecks of salt to the 100 pounds of fish, in the same manner as has been heretofore noted in preparing dry-salted cod. When orders are received, the fish are removed from the butts, cleaned with brushes, and placed in tight barrels, 200 pounds to the barrel, face side up, except the top layer, which is placed back up, the fish being bent to follow the curve of the barrel, pressure being applied, if necessary, to place the 200 pounds in the package. It is important that the fish be not repacked until thoroughly struck through, otherwise the flesh will be marked with yellow spots caused by contact of the imperfectly cured fish with one another. Coarse Trapani salt is placed at the bottom of the barrel and over each layer of fish, about 1 peck of salt being used to each barrel of fish. The barrel is then headed and strong brine is added through the bunghole, when the package is ready for shipment. The gross weight of a barrel of codfish, including barrel and pickle, approximates 325 pounds.

It requires about 430 pounds of round cod or 290 pounds of split fish to make a barrel of 200 pounds pickled. If the green fish cost 40 cents per 100 pounds, the cost of preparing a barrel of pickled cod approximates $3.05, divided as follows:

Fish, 130 pounds, at 40 cents per 100 pounds	$1.72
Labor, dressing, splitting, and pickling	.14
Salt used in pickling	.32
Barrel used in packing	.50
Labor of repacking	.07
Cooperage	.05
Wear and tear, loss, etc	.25
Total	3.05

The average price of pickled cod is about $4 per barrel, while haddock are usually worth from 25 to 50 cents less. The market is principally in New York and the West. The annual product on the New England coast ranges from 2,000 to 3,000 barrels.

It appears that there is scope for enlargement of the trade in pickled cod, especially if prepared with great care. Considerable quantities are pickled in Holland, Scotland, Sweden, Belgium, France, and the British North American Provinces. The process employed in each country differs somewhat from that in the United States; the business is conducted more systematically, and the output is much more extensive.

It is generally conceded that the choicest pickled cod are prepared in Holland, those fish selling on the European markets at an equivalent of $11 to $16, and sometimes as high as $25, per barrel of 250 pounds, compared with which the average price of our pickled cod ($4 per barrel) seems very small. Following is the usual process:

As soon as caught each fish is bled by cutting the throat and is then split down the belly from the throat to the tail, the knife running somewhat on the side of the ventral line so as to have the flesh on one side of the dorsal line much larger than the other. The head and three-fourths of the backbone are removed, and the fish immediately washed. The abdominal cavity is well brushed, and to thoroughly cleanse the parts about the remaining portion of the backbone the tail is twisted from left to right and from right to left, and also bent up and down during the process of washing. After being cleansed the fish are packed with dry salt in butts and allowed to make their own pickle. When well struck, usually in five or six days, the fish are repacked in market barrels with some fresh salt between them and with the old pickle poured over all. In packing, the tail of each fish is held in the right hand and the upper portion in the left hand, and the fish so folded that about one-half of the left side is underneath the right side, the body of the fish being bent to follow the curve of the barrel, each layer in the barrel being formed by two fish. The fish should be firm and free from a sodden or flabby condition.

In Scotland the fish are usually bled as soon as caught, and after being split and washed, as in the United States, are placed in butts or barrels with about 75 pounds of Liverpool salt to each 250 pounds of split fish. After remaining there two or three days they are removed, cleansed with brushes, and packed in shipping barrels with about 50 pounds of salt scattered among the fish in each barrel, and strong pickle is then added. Most of these fish are sold in London at from £2 to £3 per barrel.

The process used by the Swedes in pickling codfish differs little from that employed in Scotland, except that Lisbon salt is generally used, and the fish are subjected to considerable compression during the first salting. In Belgium St. Ybes salt is used, and in repacking, the old pickle from the first salting is added through the bunghole, it being claimed that this old pickle is better than new brine, because it prevents the fish from turning yellow and also gives it a better flavor.

A century ago quantities of codfish were salted in barrels provided with holes near the bottom to permit the brine to leak away. The product was not generally considered so delicately flavored as cod retained in the brine, but in dressing it for the table it swelled, whereas the latter shrinks.

BRINE-SALTED SALMON.

During the last century and the early part of the present a large portion of the salmon taken in the rivers of New England were salted in barrels for local use during the winter and for distant markets. At present, however, practically the entire catch on the Atlantic coast is marketed fresh. Many salmon are salted on the Pacific coast of the United States, especially in Alaska, where the business originated ten years ago, and at one or two points on the coast of Oregon and California, the business in the latter State dating from 1853. In Alaska the red (*Oncorhynchus nerka*), the humpback (*O. gorbuscha*), and the king or chinook salmon (*O. tschawytscha*) are salted, while lower down the coast the silver salmon (*O. kisutch*) is the species generally used, but some chinook are also salted. The annual product is about 25,000 barrels, valued in San Francisco at about $10 per barrel.

Quantities of salmon are also brine salted in the British North American Provinces, especially on the coast of Labrador and Newfoundland, as well as in the Hudson Bay territory. These fish are known in the United States as "Halifax salmon." The trade began early in the present century, and since 1840 has ranged between 3,000 and 10,000 barrels annually, the present annual receipts averaging 5,500 barrels, valued at about $15 per barrel. In the fisheries of northern Europe and Asia salmon are also salted, but it is unusual for any of the product to be received in this country.

In dressing salmon for pickling on the Pacific coast, the heads are removed and the fish split along the belly, the cut ending with a downward curve on the tail. The viscera and two-thirds of the backbone are removed, and the blood, gurry, and black stomach membrane scraped away. The fish are then thrown into washing tubs, the red-fleshed and the pale-fleshed fish being placed in separate tubs and soaked sufficiently to make them perfectly free from blood, and thoroughly cleaned with a brush or broom. They are next placed in pickling butts with about 15 pounds of salt to every 100 pounds of fish, and sometimes a little saltpeter is used to increase and set the pink color. The fish remain in the salting butts about one week, when they are removed, rubbed clean with a scrub brush, and repacked in market barrels, one sack of salt being used to every three barrels of 200 pounds each. At some of the salting establishments the fish are salted in the barrels without being first placed in butts, but these are usually repacked in San Francisco. The barrels used in packing salted salmon in Alaska are generally made of native woods at the salteries, a stock being prepared before the salmon season.

The following notes on salting salmon in Alaska are furnished by Mr. A. B. Alexander:

The demand for salt salmon is yearly increasing. A few years ago there was but little call for it, probably owing to the fact that little effort was made on the part of those engaged in the business to introduce it in the East. Seeing the absolute necessity of taking steps to place their products on the Eastern market in order to increase the demand and establish a trade for salt-cured salmon, efforts have been pushed in that direction, and the encouragement met with has induced many who had not the means or desire to enter into the expensive business of canning salmon to establish salmon salteries in various parts of Alaska. The amount of capital required to start on a small scale in this business is not large. One or two boats fitted with drag seines, a cabin on shore for living quarters, a rough shed or fish house in which to dress and salt the fish and for performing such general work as may be required in a limited business of this kind will suffice for all purposes. Many of the well-established salteries were first started in this manner and have since grown to be of considerable importance. Two or three men with only a small amount of capital, if they are fortunate in selecting a good locality where the run of salmon can be relied upon—for the success of the entire business depends upon the location—can, if they display the required amount of energy, build up a paying business. They of course must appreciate the fact that for at least seven months out of the year they must content themselves with being cut off and isolated from civilization; but the class of men who seek a livelihood in this remote part of the world care little for social life, or, if so, the prospect which looms up before them for making money is fully equivalent to any hardships of this nature they may undergo. Several small vessels manned by men of small means have, during the past few years, made annual voyages to Alaska, spent the fishing season there, and in the fall brought back the summer's catch. At first they temporarily located themselves by way of an experiment where it was thought to be a good position for carrying on the business. If the experiment proved a success, the next year greater preparations were made, and in this way from a small beginning quite a number of valuable plants have been established. The greater part of the salmon put up at the salteries are caught in drag seines, although a few are taken in gill nets and traps, but at most places where salteries are situated the drag seine has been found to be the most profitable apparatus of capture, owing to the great number of smooth beaches where the fish can be easily taken.

All barrels used for putting up salmon in southeastern Alaska are manufactured at the salteries. Suitable wood being abundant, they can be made at a reasonable price. During the winter months enough barrels are made to meet the demand for the coming season. A cooper is an indispensable person about a salmon saltery, for, besides performing his regular duties as a cooper, he is often called upon to assist in various mechanical jobs, and is paid by the piece, or so much per barrel—85 cents for making a whole barrel and 65 cents for a half barrel. At this price he can earn good wages, for he is under no expense for board. It being the object of every man owning a saltery to enlarge on the plant and increase his business as rapidly as possible, several weeks of each year, before and after the fishing season, are spent in building wharves, if needed, erecting buildings, and making such improvements as are required to keep a place of this kind in good order. Many salmon salters have gained a firmer foothold in Alaska than the mere business of salting salmon would give them. They have branched out into general trade and have stores well stocked with goods of all kinds. In this way they have drawn around them the neighboring tribes of Indians, who are ever ready to buy and trade for such commodities as they require.

In Sweden the choicest pickled salmon are the "ice-house salted salmon." These are killed as soon as caught, split and eviscerated, and the head and larger portion of the backbone removed. Each fish is then rubbed with a mixture of salt and sugar and carefully placed, skin upward, except the bottom layer, in barrels and covered with brine. These barrels are stored in ice-houses or cool cellars and kept at a low temperature.

The following method of brine-salting salmon was practiced in Scotland a hundred years ago:[*]

The Scotch salmon is not too fat, a circumstance which contributes much to its preservation. As soon as possible after they return from fishing, they split the salmon in the same manner as is done with flat cod, except that cod is cut along the belly, and salmon along the back from the head to where the fin of the tail begins, and often leave the large bone sticking to the flesh of one of the sides. Having cut the fish in this manner, taken out the gills, emptied it, and, sometimes, taken off a part of the large bone, they wash it in sea water, if they have it convenient to them, or, if not, in fresh water,

---

[*] A treatise on fishing for herring, cod, and salmon and of curing or preserving them, published by order of the Dublin Society, Dublin, 1800. pp. 110–111.

to take out all the blood, which has a great tendency to putrefaction. This seems to be a much better method than that of washing the fish in its own blood, as is used in the North, probably for the purpose of making the flesh redder. When the fish has dripped they put it into large tubs, with French or Spanish salt over it and under it. It is supposed that Spanish salt gives it a more reddish color than French salt, but that French salt gives it a less sharp taste. Some leave the fish in salt for eight or ten days and then barrel them. Others put them without salt into large tubs, filled with strong brine, and leave them there for a month or six weeks and sometimes longer, waiting until there may be a call for them, for it is thought that salmon keeps better in those large vessels than in barrels, but care must be taken to have it constantly covered with brine. Lastly, it is to be taken out of the tubs and barreled. In the bottom of the barrel they put four or five small salmon and then lay on the large, good salmon, pressing them together as much as possible and putting a little salt between them. In the top of the barrel, likewise, they put some small salmon. When the barrel is full they pour in a small quantity of strong brine and immediately close it up, for it is necessary to guard the fish against the contact of the air and to prevent the brine from being lost. Without these precautions the salmon would grow yellow and rusty and would contract a bad smell. Large salmon is more liable to these inconveniences than the smaller sort, and, therefore, requires more salt in the barreling of it. It is more difficult to preserve salmon than cod. The Scotch take care not to mix salmon of different sorts and qualities in the same barrels, and not to export such as are bad. There is a bounty on every barrel exported, and there are inspectors in every port of Scotland whose business it is to inquire into and certify the good quality, species, etc., of the fish. When the barrels arrive at their place of destination, they ought to be filled again with fresh brine.

BRINE-SALTED MULLET.

Mullet is the most important fish brine-salted in the Southern States, more of this species being pickled than all others combined, the product being especially large on the coasts of North Carolina and Florida, where about 6,000 barrels are prepared annually between the middle of August and the end of November.

As soon as removed from the seines and carried ashore the mullet are dressed. This consists in splitting them down the back and underneath the backbone from the head to the tail, so that the fish will lay out flat, and removing the viscera, stomach membrane, and gills. On the coast of Florida, where the mullet are very large, the heads are removed, and sometimes the backbones, but this is not the case on the Carolina coasts. A horizontal gash is sometimes cut in the thick portion of the flesh on the side in which the backbone is left, in order that the brine may easily penetrate it. The fish are next immersed in tubs or barrels of clean salt water and soaked for about half an hour and the blood and slime washed off. They are removed one at a time and salted with the hand, the salt being rubbed both inside and outside. Then they are usually placed, flesh side up, in old boxes or barrels of any description that are clean, with salt sprinkled over each layer of fish. Generally this work is done by the fishermen and their assistants, and on the North Carolina coast they take them to market in two or three days and sell them to the packers, who are usually wholesale grocers or dealers in fresh and salt fish, by whom the mullet are at once repacked. In other localities, and especially on the west Florida coast, the repacking is generally done by the fishermen.

In repacking the mullet are removed from the first package and placed in layers, with the face or inside of the fish up, in new white pine barrels, 100 pounds being put in each package. In order to permit the brine to easily permeate the contents of the barrel, the fish of one layer are sometimes placed at right angles to those in the layer below. Strong brine of not less than 95° test, or, as usually determined by the fishermen, strong enough to float a mullet, is then poured in until the barrel is full, when the barrel is coopered and set aside and sold to the trade. In some cases, instead of making new brine, the pickle resulting from the first salting is boiled in large kettles, strained, cooled, and poured over the fish, and dry salt is frequently sprinkled over each layer of fish as they are placed in the barrel. It requires half a bushel of salt to strike and pickle 100 pounds of mullet. The decrease in weight by dressing, when only the viscera and gills are removed, approximates 15 per cent, and the decrease in weight by curing is about 10 per cent.

The fishermen sell the partly salted mullet to the dealers at prices ranging from $1.50 to $3.50 per 100 pounds, and after pickling them the dealers usually sell them for from $2.50 to $4.50 per barrel of 100 pounds, the quality and full weight of the fish being guaranteed by the dealer who puts them up. According to the inspection laws of North Carolina, mullet are divided into three grades—those taken in gill nets of 2-inch mesh being called 2-inch mullet and branded as "number one"; 1½-inch mullet, "number two"; 1-inch and under, "number three"; and fish of different lengths and kinds are designated "mixed."

In North Carolina it is required, by an enactment of 1879, that barrels used in packing mullet shall have staves 25 inches in length and heads 13 inches in diameter. They are made generally of Maine white pine, and cost from 45 to 60 cents each. Packages made from the long-leaf pine grown in the Southern States should never be used, since the fish are liable to be flavored with the turpentine. Mullet are also placed in quarter barrels containing 50 pounds, in full barrels of 200 pounds capacity, and in kits of 10 and 15 pounds each.

If the fish are kept on hand long they are examined from time to time by removing the barrel heads, and if the pickle has leaked out more is added, for the fish must be kept under pickle to prevent their rusting and spoiling. They are also liable to rust if kept in the first salting longer than one week. Pickled mullet are at their best after they have been pickled from one to six weeks; after that they begin to deteriorate in quality, and after six months they become so strong that they are not very palatable and few are then sold.

The full value of pickled mullet is scarcely appreciated on our South Atlantic and Gulf of Mexico coasts, and there are stretches hundreds of miles in extent where none whatever are prepared, notwithstanding the fact that the fish are abundant and the industry would yield remunerative employment to the fishermen of the locality. Even where mullet are prepared many of the fishermen are unfamiliar with the best methods of cure, and some mullet are put up in so crude a manner as to injure the trade by prejudicing the public against eating these fish. Pickled mullet properly cured are among the choicest of our Southern fishery products, and if careful attention be given to their preparation, with suitable restrictions against marketing inferior products, a large trade in them could be established, and, because of their great abundance, without in any way conflicting with the supply for the fresh-fish markets.

BRINE-SALTED SHAD.

During the early part of the present century pickled shad was an important fishery product, large quantities being salted in barrels, either for local use during the winter or for shipment to distant markets. It was a staple winter food for the people living near the shad streams, most of the families who could afford it laying in from 1 to 5 or 6 barrels. People living 50 miles or more inland came to the streams to obtain their winter supply of fish, bringing their products to exchange, such as maple sugar and salt, or cider and whisky; and sometimes the fishermen sold to traders, who carted the fish inland, exchanging them for what they could get.

Shad are yet salted to some extent on Kennebec River, in Casco Bay, on Delaware River, the tributaries of the Chesapeake, and in the Carolinas. The bulk of those on the Boston market come from Canada, while of the domestic product the coast and rivers of Maine and the Chesapeake region furnish the greater number. Those salted in the Southern States are usually eaten in the homes of the fishermen or in the immediate neighborhood. There is no uniform method of preparation, the dressing of the fish, the salting, and the packing varying according to the experience or fancies of the different curers; but the following are the most general processes when the fish are to be placed on the market.

KENNEBEC RIVER PROCESS.

The shad are first beheaded and split along the belly, eviscerated, and about 6 inches of the upper portion of the backbone removed. They are next washed thoroughly, some curers washing them in two waters, allowing them to soak five or six hours in the second washing. After the soaking the end of the tail is sometimes cut off. The shad are then ready for salting. In this operation a layer of salt is placed in the bottom of a barrel or butt, and this is followed by successive layers of fish and salt, the former with the backs down, about a bushel of salt to each 200 pounds of split fish. It is desirable to rub the salt over the face or flesh side of the shad before placing in the barrel. In a few days the fish in the top layer are turned backs up and a weight is put on them to keep them beneath the pickle, and a small quantity of salt placed over all to strengthen the weak pickle floating at the top. The shad may remain in the pickle a month or more, but usually two weeks or even less is sufficient time for the curing. On removal they are rinsed off in the pickle, culled if the quantity warrants, weighed in lots of 200 pounds each, and packed backs down in tight barrels, with salt scattered at the bottom of the barrel and over each layer of fish, about half a bushel of salt being used for each barrel.

Liverpool salt is used almost exclusively for striking or curing and for repacking. Trapani salt is objectionable, as its coarse grains lacerate the smooth surface of the shad. After 200 pounds of fish have been placed in the barrel, the latter is filled with strained pickle from the curing or first packing and the head is put on, when the barrel is ready for branding and shipment; or, better still, after the barrel is headed and its contents have settled somewhat it is turned on its side and additional pickle added through the bung. When properly prepared, pickled shad should keep from 12 to 24 months. The shrinkage from dressing and salting is about 50 per cent, 400 pounds of round shad being required to make a barrel of 200 pounds salted, the number of fish to the barrel ranging from 75 to 120. The price received is usually $8 or $10 per barrel.

THE CHESAPEAKE PROCESS.

On the tributaries of Chesapeake Bay the roe shad are rarely salted, on account of the demand for them in the fresh-fish markets, and as a rule it is only during a glut in those markets that the bucks or males are salted. In preparing the fish the heads and tails are cut off and the fish cut down the back to the tail and thrown in tubs or vats of water where they soak for an hour or so, the blood within the backbone being scraped out with a knife in the meantime, when the water should be renewed or the fish placed in other tubs. Upon completion of the washing and soaking, the fish are drained and put in vats with dry Liverpool salt at the bottom, and over each layer of fish, and on top. Every 12 hours thereafter for 7 days the fish are stirred with a pole, to separate them from each other and to have all portions uniformly salted, thus avoiding spots caused by salt burning.

On the eighth day the fish should be removed, drained, and packed in barrels. First is placed a sprinkling of Turks Island salt, then a layer of shad, backs down, then a sprinkling of salt and another layer of fish, backs up, and so on until the barrel is full: and after the fish have settled for a day or so the barrel is topped up with other fish and then filled with strong pickle made of Liverpool salt, when it is coopered and stored ready for market. The usual wholesale price for salted buck shad is $7.50 per barrel of 180 pounds.

The salting of shad was once an important industry on the Connecticut River and on Long Island Sound near the mouth of that stream, and there is a provision among the laws of Connecticut requiring that—

Pickled shad intended for market shall be split and well cleansed and pickled in strong brine, and shall remain in such brine at least 15 days before they shall be put up for market, and shall be put up in barrels or half-barrels, the barrels containing 200 pounds each, and the half-barrels 100 pounds each, of fish well packed, with a sufficient quantity of salt, and filled with strong brine; and shad so put up shall be of three denominations, to wit: Shad "number one," to consist wholly of shad well saved, free from rust, or any defect, with the head and tail cut off and the backbone cut out, each barrel to contain not more than 80 shad, and each half-barrel not more than 40. The second denomination shall be shad "number two," to consist wholly of those well saved, trimmed, pickled, and prepared for packing in the same manner as shad number one, each barrel to contain not more than 90 shad, and each half-barrel not more than 45. The third denomination shall be shad "number three," to consist of such as will not answer for either of the two former numbers, well saved, with the heads taken off.

The legislature of Maine, in 1828, required that shad pickled in that State should be branded as follows:

Those of the best quality, caught in the right season, to be most approved and free from damage, having their tails cut off and backbones out, shall be branded "cargo mess"; those which remain after the best have been selected, being sweet and free from taint, rust, or damage, with their backbones in and tails on, shall be branded "cargo No. 1"; and there shall be a third quality, which shall consist of the thinnest and poorest of those that are sweet and wholesome, which shall be branded "cargo No. 2."

BRINE-SALTED SWORDFISH.

Most of the swordfish captured on the New England coast are sold fresh, yet sometimes a glut in the market or the exigencies of the fishery make it desirable that they be preserved in more permanent form, and pickling in brine is the process usually adopted. In dressing, the swords are sawed off and discarded and the heads removed with a large knife and saved for the oil factories. The fish is then split down the belly and the viscera removed. The splitting is continued down to the tail and around the back, the backbone removed, and the fins and tail cut off. The fish is then cut into pieces weighing 4 or 5 pounds each and placed in butts with a heavy sprinkling of salt, about 2½ bushels of Trapani salt being used to each 1,000 pounds of fish. Any time after 10 days or 2 weeks the fish are repacked in shipping barrels—200 pounds to the barrel—with a small quantity of salt sprinkled among them. The barrel is then filled with pickle and headed up.

Because of its being so fat, the shrinkage of swordfish in pickling is very great, amounting sometimes to 30 per cent of the weight after it has been beheaded and eviscerated. When the pickling is done on board vessel the fish are placed with salt in barrels in the same manner as in the butts, and are repacked ashore in order to insure the proper weight in the barrel, a small quantity of salt being added usually in the repacking, or the fish may be repacked in the original pickle.

The market for the product exists principally in the interior of New England, and especially in Connecticut, where many persons consider it more palatable than salted mackerel. The wholesale price is generally about the same as for No. 3 mackerel, averaging from $6 to $8 per barrel.

A fair idea of the cost and profit in pickling swordfish ashore may be obtained from the following figures, representing the handling of 4,043 pounds of pickled fish, the shrinkage from dressing and pickling amounting to 1,787 pounds:

Swordfish (5,830 pounds, at 1¼ cents)..	$72.87	Pickled swordfish:	
Salt	3.07	15 barrels, at $6.50	$97.50
Labor, cutting and pickling	5.81	10 half-barrels, at $3.45...........	34.50
Barrels, 15 at 25 cents	3.75	43 pounds, at 3¼ cents.............	1.40
Half-barrels, 10 at 16 cents	1.60		
Cartage, 15 barrels, at 7 cents........	1.05		
10 half-barrels, at 3½ cents...	.35		
Interest, wear and tear, and profit....	44.87		
	133.40		133.40

BRINE-SALTING FISH ON THE GREAT LAKES.

Large quantities of trout, whitefish, herring, pike, pickerel, saugers, suckers and other species of Great Lakes fish, were formerly salted each year, but the increased trade in fresh fish and the development of the frozen-fish business have resulted in a large decrease in the product of salt fish. The most profitable disposition of fish on the Great Lakes is in the fresh-fish markets, and when the supply is in excess of the demand the surplus is generally frozen, the salters receiving only the surplus after both the fresh and frozen trades have been supplied, the fish that are salted being only such as can not be sold with profit, either fresh or frozen. Many of these fish are salted immediately after removal from the nets; others are sent to the large ports packed in ice for the fresh trade, but on a glut developing in the market they are salted, and a small quantity consists of fish which have been frozen, but are finally salted because of exigencies in the frozen-fish trade or because of fault in the freezing. Frozen fish are not so satisfactory for pickling as are fresh fish, because of their great tendency to rust, but they are equally good for smoking.

The methods of salting fish on the Great Lakes are essentially the same for the various species, differing only slightly in the manner of splitting. They are laid open flat by splitting down the back or down the belly to the tail, or in case of ciscoes they are split down the belly only sufficiently to remove the viscera, similar to the split herring on the New England coast. The fish bring a higher price if split down the back, and when salted directly from the nets that is the usual method of dressing, but many of the large ones have already been split down the belly to the vent in dressing for the fresh fish trade, and in that case the splitting is continued down to the tail, so that the fish may be laid out flat. Trout are generally split down the back if salted by the fishermen, but most of those on the market are from the fresh fish

houses and consequently have been split down the belly, and the same is to some
extent true in regard to whitefish, but the proportion of salted whitefish split down the
back is greater than in case of trout. Blue pike, yellow pike, mullet, sheepshead,
perch, and carp are usually split down the back, even though prepared as surplus
from the fresh-fish trade, since they are usually sold round in the fresh fish-markets.
Herring split down the belly to the vent sufficiently to remove the viscera are commonly called ciscoes, but many herring are split down the back and sold under the
trade name of "family whitefish." The difference in value of fish when split down the
back and when split down the belly is shown in that species. Although the ciscoes
and the family whitefish are prepared from the same grade of fish, yet the latter
usually sells for about 50 cents per 100 pounds more than the ciscoes. It is much
easier and quicker to split herring for ciscoes than for family whitefish, 500 pounds
of ciscoes being readily split in one hour, whereas twice that length of time is required
for splitting an equal quantity of family whitefish. Not so many ciscoes are prepared
now as a few years ago, since the increased value of the fish makes the better method
of cure profitable.

On arrival at the salting house, if the fish have already been eviscerated, the
heads are cut off and the splitting continued down to the tail, so that the fish will lie
flat. In case the fish are round when received, they are beheaded and cut down the
back along the left side of the backbone, so as to lie flat, except in case of ciscoes, as
above noted, and the rough edges of the backbone are cut off. The flat edge of the
knife is run around the abdominal cavity to scrape away the blood, etc., and if the fish
are large, one or two horizontal deep cuts are made in the thick flesh of the back. The
fish are then thrown into a trough containing fresh water, where they soak for a few
minutes and are removed with a pew or fork and thrown on a draining and salting
table, three-fourths of the top of which consists of strips on edge, on which the fish lie
to drain, and the remaining one-fourth of the width is solid for holding the salt. After
draining a few moments, each fish is taken separately, laid back down on the salt
if large, and a quantity of salt spread evenly over the face, and the fish carefully
placed face up in a tight barrel. For protection from dust, etc., the top layer is placed
skin up. In salting small fish, one is taken in each hand and rubbed in the salt, as in
salting mackerel.

In case of ciscoes, the stomach cavities are scraped full of salt and closed and the
fish are thrown into the pickling barrel.

In order to avoid lacerating the flesh of the fish it is necessary to use fine salt,
either Syracuse or Warsaw being preferred. The former is a solar salt and weighs
about 336 pounds per barrel, and the latter, a pan salt, weighs about 290 pounds per
barrel, and the cost of each ranges from 90 cents to $1 per barrel, the Warsaw being
usually a few cents cheaper than the Syracuse. Cleveland salt is also used to some
extent in striking. Some salt is sprinkled in the bottom of the barrel and several
handfuls placed on top of the fish, and weights placed thereon to keep the fish down
in the pickle, these weights consisting usually of stones on top of boards. No brine
is added, the fish making their own pickle. The entire quantity of salt used for each
100 pounds of green fish ranges from 12 to 15 pounds, according to the size and
condition of the fish and the season. Fish that have been frozen do not require so
much salt as fresh fish, since they are somewhat drier and the texture is to some
extent disintegrated, permitting the salt to strike through the fish more readily.
Within four or five days the fish are struck through, depending on the grade of salt.

size of the fish, and the temperature, and at any time thereafter they may be removed and repacked. This should be done at the first convenient opportunity, for the longer the fish remain in the pickling barrel or vat after being cured the darker they will be, which detracts from their value.

On removal from the pickling barrel the fish are rinsed in the pickle to get rid of the surplus salt and at once weighed and repacked in the shipping packages, which consist almost entirely of half-barrels with capacity for 100 pounds. These are made mostly in Sandusky, and cost from 40 to 45 cents each. The fish are carefully placed face up, except the two top layers, which are placed with the skin side up as a protection from the head of the barrel. Salt is sprinkled in the bottom of the barrel, at the top, and at intervals among the layers of fish, about 8 pounds being used for each 100 pounds of fish. Syracuse salt No. 2 is usually preferred for packing, even though Warsaw or Cleveland salt has been used in striking. When the package is full of fish strong brine is poured in to fill the interstices between the fish. This brine is made by permitting water to percolate through a box or tank, the lower part of which is filled with some filtering substance, such as straw or plane shavings, and the upper part filled with salt; or the filtering box may have a false bottom covered with burlap, the salt resting above the burlap and the brine percolating through and remaining in a tank below. In case the salting establishment is connected with an ice-and-salt cold storage the surplus brine from the ice-and-salt receptacles may be used with excellent result, this brine being permitted to flow from the receptacles into a large filtering tank sunk in the ground, from which it may be pumped as required. The strength of the brine usually depends on the season of the year and the grade of fish being packed. In the summer packing of whitefish or trout the brine should be of 100° salinometer test. But in October and November packing of herring, brine of even 60° test is frequently used, this being made by weakening stronger brine with fresh water. This use of diluted or weak brine is satisfactory when the packer is assured that the fish will be used before spring; but in packing fish for the general trade, where they may not be used until the following summer, the brine should not be weaker than 95°, and 100° test is much better.

When the package is filled with brine the top is coopered on and additional pickle admitted through a hole in the head of the barrel by means of a funnel watering-pot, the barrel being overfilled to permit the pickle to soak in. In a few hours a plug is driven in the hole and the tightness of the head is tested by pressing on it in the center.

The decrease in dressing fish ranges from 15 to 35 per cent of the round weight, according to the species of fish and the season of the year. The decrease is least in case of herring and blue pike and is greatest with mullet and carp, but it varies in different seasons of the year, according to the development of the ovaries. The decrease in weight of Great Lakes fish in pickling ranges from 8 to 12 per cent of the dressed weight, according to the fatness of the fish and the extent of the salting. The decrease in herring is about 9 per cent of the dressed weight, 110 pounds of split fish being necessary to make a 100-pound package of salted fish. Whitefish, being fatter than herring, decrease more in weight in salting and consequently are drier, the salt absorbing the fat. Generally, in case of whitefish, trout, and herring, about 132 pounds of round fish are required for each 100 pounds of pickled fish. In brine-salting trout, 130 pounds round, 115 pounds from the ice, or 105 pounds from the knife are required for each 100-pound package.

Aside from the first cost of the fish and the cost of plant, superintendence, etc., the expense in preparing pickled Great Lakes fish is about the same for the various species and approximates 69 cents per package of 100 pounds, divided as follows:

Labor in dressing and salting.. $0.12
Salt in striking and packing... .07
Labor in packing, coopering, etc.. .08
Barrel.. .42

Whitefish are generally divided into three grades—Nos. 1, 2, and 3. In the first class are placed all weighing 2 pounds and over; No. 2 includes all weighing between 1 and 2, and No. 3 includes all under 1 pound in weight.

Prior to 1891 there was only one grade of trout, but it has since been customary to brand trout weighing 1½ pounds or more as No. 1 and all under that weight as No. 2.

BRINE-SALTED HALIBUT FINS.

The strips of flesh attached to the inner bones of the dorsal and ventral fins of the halibut are cut off in dressing these fish for use by the smokers, and are subsequently pickled; but in case of the fresh-halibut trade the fins are not removed, but are shipped with the fish. In dressing halibut for the smokehouses, the "feathers" or "fly" of the fins are first cut away, then with the point of the fletching knife the skin is cut on each side of the fin about 2 inches from the edge, and by a sharp stroke near the tail that end is separated, and catching hold thereof the whole fin is pulled off, the two fins representing about 3 or 4 per cent of the weight of the round halibut. They are placed in tiers in tight barrels, with salt sprinkled in the bottom and over each layer, about 1 bushel of salt being used to each 200 pounds. On being landed from the vessel the pickled fins are frequently repacked in half-barrels and other small packages, and they are used mostly as ship stores.

Several years ago pickled-halibut fins sold at $8 to $10 per barrel of 200 pounds, but the price gradually decreased to about half that amount. In 1898 they sold at about $8. This fluctuation is due to the fact that during the Iceland fishery a large percentage of the halibut were too small to warrant saving the fins, and the demand for ship stores being good, the market was not overstocked. When the Iceland fishery was abandoned for Bacalieu and other western banks the average size of the fish caught was much increased, so that many more fins were salted. This, together with a decreasing demand, greatly overstocked the market and ran the price down very low, so that in 1897 and 1898 very few vessels saved the fins. The small product resulted in equalizing the supply and demand, and in 1898 the price was advanced to nearly its former standing.

MISCELLANEOUS BRINE-SALTING.

All along the coast of the United States a small local business is carried on in pickling fish for use during the winter in the homes of fishermen and their neighbors. Among the species thus prepared are bluefish, squetague or sea trout, channel bass, croakers, perch, sheepshead, Spanish mackerel, striped bass, black bass, hogfish, etc. There is no uniform method of pickling, the fish being dressed, salted, and packed according to the fancies and convenience of the curers, and the product rarely goes on the general market. In general, the fish are dressed by removing the head and viscera, and are split down the back or sometimes the belly, so as to lie out flat. They are next washed and soaked until the blood is removed and then covered with salt and placed in barrels, first a sprinkling of salt and then a layer of fish, and so on until the barrel is filled. Then brine is poured in to fill the interstices and the barrel is headed and coopered.

In Europe a large variety of marine products are brine-salted, most of them being prepared from species of the herring family. The following descriptions apply to the method of preparing a number of them:

PRESSED SARDINES.

The "pressed sardines" of Sweden are prepared in the following manner:

As soon as the sardines are caught they are thoroughly eviscerated, cleansed, and salted in layers in large vats, 65 pounds of salt being used to 100 pounds of fish, this salt being thoroughly sprinkled between the layers of fish. Within two or three days brine forms and covers the fish, and there they remain for one or two, and sometimes three months—the longer the better. When ready for repacking, the sardines are laid flat in the barrel with their tails pointing toward the center, while before they were placed in layers. The barrel is thus filled to within 4 inches of the top, and over the fish is placed a sheet of paper, and upon that a thin board which is smaller than the opening of the barrel, and lastly a wooden block which measures one-fifth the height of the barrel. By means of a screw the wooden block and the fish underneath are slowly pressed down; then the block is removed and the space filled with more sardines, until the barrel is overfull. A sheet of paper and another thin board are put on the fish and pressed down like the first, when the barrel hoops are loosened, the cover placed on, and the barrel tightly sealed. The barrels are provided with small holes, so that the oil and moisture may run off. A barrel of 40 gallons capacity will hold from 3,000 to 8,000 pressed sardines.

SALTED PILCHARDS OR FUMADOES.

Somewhat similar to the above is the English process of preparing pilchards (*Clupea pilchardus*) in the form of fumadoes* for the Italian markets, which is thus described in Holdsworth's Sea Fisheries:

The curing is the especial work of the women, who pack the pilchards in alternate layers of coarse salt and fish on the stone floor of the curing house until the "bulk" has reached a height of 5 or 6 feet. Here the fish remain for a month, and the oil and brine draining from them are carried off by gutters in the floor to a cistern. When the fish have been sufficiently salted, they are washed and packed in hogsheads, each layer of fish being placed with their heads outward and with a "rose" of fish in the center. A circular piece of wood called a "buckler," and rather smaller than the head of the cask, is then placed on the top of the fish and strong but gradual pressure is applied by means of a lever until the mass of fish is reduced one-third in bulk and a great quantity of oil squeezed from them. This drains through the sides and bottom of the cask, the hoops of which are not at that time very tightly driven, and is collected as before. The quantity of oil obtained from the pilchards depends on the season, but at least 2 gallons of oil are expected from each hogshead. It is principally used by the leather-dressers. The cask is filled up three times before the pressing is finished, which is not until after eight or nine days, and then the hogshead (50 gallons) of fish should weigh 4 cwt. gross. The average number of fish packed in a hogshead is about 2,500. The pilchards cured at St. Ives in the early part of the season are mostly taken by drift nets, but the seine fishery at a later period is mainly depended on to provide the fish for exportation.

A large trade in pickled pilchards is carried on between Cornwall and the Italian ports, according to Francis Day, the idea having originated at Mevagissey, as follows:

In 1876 [*Land and Water*, November 18, 1882], a fish-curer here found there was a demand in the Mediterranean fish markets for bright salted pilchards. He first thought the matter out and then cured several tons of pilchards by throwing them, with salt, into barrels, and allowing the brine to

---

* As may be inferred from the name, these fish were formerly smoked. William Borlase noted, in 1758, "fuming them being for many years laid aside."

rise over them. After keeping them steeped for some weeks they were washed, packed, and pressed into clean barrels, just as was formerly done to the old-fashioned fumadoes. On their being put on the market it was at once seen they were the article wanted; for these fish, instead of having the dirty yellow hue of the fumado, had the desired bright and clean silvery color, hence they have been in demand ever since. The fish-curer in question took out no patent rights, but allowed all to use his discovery; so much so that for some seasons past not less than 1,000 hogsheads of fish yearly have been shipped for the Mediterranean from Mevagissey alone. The barrels first used have been superseded by large steeping vats, one of which here will hold over 500,000 fish. Since the business in question has been progressing, it has been discovered that the Spaniards cure sardines much after the same manner.

Pickled pilchards are not so well flavored as salted pilchards, or fumadoes, but they will keep a much longer time, it being necessary to dispose of the latter within a short time after curing.

ITALIAN SARDELS.

The method of preparing the celebrated and deliciously-flavored sardels of Italy is as follows:

After the freshly caught sardels or anchovies (*Engraulis encrasicholus*) have been well salted and washed they are cleaned and the lower jawbone is removed and the fish strongly salted in a barrel with 50 pounds of salt to 100 pounds of fish. There the fish remain for two or three months, when they are removed and loosely packed in the market barrel or package, being resalted at the same time, 25 pounds of salt being used to each 100 pounds of fish. The barrel is set upright, and after three or four months the blood pickle is poured off through a bunghole. During this time the barrel is placed in the sun, so that the pickle has become quite strong, and sometimes a little ocher is added to the pickle to give it a dark-red color. The fish may then be used within a few weeks, but to acquire its best flavor about three years are required.

GÄHRFISCH OR FERMENTATION FISH.

In some of the districts of northern Sweden there is a unique method of preserving fish, the product being known as "gährfisch" or fermentation fish. Various species are used, but mostly the stromling or Swedish anchovy. The freshly caught anchovies, after being dressed and thoroughly washed, are lightly salted and loosely packed in tight wooden barrels. A blood pickle made from the dressings of the fish is then poured over them until all the fish are covered, when the barrel is tightly sealed. It is then placed where the sun's rays can reach it, and there it remains four or five weeks, the fish undergoing fermentation. If this fermentation be too rapid the barrel is removed to a cooler place, and as soon as the fermentation has taken place the barrel is opened and its contents repacked in smaller packages, which must be kept securely sealed, otherwise putrefaction quickly ensues. These fermentation fish are eaten either raw or cooked, but the market is limited to northern Sweden. The odor is very strong and excites a feeling of disgust among persons unaccustomed to eating them, but when a taste for the fish has been acquired they are highly relished.

SAHLSTRÖM PROCESS OF BRINE SALTING FISH.

A method of pickling fish, intended especially for herring, was devised about fifteen years ago by Carl A. Sahlström, and has been used to some extent in Norway and Scotland. It is somewhat similar to the Roosen process of preserving fish fresh by means of an antiseptic, and consists, first, in placing the dressed fish in a closed cylinder, into which brine is introduced until the cylinder is full. Additional brine is then forced into this cylinder under a pressure of from 60 to 100 pounds to the square inch, thus forcing it into the tissues of the fish. The fish can be salted sufficiently to suit the necessities of the market, and the operation requires a less number of hours than the ordinary process requires weeks. The tissues are thoroughly permeated by the preserving liquid and are quite incapable of supporting organisms of putrefaction.

PICKLING WITH VINEGAR AND SPICES.

Pickling with vinegar and spices is one of the ancient forms of preserving fishery products, probably antedating even the pickling with salt. It was well known to the Greeks and Romans, the latter applying it especially to preserving mullet, swordfish, tunny, etc. The most costly spices were used, and the products frequently sold at fabulous prices. At present comparatively few fish are preserved in this manner and the business is done on a small scale. Many small herring are compounded with vinegar and spices and marketed as Russian sardines, and there is some importation from Europe of herring somewhat similarly prepared, which are sold as Christiania anchovies, marinated herring, spiced herring, etc. A small business is done in pickling eels, sturgeon, and one or two other species with vinegar; and oysters, clams, and mussels are frequently put up with this antiseptic in glass jars, but the business is of small extent.

RUSSIAN SARDINES.

The preparation of Russian sardines, or small pickled herring in vinegar and spices, is of comparatively recent origin in this country, being first undertaken by Mr. Henry Sellman in 1874, at about the time of the beginning of the Maine sardine industry. The business, which is not very extensive, is carried on in connection with the preparation of sardines on the coast of Maine. The fish used are similar to those canned in oil, varying in length from 7 to 12 inches alive and from 5 to 9 inches when dressed. As the herring are more valuable when canned they are so prepared when practicable; but when more fish are received than the canneries can handle the surplus is salted and prepared under the trade name, Russian sardines.

For many years previous to 1860 Russian sardines were prepared at various points in Europe, and especially at Hamburg, Germany, and the trade extended to this country. By 1870 the importation of this product into the United States amounted to 50,000 kegs per annum, nearly all of which came from Hamburg. In consequence of the blockading of the German ports in the early part of the Franco-German war the importation was necessarily abandoned for a time, and an effort was made to supply the deficiency with a domestic product, with such good results that at present few foreign-prepared Russian sardines enter into the United States trade.

The present method of preparation is as follows:

As soon as practicable after being removed from the water the fish are placed in strong brine contained in suitable casks. It is desirable that this be done while the fish are yet alive, so as to remove any possibility of putrefaction starting in. There they remain for about ten days, depending on the size of the fish and state of the weather, or until thoroughly struck. The fish are then scaled, beheaded, eviscerated, and cleansed in clear water, after which they are placed on sieves or other suitable receptacles for draining. After draining for several hours the fish are spread upon packing tables and assorted according to their size, each size being packed separately in kegs, with a mixture of certain preservatives combined with flavoring substances. The preservative substances are vinegar, allspice, and chile pepper, or their equivalents. The flavoring substances are sliced onions, bay leaves, horse-radish, cloves, ginger, coriander seed, and capers or their equivalents. When all are used, the following (according to Mr. Henry Sellman) is about the usual proportion for every 120 pounds of fish: Vinegar, 2 gallons; allspice, 1½ pounds; sliced onions, 4 pounds; sliced horse-radish, 2 pounds; bay leaves, 1 pound; cloves, ½ pound; ginger, ¼ pound; chile pepper, ¼ pound; coriander seed, ¼ pound; capers, 2½ ounces.

The fish are packed in kegs of uniform size, containing about 7 pounds. A small quantity of vinegar and a thin layer of the other ingredients are placed in the bottom of the keg, and a layer of fish, placed back upward, are put in and gently pressed down. Another small quantity of vinegar and thin layer of the other ingredients are then put in and another layer of fish, and so on until the keg is full, when a small quantity of vinegar is poured over the whole and the keg headed up. In order that the fish may be well flavored they should be prepared some days before being placed on the market. This length of time varies, according to the temperature, from about four days in summer to three or four weeks in winter. When properly prepared the fish will readily keep a year or longer without spoiling.

The preparation of herring in this manner was begun at Eastport, Maine, in 1874, and since then has been confined to the eastern portion of that State. Later it was found convenient to simply salt and dress the herring at Eastport and ship them in barrels to New York City dealers, who pack them in kegs for the market, and this is the way in which the business is generally conducted at present.

The fish are either shipped in the barrel in which they are being salted, or, as is more commonly the case, they are packed in shipping barrels after being dressed. In packing, each layer of fish is sprinkled with dry salt. On arrival in New York City they are subjected to the same treatment described above. The quantity of Russian sardines prepared annually in this country amounts to about 60,000 7-pound kegs, worth about $27,000.

The following is a popular method of preparing these herring on the shores of the Baltic Sea in Norway:

The fish are placed in vinegar weakened by the addition of 25 per cent of water, and to which a particle of salt has been added. In this bath the fish remain for about 24 hours, when they are removed and the vinegar drained off. Some persons place the fish for 12 hours in vinegar which has not been weakened with water, the important point being that they must be taken out before the skin becomes flabby. The fish are then carefully packed in kegs or jars with the following preservatives and spices, the quantities given being sufficient for 80 herring: Fine salt, 1 pound; powdered sugar or sometimes brown sugar, 1 pound; black pepper, ¼ ounce; bay leaves, ¼ ounce; saltpeter, ¼ ounce; sandal, ¼ ounce; cloves, ¼ ounce; ginger, ¼ ounce; spanish hops, ¼ ounce. Others use the following spices: Salt, 1 pound; sugar, ¼ pound; allspice, 1 ounce; pepper, 1 ounce; cloves, ¼ ounce; Spanish hops, ¼ ounce. In this mixture the herring should be left for at least two months before using, and if the brine should leak off, additional brine made of Lüneburg salt should be added; and under favorable conditions the product will keep for years.

CHRISTIANIA ANCHOVIES, ETC.

In the preparation of Christiania anchovies many methods and flavoring ingredients are used, depending on the skill and ideas of the curer and the markets for which the preparation is intended. The following is one of the most popular processes:

The fresh sprat or anchovies are immersed in brine for 12 or 18 hours, 15 pounds of Liverpool salt being used for each 100 pounds of fish. On removal, the fish are drained in a sieve and then loosely packed in a barrel, with the following ingredients, which have previously been finely crushed and well mixed: 4 pounds of Lüneburg salt, 6 units of pepper, 6 units of sugar, 6 units of English spices, 1 unit of cloves, 1 unit of nutmeg, and 1 unit of Spanish pepper. The anchovies remain saturated with these ingredients for 2 weeks, when they are repacked tightly in kegs or barrels, being carefully arranged in layers, with the backs downward. A quantity of the ingredients above mentioned is sprinkled over each layer, with the addition of a few cut bay leaves or cherry leaves. At the bottom and the top of the package is placed two whole bay leaves, but before the top leaves are laid on, brine is poured over the fish. The barrels or kegs are then coopered and rotated daily for the first few days, and after that every other day for two or three weeks.

The following process is also used to some extent:

The fish are salted for 24 hours and next immersed in sweetened water, 20 parts of water to 1 part of sugar being used. The fish are then packed with a mixture of Lüneburg salt with 90 units or parts of allspice, 60 units of pulverized sugar, 19 units of whole peppers, 15 units of cloves, an equal quantity of nutmeg or mace and of hops (*Origanum creticum*), and some bay leaves.

The following is a choice method of preparing "Mätjeshering" in Germany:

Fresh full herring, both spawners and melters, are well washed, and the gills, stomach, and intestines are removed in such a way as not to necessitate cutting the throat or abdomen, this being accomplished by pulling them through the gill flap. The fish are next immersed for 12 or 18 hours in a 7 per cent solution of white-wine vinegar, from which they must be removed before the skin becomes flabby and be wiped dry and covered with a preparation composed of 2 pounds of salt, 1 pound of powdered sugar, and a small quantity of saltpeter, this quantity being sufficient for 75 herring. The fish are then packed in a barrel as upright as possible, in layers, with a sprinkling of salt over each. The following day the fish are returned with the original brine to the barrel, which is sealed. When there is not sufficient brine to fill the barrel, additional should be made of 1 part of the above mixture and 4 parts of water which has been boiled.

Spiced herring (*Gewürzhering*) are prepared in Germany in the manner above described, with the addition of spices mixed with the salt. The spices commonly used consist of 1 part of Spanish pepper, 5 parts of white pepper, 4 parts of cloves, 2½ parts of ginger, an equal quantity of mustard, and a particle of mace and of Spanish marjoram, with a few bay leaves scattered between the layers.

PICKLED STURGEON.

In the early history of New England pickled sturgeon was an article of home traffic, and considerable quantities of it were exported to the West Indies. During the early half of the present century comparatively little use was made of sturgeon, either fresh or otherwise, but since 1860 there has developed a considerable demand for the flesh, especially when smoked.

A small quantity of sturgeon is brine-salted along the Southern coast and on the Great Lakes in the manner described for swordfish, but the smokers take nearly all the surplus from the fresh-fish market.

It is probable that the pickled sturgeon referred to in the early New England history was prepared in practically the same manner as is still in vogue among the Germans in the West, i. e., by boiling the meat and preserving it in weak vinegar flavored with suitable spices.

In 1629 Governor Endicott, of the Massachusetts Colony, was "ordered to send home to the company in London two or three hundred firkins of sturgeon and other fish"; and by 1633 a considerable export trade existed in pickled sturgeon, most of which were caught in the Merrimac River. An early description of the town of Newburyport, Mass., says: "At the mouth of the river stands Newbury, pleasantly situated, where abundance of sturgeon are taken, and pickled after the manner used in the Baltick." The Indians called that river Monomack, signifying sturgeon. In 1656, "a keg of sturgeon, ten shillings," was among the charges for entertaining an ecclesiastical council at Salisbury.

In 1680 the court licensed Thomas Rogers "to make sturgeon, provided he shall present the court a bowl of good sturgeon every Michaelmas court." The business was quite extensively carried on along the Merrimac River as late as 1733, and quite a large trade was established with England and the West Indies. They sold for 10 or 12 shillings per keg, and one sale is recorded of "15 kegs of sturgeon for a small cask of rum and a cask of molasses."

The general court of Massachusetts, at Boston May 7, 1673, was petitioned by William Thomas for protection in putting up sturgeon. The petitioner stated:

After sundry experiments and travels into forreigne countries upon great expense to his estate hath through ye blessing of God upon his industry therein, attained unto the art of boyling and pickling of sturgeon by means whereof it is a commodity not only in this country but in England and other parts for transportation and purposes of traffic for the procuring of goods more useful and needful to this country. (Massachusetts Maritime Manuscripts, II, p. 3.)

At the same session of the general court the following law was enacted:

Forasmuch as sundry unskilfull persons have of late yeares taken upon them to boyle, pickle, & sell sturgeon for transportation, of which sundry keggs & other caske have prooved corrupt & wholly unserviceable, to the disappointment & damage of sundry merchants & others, as also to the debasement of that comodity, & reproach of the country, wch, if duely ordered, might be beneficiall to the inhabitants for transportation & otherwise, it is therefore ordered & enacted by the authority of this Court, and be it hereby ordered & enacted, that no person whatsoever shall henceforth boyle, pickle, or packe up any sturgeon for sale in this jurisdiction but such as shall be licensed thereunto by the County Court where such persons inhabit, on poenalty of forfeiture thereof, one halfe to the informer, and the other halfe to the county. And to the end there may be no fraud or abuse in the said comodity, every such licensed person shall brand marke all caske wherein it is packed wth the letters of his name; and that there be searchers appointed & sworne to view all sturgeon made beere, or imported, before it be sold or in kinde passed away, who shall sett their marke on such as they finde sound & sufficient in all respects, both as to the quallity of the sturgeon & gage of the caske; and that only such so marked as above shall be exported, on penalty of forfeiture of the whole value thereof; ffor whose care & labour the sturgeon boyler or importer shall pay, for the veiwing & heading thereof, after three shillings fower pence per score for all kegs & firkins, from time to time. And if any shall counterfeit the sturgeon boylers or packers marke, they or he shall forfeit five pounds to the country for every such defect. And it is referred to the respective County Courts to license able & fitt persons to boyle & pickle sturgeon for sale, as likewise to appoint searchers to view & marke the same as aforesaid. (Records of Massachusetts, vol. IV, part II, page 553.)

PICKLED EELS.

Notwithstanding the abundance of eels in the United States, comparatively few are marketed except in a fresh condition, and even the demand in the fresh-fish markets is rather small in many localities, owing to their snake-like appearance. In New York City and a few other points some are pickled, and at various places they are smoked to a small extent. In Europe there are a number of valuable eel fisheries, the most celebrated of which is that of Commachio, near Venice, where pickled eels are prepared in large quantities, as follows:

The fresh eels are dressed and well cleansed with a brush; they are placed in salt brine for 2 or 3 hours, and on removal are dried thoroughly with a towel, cut in pieces of suitable length, immersed in Provence oil, and cooked in a frying pan. On the cooking being completed, the eels are removed from the oil and allowed to cool upon blotting paper, and to the oil in the pan are added some white peppercorns, whole mace, bay leaves and lemon, and a quantity of weakened vinegar, this mixture being cooked for 15 or 20 minutes. The pieces of eel are laid in glass jars or stone jugs, and over them is poured the above mixture after it has cooled, the quantity of which must be sufficient to cover all the pieces and half an inch more. The jars are then carefully sealed and put away in a cool place.

Another method of pickling eels is the following:

The dressed eels are sprinkled with salt, which is soon rubbed or wiped off; then the eels, cut in pieces of suitable length, are spread with butter and broiled brown upon a gridiron. The pieces are next placed in suitable receptacles, such as jars, kegs, etc., and among them is spread a mixture of bay leaves, whole cloves, pepper, English spices, and a little mace. A weight is placed on the eels to keep them compressed and the receptacle covered. After 24 hours the weight is removed, vinegar added to cover the pieces, and the receptacle tightly sealed.

PICKLED SALMON, ETC.

The following description of an old method of pickling salmon, in use in northern Europe to some extent at the present time, is from "A treatise on fishing for herrings, cod, and salmon, and of curing and preserving them," published in Dublin in 1800:

As soon as the salmon is caught they cut off the jowl, which they split into two parts, and cut the rest of it (as far down as below the anus) into pieces about 3 inches thick. The tail may be left long at pleasure. All these pieces are put into a large vessel full of fresh water, in which they are washed with care; the water is changed three times, so as to take out all the blood. Each piece is fastened separately to small laths, to prevent their touching one another. They then boil, apart, as much water as may serve to cover all the fish, to which they add two bottles of Rhenish wine, a bottle of good vinegar, some mace, cloves, pepper in grain, or long Portuguese pepper, coriander seed, thyme, leaves of laurel, a clove of garlic, and more or less salt, according as they intend that the salmon should keep for a longer or shorter time. When this water boils they put the pieces of salmon into it, placing the jowls uppermost, as they boil sooner than the rest, and when the whole is boiled they take it out and let it drip on a linen cloth. When the water has cooled they pass it through a searce, or sieve; then they put the salmon, in pieces, into an earthen jar or pot, upon a bed of leaves of laurel, and throw between the pieces a little salt and some slices of lemon. They then pour upon it the sauce in which the salmon was boiled, until it is quite covered, and fill up the vessel with the jowls and tails; after which they pour good oil upon it and close the vessel. Salmon prepared in this manner will keep a considerable time.

Among the proprietary compositions for preserving fish in vinegar, spices, etc., was one patented* in 1881 by Paul Brick, of Cape Elizabeth, Me. This method was intended particularly for mackerel, but it is claimed to be equally applicable to other species of fresh fish. Brick's process is as follows:

The fish after being scaled and dressed are cut into pieces of about 2 inches in length, cleansed and placed for about 12 hours in a cold sauce or pickle made of 1 gallon of strong cider vinegar, one-half ounce of green parsley, eight bay leaves, 4 ounces of onions, one-half pound of salt, and 1 ounce of the following spices mixed in equal parts: Mustard seed, cloves, allspice, mace, cinnamon, and pepper. At the end of 12 hours the fish are removed from this pickle and placed in air-tight pots or jars with another sauce or pickle composed of similar ingredients to the first and in the same proportion, and to which have been added 1 gill of capers, a half-pint of olive oil, 1 gill of Worcester sauce, 2 lemons, and a small quantity of extract of anchovy, and allowed to simmer for 5 or 6 hours at a temperature of about 110° F., when the jars are sealed.

From *Bayerische Fischerei-Zeitung*, No. 30, Munich, 1885, is taken the following method of pickling fish, applicable to sturgeon, salmon, and other large species:

The fish is cut into pieces, strongly salted down, fried rather quickly in butter and oil, then laid upon a plate, each piece by itself. Before the pieces have cooled off they are put in layers in a porcelain or glass vessel, with some "tan liquor" (beize). This covers all the pieces. To 1 kilo of fish a sauce is made from 3 deca of the finest olive oil and finely sliced onions. This is cooked until the onions turn yellow. To this is then added heated strong vinegar, whole spices (white pepper, cloves, and Jamaica pepper), a few bay leaves, and shalot. All this is cooked together and then put away to cool. Enough vinegar is used to make sufficient pickle and also to cover well the pieces of

* See Letters Patent No. 241187, dated May 10, 1881.

fish. The vessel has either a tight cover or it is tied up with paper. When some of the pieces are taken out and there is not enough "tan liquor" in the vessel to cover the remaining pieces, either some more of this liquor is added or a sufficient amount of vinegar is used.

With a view to softening the bones of small pickled fish so that they may be freely masticated, a method was introduced about thirty years ago in which the dressed fish are placed in a suitable receptacle with a mixture of vinegar, salt, cloves, and cayenne pepper. The receptacle is then closed and the contents subjected to a temperature of about 170° F. for 24 hours.* The acid of the vinegar neutralizes or dissolves the phosphate of lime and the heat serves to reduce the coherence of the particles of animal matter contained in the bones, so that the latter may be masticated and swallowed without inconvenience. This process was used to a limited extent only and is no longer in vogue.

PICKLED OYSTERS.

In pickling oysters, clams, mussels, etc., the mollusks are usually cooked a short period either before or after removal from the shell, cooled, and placed in glass jars or other receptacles with vinegar, spices, etc., according to the ideas of the preparer.

A generation or two ago large quantities of pickled oysters were put up for use in and about New York City, the consumption being especially large during the Christmas holidays; but their popularity has greatly decreased, and during recent years probably not over 50,000 have been pickled annually, mostly in New York City, to fill special orders. The choicest oysters are generally used. The process is as follows:

The raw oysters are removed from the shells in the usual manner, as much as practicable of the liquor being saved. The oysters, with their liquor, are boiled in an open kettle for 5 to 30 minutes, according to the length of time that will elapse before they are to be used. It is important that the boiling be reduced as much as practicable for good keeping, since the longer they are boiled the smaller and harder they become and the more difficult to flavor. When boiled sufficiently the oysters and liquor are separated, the former spread on shelves to cool and the latter strained and mixed with sufficient vinegar to impart the flavor desired, to which may be added mace, lemon, and other flavoring ingredients, if desired. When both the oysters and liquor are quite cool, they are combined and sealed up in glass jars or other suitable receptacles and stored in a cool place. If the cooling of the oysters and liquor after boiling be not done separately, the oysters become soft and disintegrate, affecting both the appearance and keeping qualities.

An objection to the above method of pickling is that when the oysters are cooked they become shriveled and somewhat unsightly, and if merely scalded the vinegar soon acts upon the tissues, rendering them soft and equally unsightly.

In 1879 the following process of preparing "jellied oysters" was patented[†] by Katherine L. Jewell, of New York:

A quantity of freshly opened oysters are slightly cooked so as to plump them. They are immediately placed in the vessel in which they are to be marketed and covered with a liquid prepared in the following manner: A suitable quantity of oyster liquor containing a few fresh oysters is boiled until the liquor is so far inspissated as to form a jelly (solidify) when cooled. This liquor is strained and while warm is mixed with its weight of heated vinegar, to which spices are added to suit the taste. This liquor is poured over the plumped oysters so as to cover them, and it will, when cooled, form a jelly sufficiently firm to support the oysters and form with them a semisolid mass impervious to air.

The small oyster crabs (*Pinnotheres ostreum*) found at times in the oysters are sometimes pickled at Chesapeake ports in a manner similar to that applied to oysters, but they are so scarce and the price for them fresh is so high that the quantity pickled is very small.

* Letters Patent No. 70435, November 5, 1867.   † Letters Patent No. 215628, May 20, 1879.

PICKLED CLAMS, MUSSELS, SCALLOPS, ETC.

The process of pickling clams, mussels, and scallops is quite similar to that employed in preserving oysters, differing principally in the manner of removal from the shell and in cooking. The trade in these products is very limited and is centered about New York City.

Clams or quahogs are generally steamed in the shell, a basketful being placed in the steam box at a time, where they remain for 10 to 30 minutes, according to the time for which they are to be kept. On removal the clams and liquor are cooled separately, the latter being first strained and flavored with vinegar, lemon, mace, etc., and then combined and sealed up in suitable receptacles. The object in steaming the clams is to avoid cutting and bruising the meats, which would result if they were opened raw.

The shells of mussels are usually covered with dirt, which should be thoroughly rinsed off. The mollusks are then generally scalded in brackish water in the shell for 10 or 15 minutes and on removal therefrom the dark filament or beard is pulled off, when the meats and liquor are cooled separately and treated similarly to the process of pickling oysters or clams, the flavoring ingredients being selected according to the individual fancies, but consisting usually of mace and cloves in addition to vinegar.

The quantity of ingredients suitable for 100 mussels is about a pint of white vinegar, an ounce of mixed cloves and allspice, with a large red pepper and a few blades of mace. These should be boiled with the liquor from the mussels, and when cooled the whole is poured over the meats. The quantity of vinegar used is small, only sufficient for flavoring. While almost any vinegar might be used, white wine or cider vinegar is preferred. Pickled mussels should be kept in a cool, dark place, for if not well excluded from the light they will turn dark.

PICKLED LOBSTERS.

When a lobster dealer is overstocked with boiled lobsters or with lobsters so weak that they must be boiled to save them, or less frequently when a fisherman desires to dispose of short lobsters caught contrary to local regulations, the usual method is to pickle them. For this purpose the live lobsters are first boiled and the meats extracted, 100 pounds of round lobsters yielding about 25 pounds of meat. The meat in the tail and the large part of the claw is the only portion used, that in the knuckle being discarded, since the quantity is so small that it does not pay for the work of removing it. One man can usually pick out 100 pounds of meat in three hours. The meat is immersed in vinegar for two or three days, then replaced in fresh vinegar and placed in suitable packages, which are usually glass jars with cork stoppers when prepared by the dealers, and barrels, kegs, or stone jars when the pickling is done by the fishermen. White-wine vinegar is preferred to cider vinegar, since the latter has a tendency to make the meat dark, and the vinegar may be weakened to suit the taste desired unless it is proposed to keep the lobsters a long time. If carefully protected in tight packages, the pickled meat will keep several months under ordinary conditions. It sells for about 12 or 15 cents per pound, representing an equivalent of 2 or 3 cents per pound for the live lobsters.

PRESERVATION OF FISHERY PRODUCTS BY SMOKING.

Fish and other food products have been preserved by smoking from time immemorial. The process was well known in Europe during the fourteenth century, and it appears to be used by savage tribes of many different localities. It consists in exposing the articles fresh or, as is more frequently the case, slightly salted, to the action of smoke 'produced by smoldering wood, bark, or sawdust. Its efficiency depends upon the drying as well as the action upon the texture of the fish of the pyroligneous acid produced by the smoldering, which at the same time imparts an agreeable flavor to the product. Smoking is practiced to some extent by nearly all nations, especially in curing oily species of fish, such as herring, haddock, halibut, salmon, etc.

In the United States smoked fish are cured either round, eviscerated, split and beheaded, or cut into small pieces with or without the skin removed, according to the species. Small sea herring, cured as hard herring, and bückling, alewives, fresh mackerel, etc., are usually not dressed at all: bloater herring, lake herring, eels, salt mackerel, flounders, etc., are usually split down the belly to the vent and eviscerated; salmon and haddock are usually split so as to lay out flat like dried codfish, and halibut, sturgeon, and sometimes catfish, are cut up into small pieces before smoking.

After being dressed the fish are at once struck with salt, the length of the salting differing according to the species being prepared, but ranging from an hour or two to a week or more, and in case of halibut, salmon, mackerel, etc., they may be smoked after being salted a year or two, the excess of salt being removed by soaking in water. On removal from the pickle the fish are cleansed and attached to smoking-sticks and after drying for a few hours are placed in the smokehouse, or, in case of halibut, they may be dried on cod flakes for a day or so and then strung on sticks and placed in the smokehouse. All fish cure better and present a neater appearance when cured, if dried in the open air a few hours before being placed in the smokehouse.

Both cold-smoking and hot-smoking are employed, the result of these two processes being quite different. In the former, the fish are suspended at a distance from the fire and smoked at a temperature less than 80° F.; in the latter process the fish are partly or entirely cooked while smoking, being hung near the fire. In cold-smoking the exposure may be only a few hours, as in the case of salmon, or it may continue for weeks, as in curing hard herring, the length of exposure depending on the article prepared and the time that will probably elapse before it is consumed, whereas hot-smoking is always completed within a few hours, usually within three or four. Cold-smoking is used principally in the United States, England, Norway, Holland, Russia, and Scotland. It is applied to herring, alewives, halibut, haddock, salmon, salt mackerel, flounders, butter fish, etc. In Germany and Sweden hot-smoking is the more important, but it is not extensively employed in the United States, being confined mainly to New York, Chicago, Milwaukee, and other centers of foreign population, the species so treated being sturgeon, lake herring, whitefish, eels, catfish, fresh mackerel, etc.

The style of the smokehouse depends on the particular product for which it is intended. The large houses used for smoking herring, halibut, and Finnan haddie are described in the paragraphs relating to the preparation of those respective products. The houses for smoking sturgeon, lake herring, eels, etc., are generally much smaller. Three or four smoking chambers are usually ranged side by side and are commonly built of brick with 8-inch walls with a ceiling of tin or zinc. The height ranges from 6 to 14 feet, inside measurement, the width 3½ to 5 feet, and the depth or length 6 to 12 feet. In most cases provision is made for smoking only three rows of fish, the lowest of which is from 3½ to 6½ feet above the floor, and the others at intervals of 13 to 18 inches above that, the uppermost one being from 8 to 18 inches below the ceiling, so that the fish will be removed somewhat from the body of hot air which accumulates at the top. In the ceiling there are eight or ten small holes, an inch or so in diameter, leading to the flue or chimney.

Most of the houses are of the larger size above given, and when smaller sizes are used it is sometimes necessary to protect the fish from the heat, or they are liable to become too hot. This may be done by placing two small stands of bricks, about 14 inches high, on the floor and building the fire between these, and when necessary to protect the fish from the heat a metallic pan is placed over the fire, the ends resting on the brick piles. In places where the smoking is of small extent the houses are generally cheaply constructed, and some curers do their smoking in an old dry-goods box, the top of which is covered with boards, mats, or sacking. The fish are placed on sticks, and these are placed crosswise inside the box. This is suitable only for hot-smoking, and to adapt the box to cold-smoking the smoke is admitted to the box at its lower end through a trough or channel of boards. Years ago the old-fashioned open kitchen chimneys were used for hot-smoking by arranging the sticks of fish 3 or 4 feet above the fireplace. This form of chimney is being gradually discarded, but a few are still used in smoking fish for home consumption.

The hogshead smokehouses used in a few localities for hot-smoking sturgeon, eels, herring, etc., are quickly and cheaply constructed and well adapted to the purpose. An old sugar or molasses hogshead, with the head removed, is placed on end on the ground, 2 or 3 bushels of earth being first removed so as to form a pit for the fire. For convenience in placing the fuel and in making the fire, 12 or 14 inches of the lower end of one or two of the staves are removed. Across the top of the hogshead in the middle is an iron rod or a piece of old gas pipe, on which rests one end of the smokesticks containing the fish, the other end of the sticks resting on the chime of the hogshead. After the fires are built and the fish placed in the hogshead, the latter is covered with old sacking, such as discarded salt sacks, to confine the smoke. The capacity of each of these hogshead smokehouses is 200 pounds of fish at one time. They are intended especially for hot-smoking, and a mixture of oak and hickory wood is used for fuel. Usually two or three hogsheads are ranged side by side, and for protection against the weather they should be inclosed within a shed or house.

. The foreign smokehouses are quite similar to our own. The following is a description of two, the first being situated at Masnedsund, and the other on the island of Bornholm, in Denmark:

The oven, with the fireplace below, is 6 feet broad, 5½ feet high, and 3 feet deep. In front there are iron doors. There is room in the oven for three rows of poles; the distance from the fire to the lowest row is 3½ feet, and the distance between the rows 14 inches. At the top the oven can be closed by a lid, which opens outside, toward the back wall of the chimney. The chimney projects about a foot beyond the front of the oven, and therefore forms an opening for the escape of the superfluous smoke. The oven is about 6 feet high, and grows narrower toward the top, which is about 1 foot in diameter. The chimney is held together by a strong iron bar. When the fish have been dried in the air, smoking may be done on all three irons, therefore in three rows. The lid at the top is then kept closed. If, however, the oven is to be used for drying, the two upper rows are used for this purpose, and in that case the lid must remain open, and the opening is covered by bags or pieces of board. Gradually, as the two lower rows have been smoked, the two upper ones are put a row farther down, and a new row is hung on the upper iron.

A larger smokehouse in Svanike, on the island of Bornholm, is about 18 yards square and 4 yards high, while the chimney is 6 feet high and 4 feet broad. There are seven smoke rooms, or ovens, for hot-smoking, and one for cold-smoking. The herring are hung in pairs over poles 3 feet long, one herring's head being stuck through the gills of the other and coming out at the mouth. If necessary, a thin stick of wood serves as a skewer. On each pole about 40 herring can be hung, which must not touch each other. The poles are arranged crosswise over square frames, 3 feet broad and 7 feet long, which are run into the oven on ledges. Each frame contains 26 poles, and about 1,040 herring can be smoked in it at the same time. The entire smokehouse can contain 22,400 herring, which are smoked by the hot method. The lowest frame is about 3 feet above the fireplace. In the cold-smoke chimney about 12,000 herring can be smoked.

A few smokehouses, which are devoted principally to smoking river herring or alewives, are constructed with the fire-box outside of the house containing the fish, to avoid heating or burning the fish and to more carefully regulate the smoking. One of these is constructed as follows:

A foundation is made of brick, 9 feet square, 2 feet deep, and 12 inches thick, on which rest brick walls 8 inches thick and 15 feet high on the rear or furnace side, and 16 feet on the front or entrance side, giving the roof a pitch of one foot in eight. About 9 feet from the floor there is set into the walls, on the inside, a ledge of iron, on the front as well as on the rear wall, on which rest pieces of scantling for holding the herring sticks. These are followed by other ledges 12 inches apart until within a few inches of the top of the rear wall. The house is ventilated by a door in the roof, 12 by 15 inches in area, which may be opened or closed by means of a long rod. The furnaces are constructed in the rear of, and adjacent to, the smokehouse, and are 3 feet high, the end and division walls 4½ inches thick, the four grates 20 inches wide and 28 inches deep, and the doors of cast iron 11 by 12 inches in area. The smoke generated passes into four inclined flues, 8 inches square, connecting with the smoke or fish room. These smoke flues are 6 feet long and project two-thirds across the width of the house. In the top of each there are two openings which may be stopped with caps when but little smoke is needed, or each may be covered with a smoke spreader, which consists of a circular piece of tin or iron supported by wires attached to a rim made to fit the openings, and is 12 or 15 inches in diameter and set about 18 inches above the tin to which it is attached. In such a building 5,000 river herring may be smoked in 3 days.

The material which is used for producing the smoke consists of some hard wood or hard-wood sawdust. Oak or hickory mixed with sawdust is the most common in this country, but a variety of other woods are used, depending on the facilities for obtaining it as well as its suitableness for the purpose. In the extensive herring smokehouses at Eastport, Maine, white birch is generally preferred, but driftwood which has been soaked with salt water is used to a considerable extent. At Gloucester and Boston ship carpenter's chips of oak or oak edgings, with sawdust to smother the flames, are used principally. In New York City mahogany and cedar sawdust are used extensively, and at Buffalo maple wood is used exclusively. At Sandusky and Detroit the smokers use hickory wood and sawdust. Shavings and sawdust of pine wood are not very desirable, as they are apt to impart a resinous flavor to the fish. Dry chips of oak are used in Holland, and when those are not readily obtained, poplar, birch, or ash are used. In Denmark the fuel used is alder wood slightly moistened so as to make more smoke, and oak and beech sawdust is used to keep the flames down when they blaze up too high. The smokehouse should always be warm and dry before the fish are put in, as the development of steam is apt to injure the fish. Even when using the same kind of wood, the length of time required to smoke an article of uniform grade depends largely on the condition of the weather, much longer time being required when the weather is sultry than when it is clear and windy.

The length of time that smoked fish will keep depends on the extent of the salting and smoking, and on temperature conditions. Hard herring will keep for a year or more; smoked halibut and haddock will keep only a few weeks, and those products smoked only a few hours are not likely to keep more than a week or so. If the weather be cold and dry, smoked fish keep very much longer than when it is sultry. Some curers, especially halibut smokers, prevent a liability to mold by sprinkling a small quantity of fine dry salt over the fish after smoking; others use compositions of boracic acid, salicylic acid, and other antiseptics sold under various trade names, but the best preventive is to keep the fish in a cool, dry place and dispose of them as soon as practicable after smoking.

Not content with the somewhat slow process of smoking, some dealers have introduced methods by which they reduce its extent, or else do away with it altogether, thus saving in time and in loss of weight of fish. Their process consists in coating the fish with a form or composition of pyroligneous acid to impart a smoked flavor, and a coloring substance to give the fish the appearance of having been smoked. It is gratifying to know that these devices have not been favorably received in the United States.

For the purpose of preparing a choice product especially for exportation to warm climates, the following process* of treating smoked fish has been introduced, but as yet its application in this country is of small extent:

The fish, after being smoked, are cooled off and placed in layers in wooden barrels. Between each layer of fish a layer of dry salt is placed in a quantity of about 6 pounds of salt to 100 pounds of fish. The barrels, after having been filled, are kept in a cool place until the fish have become completely hard in consequence of the salt combining with the natural fat of the fish. This process of hardening must take place through the whole body of each fish, and can be ascertained by pressing the fish with the finger, which must leave no recess or impression whatever on the surface of the fish. After the process of hardening has taken place, which will be, according to the sort and size of fish, from within 3 to 15 days, the barrels are filled up with brine and then closed by a cover fitting tightly. The preparation of the brine must be executed carefully in the following manner: Filtered water is boiled with salt to a saturated solution, which latter is allowed to cool off, after which it is skimmed and drawn off as far as it appears fully clear and pure. If the brine is not carefully prepared, as above stated, the fish will not keep for so long a time, which will likewise not be the case if the process of hardening, before described, has not completely taken place. Fish prepared in the mode described will keep for many months and can be sent to hot climates without danger of spoiling. For making such preserved fish eatable it must be taken from the barrel and placed in fresh water to remove its rigidness. This will, according to the size of fish, take place within from 3 to 8 hours, when the salt will be sufficiently removed from the fat. The fish is then dried in the open air and will now fully resemble newly smoked fish. By first taking the fish in their natural condition and smoking them the juices are retained and the fat of the fish is brought to such condition that the salt when applied will readily combine with it and make the fish perfectly hard and solid, especially on the exterior. After the fish are thus smoked and hardened with salt they are brought to a condition in which brine will simply preserve and protect them from atmospheric influences without changing their character in any material way. By thus treating the fish they are preserved without having the entire body of the fish permeated with salt, as after being smoked the dry salt in which they are packed combines chiefly with the fatty substances and forms a hard exterior surface which is not much penetrated by the brine.

---

* See Letters Patent No. 352666, dated November 16, 1886.

For use in smoking chunks or portions of large fish, such as sturgeon, previous to canning them, a wire disk-form receptacle, in which the chunks are compressed to a size adapted to the cans in which they are to be placed, is here described:

This receptacle is so arranged that it may be rotated during the smoking process, thus causing the dripping juices to pass through the mass. The product of the usual method of smoking does not remain sufficiently juicy for canning and the irregular chunks require a large amount of oil or other liquid to fill the interstices within the can. The receptacle is drum-shaped, with a cross section equal to the cross section of the can wherein the product is to be packed. It has a movable or inner head and a spring clasp for forcing the head inward, and is so suspended that it may be frequently rotated on the spring clasp. In carrying out this process the sturgeon or other fish is cut up into suitable pieces, salted in brine for the proper length of time, and then neatly placed in the drum until the latter is filled. The head and clasp is then placed in position and the drum suspended in the smoking-room. While subjected to the action of the smoke, and as the fish becomes more and more compact, it assumes the shape of a disk with comparatively flat ends. During the smoking the receptacle is turned from time to time so that the juice that settles at the bottom is frequently brought to the top and again compelled to flow through the mass. These disks may be much thinner than the height of the can in which they are placed, in which case two or three or more are superimposed until the can is filled.*

SMOKED HERRING.

The process of smoking is almost as important in the cure of herring as the use of salt in preserving codfish. This was one of the earliest marine products to which smoking was applied, and at present the various species of this family are probably smoked in greater quantities than all other species combined. By varying the process of smoking different products are obtained, almost wholly unlike in appearance, flavor, and keeping qualities, instances of which are the hard or red herring and the bloater herring, both prepared from the sea herring of the New England coast (*Clupea harengus*). The former are smoked three or four weeks, until quite dry, while the latter are exposed to the smoke for only a few hours and will keep but a limited time; the sooner they are eaten the better their flavor. Kippered herring differ from bloater herring principally in that they are split and eviscerated before being smoked. The bückling or pickling prepared in New York City from frozen Newfoundland herring are somewhat similar to the bloaters of Boston, differing principally in that they are smoked at a higher temperature and for a much shorter time. A few Labrador herring pickled in barrels are smoked in New York City, and along the Great Lakes and in the near-by localities quantities of lake herring are smoked. The smoked herring of the Southern States is made from the alewife (*Clupea cerualis*), so abundant in rivers of that region. The methods of smoking applied to each of these various species will be described in detail in the following pages.

HARD HERRING.

The original process of smoking hard herring, or red herring, as practiced in the United States, is said to have been derived from Scotland by way of Digby, Nova Scotia. In 1795 a Scotch fisherman located at the last-named place and devoted his attention to smoking herring as practiced in his native country, the product being sold in Nova Scotia and the adjacent parts of New England under the name of "Digby chickens." Others went into the business and the process gradually extended to the United States, the business being established at Eastport in 1808 and at Lubec in 1812. The trade gradually increased until the beginning of the Washington treaty in 1873, being particularly extensive during the civil war. The average annual output

* See Letters Patent No. 423545, in favor of Max Ams.

from 1845 to 1872 was not less than 500,000 boxes. The Washington treaty had a very serious effect on the smoked-herring industry, the product in Maine in 1880 being only 370,615 boxes, or 4,434,111 pounds, worth $99,973; whereas importations increased from 1,029,095 pounds, valued at $34,670, in 1874, to 10,411,355 pounds, worth $129,034, in 1885. After the abrogation of the treaty in 1885 the business again reached its former proportions and has been more extensive during the past few years than ever before. The annual product amounts to about 1,300,000 boxes, containing 6,500,000 pounds of cured fish, the wholesale value of which approximates $115,000.

The smoking of hard herring in the United States is confined principally to the State of Maine and to Washington, Hancock, and Knox counties, the business being centered at Eastport and Lubec. The mature *Clupea harengus* is used, taken almost wholly by weirs, the season extending generally from the first of September until late in December. The smoking is done principally by the persons catching the fish, who also depend partly on farming for a livelihood. Usually several of them own a weir in common, dividing the herring equally and preparing them on their separate premises.

The following description of the process of smoking hard herring at Eastport and Lubec is from an account of the industry by Mr. Ansley Hall: *

Description of smokehouses.—The smokehouse is generally only one of a number of buildings used in carrying on the smoked-herring industry. In addition to it there are sheds and shops of various kinds, in which is done a variety of work incidental to the business. There is a shed for pickling and salting herring, a shop in which the smoked-herring boxes are made and where the herring are packed, and there is sometimes a cooper shop for making herring barrels. The buildings are usually located on a wharf or near the shore for convenience in landing the fish from the boats. The frame of the smokehouse is covered with boards and made sufficiently tight to prevent the smoke from escaping. There are board windows in either end and ventilators in the roof. The latter are provided by arranging the boards on either side of the ridgepole so that they can be raised or lowered by means of cords attached to levers. The building is entered by large doors in the end. The value of the smokehouses, including the sheds and equipments, varies from $60 to $3,500 each; for an entire stand of buildings the average value is from about $200 to $500.

In the early days of the industry the smokehouses were very inexpensive, being built of slabs obtained at small cost from the sawmills in the vicinity. A very few of these primitive structures, now almost a century old, are still in use, but in most instances they have been replaced by better ones. As the business increased larger smokehouses were built, in order to make it possible to meet the greater demand for the product. The largest one now in use is at Lubec. The length of the building is 231 feet, 115 feet of which is included in the smokehouse and 116 feet in sheds of various kinds. The width is 25 feet, the length of posts 16 feet, and the height of the ridgepole 29 feet. The smokehouse is divided into three compartments, each having 10 "bays" or spaces in which to hang herring, and its capacity is about 45,000 boxes of medium or 60,000 boxes of large herring. It is as large as three smokehouses of the ordinary size. The smokehouses have no floors, as the area has to be used for the fires. The interior is arranged with a series of vertical rows of 2 by 4 inch scantlings. The spaces between the rows are termed "bays" and are 38 inches in width. The scantlings in each row begin near the ridgepole and extend horizontally crosswise of the building, each one being placed from 13 to 14 inches below the other, to within 6 or 8 feet of the ground. In smokehouses of the average size there are usually 10 "bays," and the capacity is about 15,000 boxes of medium or 20,000 boxes of large herring.

Equipment.—The only equipment used exclusively in a smokehouse are the herring sticks. A large number of these are necessary in the larger houses, as it requires on an average about two sticks to each box of herring. The sticks are prepared at the sawmills in long strips. The size of the sticks as they come from the mill is one-half inch square for medium and five-eighths inch square for large herring. After being cut into lengths of 3 feet 4 inches each, the edges taken off, and one end sharpened they are ready for use. They cost at the rate of about $3 per 1,000 at the mill, and are estimated to be worth from $4 to $5 per 1,000 after being made at the smokehouse.

* "The herring industry of the Passamaquoddy region, Maine," by Ansley Hall, United States Fish Commission Report for 1896, pp. 454–463.

The pickling and salting shed is supplied with wooden tanks for use in pickling the herring. These are from 7 to 8 feet long, 5 feet wide, and 3½ feet deep, having a capacity of about 4 hogsheads, or 20 barrels, of herring each. From 2 to 9 tanks are required in each salting shed. In many instances hogsheads are used instead of tanks. There is also a variety of other utensils, such as tubs, baskets, shovels, and "herring horses." The latter consists of an oblong wooden frame having four legs, the sides extending far enough beyond the end to serve as handles. It is used to hang the herring on to dry after they are strung on the sticks and before putting them into the smokehouse. Its capacity is from 25 to 30 sticks of herring. The cost of the whole outfit for a smokehouse and salting shed varies from $50 to $500.   *   *   *

Herring utilized.—The herring utilized for smoking and salting are derived chiefly from the weirs in Passamaquoddy Bay and its tributary waters. In 1895 the quantity obtained from the American weirs in the bay for these purposes was 5,903 hogsheads, valued at $12,121, and from the Canadian weirs 5,571 hogsheads, which cost $20,036. The average value per hogshead of the former was $2, and of the latter about $3.60. This difference is explained in a measure by the fact that a large part of the American fish was smoked by the fishermen, who carried them to the smokehouses in their own boats, while those from the Canadian weirs were collected at the rate of $1 per hogshead. A considerable quantity of herring was also obtained from other sources. From Machias Bay there were 1,296 hogsheads, costing $1,605; from Grand Manan, 935 hogsheads, $2,323; from the Magdalen Islands, 768 hogsheads, $4,669, and from Newfoundland, 174 hogsheads, $1,710. The total quantity used was 14,647 hogsheads, or 73,235 barrels, the cost of which, landed at the smokehouses, was $45,494. Of these, 12,148 hogsheads, costing $36,215, were smoked and packed in boxes, and 2,499 hogsheads, costing $9,279, were salted in barrels. The herring from Passamaquoddy Bay, Machias Bay, and Grand Manan are received in a fresh condition, while those from the Magdalen Islands and Newfoundland are cured on board the vessels and need no further salting after they arrive at the smokehouses. The Newfoundland herring are used largely in preparing the grade of smoked herring termed "bloaters," but those from the Magdalen Islands do not serve that purpose so well and are generally either packed in barrels as round herring or smoked and packed in regular boxes lengthwise.

Pickling.—When the fresh herring intended for smoking are landed at the salting sheds, they are immediately put into the pickling tanks, which have first been partially filled with a weak pickle. The pickle is made of salt water with about 1¼ bushels of Liverpool salt or a smaller quantity of Cadiz or other coarse salt in each tank. The salt is stirred in the water until it is wholly dissolved. If the fish are poor the water is sometimes used without the salt being added. The quantity of fish which is at first put into the tank is generally from 2 to 3 hogsheads, or enough to be of sufficient weight to rest or, as the fishermen term it, "ground" on the bottom. A light layer of salt, or about one-half bushel, is then distributed over them, after which another layer of fish of from 1 to 2 barrels is put in. This is again covered with a layer of salt rather heavier than the first, being from 1 to 1½ bushels. The remainder of the fish necessary to fill the tank is then put in and covered with from 3 to 5 bushels of salt. Each tank when filled contains 4 hogsheads of fish, and the quantity of salt used on them varies from 6 to 9 bushels, according to their size and fatness and the condition of the weather. It is also necessary to have the greater part of the salt at the top of the tank, so it will not work down through the fish and lodge at the bottom without being dissolved. In that case the fish at the bottom are liable to become too salt and those at the top not salt enough. For smoking purposes the fish are pickled in a round condition as they come from the water. When hogsheads are used instead of tanks the quantity of fish and salt in each layer is regulated to correspond with the capacity of the hogshead.

The small herring are generally allowed to remain submerged in the pickle from 24 to 36 hours, and the larger ones, especially if they are very fat, about 48 hours, and sometimes a longer period. If the herring are small and not fat the length of time required for them to "strike" may not exceed from 12 to 15 hours. Fish will also absorb salt more readily in warm than in cold weather, and if they have been caught a few hours before being salted they do not require so long a time in the pickle as when immediately taken from the water.

When the fish have been properly "struck" or salted, if the weather is fine, so as to afford them an opportunity to dry before being put in the smokehouse, they are taken out of the pickle; but it sometimes happens that the weather is rainy, and they have to remain in pickle much longer than would otherwise be necessary. As a result they become more or less oversalted. In such cases, when favorable weather returns, they are taken out and put in tubs of salt water to be freshened or "soaked out." Newfoundland and Magdalen herring, which are heavily salted on board the vessel when caught, invariably require to be treated in this manner before being smoked. Generally about four tubs of water are used, which are in succession filled with fish. As soon as the last tub is filled the fish are taken out of the first one and then out of the others in regular order, each tub being at once refilled with other fish, and this process is continued until all the oversalted fish have passed through the water, remaining there only long enough to secure the desired result. If the quantity of fish is large the water in the tubs is changed whenever requisite. It is customary to use salt water for nearly all purposes. The fishermen and smokers claim that fresh water has a tendency to make the gills of the herring tender and more liable to break and allow the fish to fall from the sticks after being hung in the smokehouse. They also think that the salt water makes the flesh of the fish more firm and not so apt to become soft after being smoked. The salting sheds are, therefore, sometimes furnished with steam pumps for obtaining the necessary supply of salt water. * * *

Scaling.—It was formerly customary to remove the scales from the herring intended for smoking purposes before taking them from the boat. The fishermen, with their rubber boots, walked through the mass without lifting their feet, and the contact of the fish with each other and with the legs of the men removed the greater part of the scales. This laborious process was called "treading them out." It is said to have begun in 1820 and was continued until about 1880. Another method of scaling the fish during that period was to stir them with a spudger. In recent years it has not been considered necessary to resort to these or other methods for removing the scales, since the frequent handling which the fish undergo renders them practically scaleless when they reach the smokehouse. The scales of the herring come off very easily when the fish are first taken from the water, but if allowed to dry they become set and are removed with difficulty. The methods for removing them above described insured a more thorough and uniform scaling of the fish than would otherwise be effected, but if the work was not carefully performed it was liable to result in bruising the fish and in an increased loss in "broken-bellied" herring.

Stringing.—When properly salted the fish are taken out of the pickle to be strung on herring sticks preparatory to being hung in the smokehouse. This is done with ordinary dip nets, or "wash nets," as they are called in this locality. As the fish are dipped out they are washed or rinsed in the brine with the nets, after which the pickle is allowed to run off of them and they are laid on the stringing tables. The dipping and stringing proceed simultaneously.

The "stringers," or persons who string the herring, are of both sexes, the females often predominating in number. In some instances the fishermen do the work themselves, but generally men and women and boys and girls are hired for this purpose. The number of stringers employed in each smokehouse varies from 2 to 8 and sometimes more, according to the amount of work to be done. They receive 20 cents per 100 sticks for stringing large herring and bloaters and 25 cents for small herring. The cost of stringing is estimated to average one-half cent per box, but is probably a little less than that. At these prices each stringer can earn from $1 to $2 per day. There are from 25 to 35 herring on each stick, and a person can string from 500 to 1,000 sticks in a day. The work is performed very rapidly. The herring is taken with its back in the palm of the right hand, the stick being held by the blunt end in the left hand; the left gill-cover is then raised by a movement of the right thumb and the pointed end of the stick is inserted and passed through the mouth, the fish being moved down to its proper place. The work is often done by reversing this order, the fish being taken in the left and the stick in the right hand, but in either case the herring when strung hang on the stick with their backs toward the stringer.

Draining and drying.—After the herring have been strung on the sticks they are washed in a trough of clean salt water and hung on the herring horses. They are then carried out into the open air, where they are allowed to remain until the water drains off of them and they have become sufficiently dry to hang in the smokehouse. The time required for drying varies according to the condition of the weather, but is usually from one to several hours. The drying not only hardens the gill-covers and prevents the fish from falling from the sticks in the smokehouse, but also improves their quality when smoked. The work of stringing and drying the herring is generally done in the fore part of the day and in the afternoon they are hung in the smokehouse. If the weather is not fine it is sometimes necessary to dry the fish in the smokehouse after leaving them in the open air long enough for the water to drain from them. When this method is resorted to, the doors and windows are opened to give a free circulation of air and fires are kept burning until the drying is completed.

Filling the smokehouse.—The smokehouse is not usually filled all at one time, and it often happens that the work occupies several weeks. The herring are taken care of as fast as they are obtained from the weirs, the time required to fill the smokehouse depending somewhat on the abundance and constancy of the supply. If the supply is steady, the work progresses as rapidly as herring can be prepared, otherwise the period may be extended to three or four weeks and perhaps longer.

When the herring have been sufficiently dried in the sun they are carried on the herring horses to the smokehouse, where the sticks are placed in the "bays," their ends resting on the scantlings or beams on either side of each "bay." The work of "hanging" the herring requires the services of at least two men, and if a larger number are engaged in it they work in pairs. One man stands in the "bay" with his feet on the beams, while the other stands on the ground or floor and hands the sticks of herring up to him, two at a time, keeping the sharp end of the stick downward so the herring will not slip off. The sticks are made long enough to reach across the "bay" and to nearly the center of the beams which support them at either end.

The lower part of the "bays" is usually filled first. The fires are then kindled and the herring smoked until they acquire a good color. When this is effected the fires are allowed to go down, the doors and ventilators are opened to let out the smoke, and the herring are shifted to a place nearer the top of the smokehouse. The lower part is then ready to receive another lot of fish. This preliminary smoking occupies from about 12 to 15 hours. The work is continued in this manner until the smokehouse is filled. Two smokehouses are very often filled at the same time. In that case, after the top of the house has been filled by shifting the herring, the lower part is completed by putting about three tiers of herring in each house on alternate days. When two houses are filled together, the work can be done in almost as short a time as would be required to fill one alone.

The object of putting the herring into the house by degrees, instead of all at one time, were that practicable, is to insure their becoming thoroughly dry before being subjected to the smoke, and also to smoke them more evenly and secure a greater uniformity of color. If a large body of fish were put into the smokehouse at once they would gather dampness and great difficulty would be met with in preventing them from spoiling. To fill a smokehouse holding 20,000 boxes of herring in a proper manner requires at least two weeks and a somewhat longer period if two such houses are filled at the same time. The length of time also varies according to the size of the smokehouses. Small houses may sometimes be filled in a few days. After the smokehouses have been filled the additional length of time required to complete smoking the herring is about three weeks. Regular herring are placed as close together on the sticks as possible without touching each other, the gill-covers generally keeping them far enough apart. The sticks, when hung, are placed about 3 inches from each other.

Fires and wood.—The fires for smoking the herring are built on the ground at equal distances apart over the entire area of the smokehouse. The wood used is of various kinds, but white birch is generally preferred; driftwood which has been soaked with salt water is also used. The main consideration is to have wood that will burn slowly and produce an abundance of smoke. The fires are kept burning very slowly, the smokehouse being visited every few hours during the night as well as the day. If too much heat is generated the herring are soon damaged and may be completely spoiled.

Previous to 1820, only two brands of smoked herring were known, namely, "number ones" and "number twos." On the introduction of scaled fish, a third brand was added, the "medium scaled," including all the best fish of medium size that were well scaled. At present there are three principal brands of hard herring, viz, "lengthwise," "medium-scaled," and "No. 1." Another brand known as "tucktails" is also prepared to some extent. The lengthwise herring are the largest of the hard herring prepared, and must be packed lengthwise with the box, hence the name. Of this grade each box contains only about 15 or 20 fish, weighing about 6 pounds, the boxes being of uniform size, 12 inches long, 6½ inches wide, and 2¾ inches deep, the thickness of the ends being five-eighths inch, and of the other parts one-fourth inch, the cost of the boxes approximating $15 per 1,000. The tucktails are also longer than the width of the box, but they are packed crosswise of the box, the tails being tucked or bent over them, as indicated by the name. The medium-scaled form the popular size and sell for the highest prices. They are packed crosswise of the box and are usually divided into two sizes, viz, large and small medium herring, 30 to 40 of the former and 40 to 50 of the latter filling a box. The "No. 1" grade is composed of the smallest fish, each box containing from 55 to 75 fish.

Several of the New England States have very extensive and precise regulations affecting the grading, packing, inspecting, and branding of smoked fish, but these regulations are rarely enforced. In Maine regulations were made from time to time beginning in 1821, affecting the smoking of herring, but in 1871 it was provided—

Hereafter no inspection of smoked herring shall be required, but all smoked herring put up in boxes or casks for sale in this State shall be branded on the cask or box inclosing them with the first letter of the Christian name and the whole of the surname of the person putting up the same, and with the name of the State and the place where such person lives, and all such fish offered for sale or shipping not thus branded shall be forfeited, one-half to the use of the town where the offense is committed, and the other half to the person libeling the same.

Early in the present century the price realized by the fishermen varied from $1 to $1.25 per box, 18 inches long, 9 inches wide, and 7 inches deep, inside measurement. From 1830 to 1850 the average price was about $1.10 for "scaled herring," 80 cents for "number ones," and 35 to 40 cents for "number twos," the size of the box being 17 inches long, 8¼ inches wide, and 6 inches deep, measured on the inside. From that date the price decreased quite rapidly for a number of years, and fish of good quality often sold as low as 7 and 8 cents per box. Later, with the revival of trade, it again improved, until in 1880 it ranged between 12 and 25 cents, according to the quality of the fish, good scaled herring averaging fully 22 cents, while lower grades usually sold at 15 or 16 cents. The boxes in 1880 were usually 15½ inches long, 7½ inches wide, and 4 inches deep, inside measurement. Since 1880 the prices have decreased considerably. In 1894 medium-scaled herring sold for 9 cents, and No. 1 for about 6 cents.

A choice method of packing smoked herring, introduced in 1878, has met with much favor. After the herring have been salted and smoked in the usual way, the skin, head, and viscera are removed and the bones extracted. The flesh is then packed eight to twelve in small wooden boxes with glass fronts or tied in bunches of about one dozen fish each, six of such bunches being packed in a neat wooden box, which also sometimes has a pane of glass introduced in one of the sides to render the contents visible without opening the box. By skinning the herring and placing them together their flesh is brought in close contact, preserving their inherent moisture and flavor, this effect being further increased by packing them in a box. The fish also present a much neater appearance when offered for sale and are more attractive to customers. This process was protected by Letters Patent No. 207980, dated September 10, 1878.

Large quantities of foreign smoked herring are imported into the United States, approximating 1,000,000 pounds annually, worth about $100,000—mostly from Nova Scotia and New Brunswick, though large quantities are received also from Newfoundland, Norway, Great Britain, and the Netherlands. The exports of smoked herring are equal in quantity to the imports, the great bulk of them being sent to Haiti, and smaller quantities to Santo Domingo, Cuba, and other tropical countries.

The following notes on the methods of smoking hard or red herring in Holland and in England are furnished by Mr. Adolph Nielsen:

SMOKING HERRING IN HOLLAND.

The greater part of the herring are caught in the North Sea and salted round on board of the vessel in barrels. After they are brought to the smoking-houses the barrels are opened and the herring put into large vessels to be steeped in fresh water. The length of time in which the herring are steeped depends upon the different markets for which they are prepared. For the local markets, Antwerp and Brussels, they are steeped for two days, while for the Italian markets they are steeped one day, and sometimes not steeped at all, but only washed. In order to liberate the herring as much as possible from scales they are stirred about several times during the day with a stirring pole. The herring which are salted heavy or have remained in salt over the ordinary time are first steeped one day, then taken up and put in baskets for 12 hours, and after this again steeped another 24 hours. After the herring are sufficiently steeped the water is drawn off and the herring sorted and put in baskets, which contain about half a barrel, and are left in these in the balcony for 18 to 21 hours. The object of this is that the herring, by their own weight, in the baskets, shall press out some of the water, and serve instead of drying, and thereby facilitate the smoking. Subsequently the herring are threaded on willow twigs, as in England, and brought into the smoking-rooms, where they in the meantime, until they can be hung up in the raftwork, are placed on stands made for that purpose. When hung up to be smoked, the fattest, and such herring as are to be smoked strongest, are placed nearest the roof. The fire is made on the floor in a dozen small heaps (according to the size of the room) in each room, and chips of oak are generally used for that purpose if they possibly can be obtained; if not, a mixture of poplar, ash, elm, and birch. Sawdust of oak is applied in order to smother the fire and keep it from flaming, also to form a good body of smoke. After the fire is kindled the small windows close to the roof and the lower part of the door are kept open in order to give a better draft, and also to give the dampness from the herring a chance to escape. The fire is renewed whenever the glowing chips are getting low. When the water after a couple of hours has evaporated from the herring, the small windows close to the roof and the lower part of the door are shut if sufficient draft can be had through the ventilators in the roof. The temperature is kept as near 65° F. as possible and is regulated by opening and closing the small windows and the doors.

Herring prepared for the two principal markets, Antwerp and Brussels, are generally smoked for 12 hours, and supposed to keep good for two weeks. These markets want the herring to be lightly smoked and of a bright bronzed color. Herring prepared for Germany, Italy, and other Belgian parts are smoked for 24 hours, and are supposed to keep good from one month to six weeks. They are dark-colored. After the herring are smoked the small windows and doors are opened and the herring left to cool, either in the smoking-rooms or in the balcony in the raftwork erected there, for a couple of hours before they are packed. The herring are packed in baskets, made of willow twigs, 28 inches long, 17 inches wide at the top, and 12 inches at the bottom, the height about 7 inches, containing 200 each, for the Belgian, German, and home markets; while for the Italian markets boxes and drums made of soft wood are used. The boxes are 21½ inches long, 12 inches high, and 9½ inches wide, and hold about 200 herring. The drums are 20 inches high and 12 inches in diameter and contain from 300 to 400 herring. The herring are packed slantways, back down, across the boxes or baskets, with the exception of the two uppermost layers, which are packed back up. The packing in the drums is just the same as the common packing in barrels. As a rule a little straw is put in the bottom of boxes and baskets. The bowed basket lid is sewed fast by the help of long needle and twine. The herring are sorted for the Belgian and German markets in full and spent, with no regard to size. For the Italian markets they are sorted in large full, medium full, and spent herring. Of the large full it takes about 300, of the medium about 400, and of the spent about 370 to fill a drum of the above-mentioned dimensions. Besides the herring caught in the North Sea and treated and smoked in the manner described, a lot of smaller herring caught in the Zuyder Zee is also smoked for local consumption. These herring are generally brought to the smoking-houses fresh, are pickled in strong pickle for about an hour, left in the baskets to dry a while and smoked for 4 to 8 hours. The smoking of herring for export to foreign countries has not been carried on in Europe to any extent, except in Great Britain and Holland, until of late years, when a lot of this article has also been exported from Norway and Sweden.

SMOKING HERRING IN ENGLAND.

Red herring are for the most part prepared of fresh herring, which are salted dry on the floor in the salting-room connected with the smoking-house, and allowed to remain in the salt 2 to 6 days, according to the length of time they are intended to keep, and according to markets for which they are prepared. After being left a sufficient time in salt they are rinsed in clean water and then threaded on sticks in the same way as the bloaters and hung up in the smoking-rooms, where they are smoked for about 4 weeks. If high dried are wanted, the time of smoking is about 6 weeks. The red herring are smoked with a small fire made of chips of oak and sawdust, and the fire only renewed once every day. The average temperature in the smoking-rooms is kept at about 62° F. Sometimes red herring are prepared from herring which have been pickled in large cisterns. These are soaked in fresh water before they are hung up to be smoked, but otherwise treated in the same manner as the dry-salted red herring. The pickle-salted are called Scotch reds, and are lower in price than the dry-salted herring. Herring which have fallen down or are headless are smoked on the same sticks of wood as the kippers, and are called "red tenters" and "plucks."

In preparing red or hard herring at Yarmouth, England, for the Mediterranean markets, the fish on arrival at the curing-house are, if previously salted on board of the vessels, rinsed to remove the incrusted salt, and then, without other preparation, are again put in salt, that from Liverpool being preferred. After remaining in salt for 10 to 14 days the herring are washed, strung on smoke-sticks, 25 fish to each stick, which is generally about 56 inches long, and placed in the smoke-room, which may be 16 or 18 feet square. A dozen or more fires are made on the floor, the fuel generally being small sticks of oak or ash. The fires are kept burning for two days, when they are permitted to go out and the fish allowed to drip for a day. Then the fires are again lighted for 2 or 3 days, and this process of alternate smoking and draining continued for 2 weeks or even longer, when, after cooling, the fish are ready for packing. For the home market Yarmouth hard herring are packed in flat boxes or in kegs 10 inches high, holding from 80 to 100 fish, and for the Mediterranean markets in barrels and half-barrels. In the latter case, when the barrel is filled to the top, by means of a screw press the fish are pressed down and an additional number placed in, 600 or 700 full-sized fish being the usual number to each barrel.

BLOATER HERRING.

Scotland has always led in the preparation of bloater herring, as in most of the smoked fishery products. It is not known when this article originated, but doubtless as early as the sixteenth century. For two centuries or more it has been an important product in Europe, but it has been prepared in the United States during the last 40 years only. The business is said to have begun at Boston in 1859, as an outgrowth of the importation of large salted herring from Bay of Islands, Newfoundland. The process was similar to that employed in Scotland, and the trade increased until in 1868 10,000 barrels of large herring were smoked and sold as "Yarmouth bloaters." During that year the business was started at Eastport, Me., and on account of the convenience of carrying it on in connection with the extensive smoking of hard herring in the vicinity the trade has largely centered at that port. The preparation of bloaters was begun at Gloucester in 1883, the fish being received salted from Newfoundland, and at present the business at that port is quite large and is carried on principally in connection with the smoking of halibut. Several ports in Maine also prepare quantities of this product, among which are Portland and Lubec. Some are also prepared in New York City.

The Washington treaty in effect from July 1, 1873, to June 30, 1885, had a very disastrous effect on the bloater-herring industry, large quantities being prepared at Grand Manan, Campobello, and other islands in the British North American Provinces, and shipped to Boston and New York. Since the abrogation of that treaty, however, the duty of ½ cent per pound has restricted the preparation of the supply for United States markets almost exclusively to this country. The present annual product approximates 5,500,000 pounds, valued at $170,000 wholesale.

Two general grades of bloater herring are prepared in this country, those from the large Newfoundland salted herring and those from herring caught along the coast of Maine. The business at Gloucester and Boston depends almost exclusively on the salted herring from Newfoundland, while Eastport, Lubec, and Portland use mainly the herring caught on the Maine coast, most of which are received in a fresh state. The Newfoundland herring are obtained from Bay of Islands, Boon Bay, Fortune Bay, Placentia Bay, St. Morris Bay, etc., being purchased of the fishermen at about 60 cents to $1 per barrel, and from 12,000 to 20,000 barrels being imported each year. The vessels engaged in that trade arrive in Newfoundland during October with from 1,200 to 1,800 bushels of salt each and are moored in some convenient cove. As the fresh herring are landed on deck, a barrel at a time, about $1\frac{1}{2}$ bushels of Trapani salt is spread among them, the mass dredged back and forth several times and then shoveled into the hold in bulk until a cargo is secured amounting to 1,000 or 1,500 barrels. There they remain until the vessel reaches port, the pickle being pumped out when necessary. The fish are then removed and stored in bulk on the floor of the warehouse, where they may remain until the following April if not required in the meantime.

There are two general processes of treating these salted herring preparatory to smoking, the difference being in the manner of soaking. At Boston the salted fish are soaked in large square tanks sufficiently to remove the extreme saltiness and all dust, slime, etc., the length of the soaking depending on the degree of saltiness and varying from 15 to 24 hours. On removal with a dip net they are thrown on a stringing table, where a gang of men stand ready to place them on small square sticks about $3\frac{1}{2}$ feet in length. In stringing them the stick is held in the left hand, the lower end resting under the left elbow; each fish is grasped with the right hand about the head, and by pressing it vertically the gills are opened, when the fish is entered on the stick at the left gill opening and out at the mouth. Usually about 15 herring are placed on each stick. Each stick with its load of herring is then dipped in water for a moment and allowed to drain, and placed in the smokehouse. At Gloucester the salted herring are soaked for only a few moments before stringing, and round sticks, $\frac{1}{2}$ inch in diameter and 2 feet long, are used. The sticks with the attached herring are then immersed in tubs or vats which are filled with fresh water, and the fish are soaked from 8 to 16 hours to freshen them. On removal they are allowed to drain for a few moments, and are then placed in the smokehouse.

When fresh herring from neighboring points are used, as at Portland and Eastport, they are immediately pickled on their receipt at the smoking establishments, about a bushel of salt being used to each barrel of fish. After remaining in the pickle for 2 or 3 days they are removed, drained, and placed on the ordinary herring sticks, and hung in the smokehouse and smoked like the Newfoundland salted herring.

In order to "bloat," the herring must be thoroughly moist, and after they have commenced to dry in the smokehouse the heat must be increased. If they are permitted to hang 10 or 12 hours without heating they will not bloat, but will become hard herring. The smoking is continued from $2\frac{1}{2}$ to 6 days, when the fish are usually sufficiently cured. They are removed from the houses, allowed to cool for a few hours, and placed in boxes holding 50 or 100 fish each, the larger size being by far the most numerous. The average weight of 100 bloaters prepared from Newfoundland herring is about 40 pounds, whereas an equal quantity prepared from Gulf of Maine fish weighs from 25 to 35 pounds, according to their size and the extent of the smoking. The Eastport bloaters weigh about 25 pounds per 100 fish, being smoked 2 or 3 days longer than the Boston bloaters, as they are intended to keep a greater length of time and in warmer climates. They are placed in boxes $18\frac{1}{2}$ inches long, $11\frac{1}{2}$ inches wide, and $7\frac{1}{2}$ inches deep, inside measurement. The thickness of the ends is generally $\frac{7}{8}$ inch and of other parts $\frac{3}{8}$ inch, and the cost of boxes approximates $12 per 100. The boxes at Boston, Gloucester, and Portland are usually considerably larger. One barrel of round fresh herring yields about 5 boxes of 100 bloaters each. Those smoked $2\frac{1}{2}$ or 3 days will keep usually 3 or 4 months under favorable conditions, while those smoked 5 or 6 days will keep until warm weather. Very few bloaters are sold after the month of May.

The market for bloaters is principally in Boston, New York, Canada, and the West, and the average wholesale price for those prepared from Newfoundland salted herring is about $1.20 per 100 fish. The Boston-cured bloaters sold in 1859 at $1.25 to $1.50, and in 1865 at $1.80, per 100. In 1880 the value of the Eastport bloaters was about 95 cents, in 1893 it was 77 cents, and in 1898 it was about 80 cents per 100.

The cost of preparing bloaters at Eastport is considerably less than at Gloucester or Boston. At Boston it approximates 84 cents per box of 100, as follows:

Salted fish (at $2.50 per barrel of 550)	$0.45
Cost of smoking and packing	.25
Box	.14
Total	.81

During the past two or three years some curers have packed bay leaves between the layers of bloaters in the boxes, but fish so packed have a tendency to mold when placed in cold storage. Choicest bloaters are very little salted, and are smoked so slightly that there is little discoloration of the skin, but prepared in that way they will keep only three or four days. These mild-cured bloaters are very popular in Great Britain, but are not prepared in this country to any great extent.

The "pickling" or "bückling" prepared in New York City are quite similar to the bloaters prepared at Boston. The large fat frozen herring from Newfoundland are used, their average weight being nearly a pound each. These are placed in cold storage, whence they are removed from time to time, as the trade requires. On removal they are thawed out and pickled round for 10 or 12 hours and placed on rods in the smokehouse, and after smoking cold for 8 or 10 hours they are placed in the smoke oven and hot-smoked or cooked for an hour or two. About 10,000 pounds of these are prepared in New York City annually, selling at about 12 cents per pound. These fish are sometimes beheaded and eviscerated before being smoked, and are then sealed in tin cans, small fish being selected for this purpose.

Labrador and Newfoundland split herring, salted in barrels, are also smoked in New York City and a few other points, but the business is not so extensive as formerly, probably not exceeding 12,000 pounds annually. These are soaked out, strung up, and cold-smoked for 8 or 10 hours, just enough to give a slight color to them. In New York they are generally tied 3 in a bunch and sold to the stores at 6 or 7 cents per bunch.

The preparation of bloaters is much more extensive in Great Britain than in the United States, Yarmouth being the principal place where they are cured. Usually they are prepared for immediate consumption and are smoked for 10 or 12 hours only. When using fresh herring, the fish are placed in strong brine for 6 or 8 hours, then washed in clean water to remove scales, slime, etc., placed on smoke-sticks by pressing the latter through the gills, dipped or rinsed in water, and suspended in smokehouses, where they are smoked for 10 or 12 hours at a temperature of about 80°. When using salted herring, they are soaked for a time to remove the excess of salt, the length of the soaking depending on the degree of saltiness of the fish.

The delegates appointed in 1889 by the Canadian government to inquire into the herring industry of Great Britain and Holland, state as follows regarding the Yarmouth bloater industry, on pages 30–31 of their report:

One of the best bloater curers in Yarmouth informed us that one reason why his fish stood so high in the market was that he was always very careful, in the first place, to select the very best fish for the manufacture of bloaters, reserving for other purposes all inferior and unsuitable fish. Then he is very careful in salting, curing, and smoking them. We saw in the fish stores in Yarmouth, also on Billingsgate market in London and on the tables in the hotels, a bloater very slightly salted, and smoked so slightly that there was no discoloration at all of the herring. This bloater so prepared is a most delicious fish. It is prepared in this way for immediate use in the nearest cities, towns, and country places, and will only keep some three or four days. Other classes of bloaters, intended for consumption at greater distances and therefore designed to keep longer, are more highly salted, smoked in various grades. The bloaters we saw were fairly fat, but very fat herring will not do for bloaters. Bloaters are salted in heaps on the stone floors of the warehouses—some for a few hours, some for one or two days or more. They are never so highly smoked as the mildest red herring. There is no difficulty in manufacturing bloaters. All that is required is intelligence, good judgment, quick observation, and honesty of purpose, together with a knowledge of the tastes of the consumers; and also whether the fish is required for immediate use near by or for exportation to places at a distance. The gentleman who gave us so much information said that first of all he required to know exactly the kind of bloater required and that he then did his best to supply the article. When the herring have been quite sufficiently salted, they are washed clean on the outside, but are not opened, gibbed, or gutted. They are then strung on rods and hung up to drip and dry, and then smoked. The fuel preferred in Britain for smoking purposes is the sawdust or the waste from the turning lathe of birch, although oak and elm are sometimes used. All agreed that the birch made the sweetest smoke. The white bloaters put up for immediate use are packed in neat light boxes, containing 50 herrings each. Those more highly salted and smoked are put up in larger packages. The bloaters we saw were considerably smaller than our own herring; they are deep from back to belly, and are an excellent fish. Too much attention can not be given to the selection of the herring used for bloaters and to the respective curing processes. The excellence of any particular curer's bloaters does not arise from any special mode of curing, but from special care and attention and that practical knowledge which close observation and experience alone can confer. At the hotel bloaters were opened and split from the belly to the backbone, the gills and viscera taken out, and the herring, without being washed, cooked with the milt and the roe. The roe furnishes pleasant eating.

In the case of bloaters for immediate use, the herring may be put, immediately after being landed and selected, into a strong pickle from six to eight hours. They are then put on the spits and washed by dipping in large tubs of salt water or very weak brine, and then hung up in the smokehouse. The fires should have been burning previously, therefore emitting only a light smoke. A few hours—six to ten—in the smoke room will suffice. They should be cooled off before being packed for the market.

The bloater business in Britain is simply enormous and uses up an immense amount of herrings, thus greatly benefiting the fishermen and the curers, who realize at once on this branch of the herring industry, while the public are supplied with herring in an agreeable and popular form.

KIPPERED HERRING.

Comparatively few kippered herring are prepared in the United States, the round bloaters being so much more popular. The kippered herring are split along the back from the head to the tail, like mackerel, eviscerated, washed, and salted in a manner similar to that applied to bloaters, except that they are not kept in the pickle so long. They are next hung up to dry for a few hours, then smoked for 6 or 8 hours at a temperature of 80° or 85°, each fish being suspended by the napes to keep its abdomen open. With the exception of splitting, the cure is similar to that of bloaters. They sell for about $2 per 100, but the trade is of very limited extent.

The Canadian delegates previously referred to reported as follows regarding the kippered-herring industry of Great Britain:

There is a very large business done in kippered herring in Britain. Herring put up in this way are in great demand everywhere and are preferred by many to the bloater. The very best herring are required for the kippering process. The herring of the west coast of Scotland are in great request for this purpose. The fish used for kippers should be had as soon as possible after they are taken out of the water. They are then carefully selected as to size and quality. Where we saw them at work an active girl stood at a bench laying the herring on its side with the back toward her; with two cuts of a sharp knife she split it from mouth to tail, and with a third motion of the knife she scraped out the stomach and gut and any loose blood inside the fish. She did her work with great rapidity. The herring were then placed carefully into vats of pickle, where, being for immediate use, they remained for 35 minutes, and were then carefully taken out and placed in baskets to drip. They were then spitted on fine rods containing from 12 to 20 herrings each, and hung up in the smoke-house and smoked for a few hours—five or six—then cooled off and packed up in small boxes and dispatched to London by train before midnight of the day on which the fish were caught. When the fish are intended to be kept longer more salt and more smoke are applied. Where circumstances are favorable kippering may be carried on to advantage either on a larger or smaller scale. Herring put up in this way are most delicious. They cost a trifle more, because of the extra labor and the greater care requisite in handling them. The same materials are used for smoking kippers as are used for smoking bloaters and the same conditions apply, only that kippers, presenting a larger surface to the smoke as they do, do not require to be so long exposed to the smoke. As in the case of bloaters and red herring, the tastes of the consumers must be ascertained and the curing as to salt and smoke regulated accordingly. The manufacture of kippers is greatly on the increase in Britain. It is an important branch of the herring industry and utilizes a large proportion of the British catch of herrings.

SMOKED ALEWIVES OR RIVER HERRING.

River herring or alewives are smoked in a number of localities, but principally in Maryland and Virginia, and to a less extent along the Delaware and Hudson rivers and in the waters of North and South Carolina. In New England smoked alewives are prepared at Taunton and at Boston, as well as on the Connecticut River; but most of the supply of these fish in the New England States is from New Brunswick. The trade is mainly during the spring and early summer months, more particularly in April, May, and June, when there are few other smoked fish on the market. The business is not concentrated, but is participated in by many small smokers located at numerous points on the Atlantic seaboard. For this reason it is difficult to estimate the quantity smoked annually with any great degree of accuracy, but it is probably not far from 5,000,000, their wholesale value being about $90,000.

In preparing these fish in the Chesapeake region they are washed in vats and scaled with a knife as soon as practicable after removal from the water. They are next immersed over night in strong brine, containing 12 to 14 pounds of Liverpool salt to each 100 pounds of fish, with some dry salt on top to strengthen the weak pickle that rises to the surface. The following morning the round fish are strung on smoke-sticks, the stick being usually entered at the left gill-opening of each fish and out at the mouth, as in case of hard herring or bloaters on the New England coast. The strings of fish attached to the stick are then dipped in fresh water to rinse them off, and after draining and drying for a few hours are suspended in the smokehouse about 6 or 8 feet above the fire, and exposed to a dense but cool smoke made of pine shavings or similar material for about 2 or 3 days. Care must be taken to prevent the fire from becoming too hot, thus causing the fish to crack at the lower end or possibly to fall from the sticks to the floor. Prepared in this manner the river herring will usually keep in good condition in the Chesapeake region for 30 days during the spring and for a somewhat less period in the summer. As the fish are not eviscerated before smoking the decrease in weight is small, 100 pounds of round fish yielding about 85 pounds smoked. The wholesale price is about 20 or 22 cents per dozen, according to the size and condition.

In Washington, Baltimore, and one or two other places the river herring are prepared in the following manner:

The fresh herring are scaled with a knife, gibbed like the pickled herring of Scotland, washed, and pickled for 3 hours in brine, about 20 pounds of Liverpool salt being used for each 100 pounds of fish. On removal from the pickle they are strung on small iron rods, the rod passing through the eye sockets of the fish, drained for an hour or so, and hung in the hogshead smokehouses, in the bottom of which a fire has been made of equal quantities of oak and hickory wood. The fish are dried for a few minutes and then the tops of the hogsheads are covered with old salt sacks or other suitable material. From time to time the fire is sprinkled with water to produce a vapor and the fish thus exposed to heat, smoke, and steam for about 3 hours, when they are removed and cooled and are then in condition to be eaten. Only oak and hickory should be used as fuel, as other materials do not produce the proper flavor. If the fire becomes too warm it should be smothered with oak or hickory sawdust.

Herring thus prepared sell for about 40 cents per dozen wholesale, and the trade is at times quite extensive. During the season 1,000 dozen are usually shipped each week from Washington to New York City.

The process of smoking alewives commonly employed in the New England States differs from the Chesapeake process in a few minor particulars. The smokers are usually not so careful about removing the scales with a knife, depending generally on the frequent handling of the fish to scale them if cured soon after removal from the water. It is also customary in salting the fish to permit them to make their own pickle, the fish remaining in the pickle for 3 to 5 days. On removal they are soaked in fresh water for 5 to 6 hours and strung on hard-wood sticks, the stick entering through the left gill-opening and out at the mouth. They are next rinsed, drained, and dried for a short while and suspended in the smokehouse, where they are exposed to a smoldering fire of hard wood and sawdust for 3 or 4 days, when, after cooling, they are ready for sale.

The wholesale price in New England is usually from $1.50 to $2 per 100.

In Massachusetts so few smoked alewives are prepared that little attention is paid to the following law respecting the methods of inspecting and packing:

Sec. 48. Alewives or herrings intended to be packed for sale or exportation shall be sufficiently salted and smoked to cure and preserve the same, and afterwards shall be closely packed in boxes in clear and dry weather.

Sec. 49. Smoked alewives or herrings shall be divided and sorted by the inspector or his deputy, and denominated, according to their quality, "number one" and "number two." Number one shall consist of all the largest and best-cured fish; number two, of the smaller but well-cured fish; and in all cases those which are belly-broken, tainted, scorched or burnt, slack-salted, or not sufficiently smoked shall be taken out as refuse.

Sec. 50. Boxes made for the purpose of packing smoked alewives or herrings, and containing the same, shall be made of good sound boards sawed and well seasoned, the sides, top, and bottom of not less than ¾-inch boards, securely nailed, and shall be 17 inches in length, 11 inches in breadth, and 6 inches in depth, in the clear, inside.

Sec. 51. Each box of alewives or herrings inspected shall be branded on the top by the inspecting officer with the first letter of his Christian name, the whole of his surname, the name of the town where it was inspected, with the addition of "Mass.," and also the quality of "number one" or "number two." Herrings taken on the coast of Nova Scotia, Newfoundland, Labrador, or Magdalen Islands, and brought into this State, shall also be branded with the name of the place or coast where taken.

Sec. 52. The fees for inspecting, packing, and branding shall be 5 cents for each box, which shall be paid by the purchaser, and the inspector-general may require from his deputies 1 cent for each box inspected, packed, and branded by them.

Sec. 54. No smoked alewives or herring shall be exported from this State unless inspected and branded as aforesaid, under a penalty of $2 for each box exported, nor shall alewives or herrings be taken from a box, inspected and branded as aforesaid, and replaced by others of an inferior quality, with intent to defraud any person in the sale of the same, under a penalty of $5 for each box so changed: *Provided*, That all smoked herrings and alewives arriving from any other State in the United States, and having been there inspected, may be exported in a vessel from this State without being reinspected. (General Statutes of Massachusetts, 1859, ch. 49.)

New Hampshire has laws somewhat on the same lines as the preceding, but very few alewives are smoked in that State.

SMOKED LAKE HERRING AND WHITEFISH.

Formerly along the shores of the Great Lakes and in the fish markets using supplies therefrom, many whitefish were smoked, but the increasing scarcity of that species gradually led to the substitution of lake herring, and during recent years very few whitefish have been prepared in this manner. The trade in smoking lake herring is quite extensive, amounting to probably 2,000,000 pounds annually, prepared principally at Chicago, Milwaukee, Detroit, Sandusky, Cleveland, Buffalo, Cincinnati, Erie, New York, Baltimore, and Washington.

The process of smoking lake herring and whitefish is identical. If the fish are frozen when received at the smokehouse, they are thawed in the open air or, better, by immersing and stirring them in a barrel of water of medium temperature. After thawing they are split down the belly to the vent, eviscerated, washed thoroughly, and pickled in butts or barrels, about 4 pounds of fine salt to 100 pounds of fish being scattered among them and sufficient brine of 90° salinity to cover them. Either dry salt or brine alone may be used, the former being preferred in warm weather and the latter during the winter. In case brine alone is used, some dry salt should be placed on top to strengthen the weak pickle floating at the surface. After remaining in the pickle from 10 to 16 hours, according to the strength of the pickle and the flavor desired, the fish are removed and strung on the smoke rods, 10 to 20 fish to each rod, according to its length and the size of the fish.

In stringing, some curers pass the rod through the body immediately below the nape bone, effectively preventing the fish from falling down in smoking, but also marring its appearance somewhat. A more usual way is to pass the stick in at the right gill-opening and out at the mouth. Others pass the rod through the head near or through the eyes, and a few pass it immediately back of the throat cartilage. The latter leaves a neat appearance, yet it permits more fish to fall in the smoking process than when the rod is passed through the head or the shoulders. In some houses the smoke-stick is not passed through the fish, but instead a stiff iron wire, curved in S shape, is used to attach the fish to the stick, one end of the wire passing through the fish at the head or beneath the nape bone and the other hung over the smoke stick. At Grand Haven, and to some extent in Chicago, Milwaukee, and one or two other places, the fish are secured by having stout smoke-sticks, about 1½ inches thick and 2½ inches wide; in the top of each, and about three-fourths of an inch from the edge, is driven a row of tacks or small wire nails at intervals of about 3 inches, projecting about one-half inch above the surface. Ordinary cotton wrapping cord is tied to the wire nail at the end of each stick, and by means of this cord passing around each nail a single herring is held in place between each two nails throughout the length of the stick, the fish being placed with the back of the neck against the stick and the cord passing from one nail around the throat of the fish, entering under the gills on each side, and then around the next nail, and so on to the end. By having the stick of sufficient width, a row of small nails may be placed on each edge, so as to attach a row of fish at each side. This removes nearly all risk of the fish falling, and their appearance is not marred by holes through which the smoke-stick has been passed.

Some markets prefer the herring well smoked on the inside, and to accomplish this the sides of the abdominal cavity are stretched open by means of small wooden sticks or toothpicks, either one or two sticks to each fish. This permits the smoke to permeate the stomach cavity better and results in a more durable article. In general, the Western trade prefers the stomach cavity stretched open, while the Eastern markets prefer them without the sticks; but there are exceptions. The smoked lake herring sold in Washington are mostly extended by means of a small stick, or, in case of large fish, by two small sticks.

The fish attached to the sticks are dipped in fresh water to remove surplus or undissolved salt, loose scales, etc., unless they have been rinsed before stringing, drained, and suspended in the smokehouse 4 to 8 feet above the floor, and subjected to a gentle smoke for 4 or 5 hours. The door or damper is then closed, the fires spread or built up and the fish cooked for 1 or 2 hours according to the amount of fire, the height of the fish, and the particular cure desired. After cooling, which is accomplished either by opening the doors of the smokehouse or by removing the fish to the outside, they are ready for the trade. 100 pounds of round fish, or 85 pounds dressed, yield about 65 pounds smoked. Ordinarily these fish keep one or two weeks, and even longer, and the wholesale price ranges from 6 to 12 cents per pound, according to the locality and the season, the former being the price for the Great Lakes and the latter for New York City. In New York about 100,000 pounds of these fish are smoked annually and they sell throughout the year, being known usually as ciscoette, competing with pickling or biickling. In Washington the smoked lake herring are usually sold by the number, averaging about 50 cents per dozen wholesale.

In some of the north European countries the sea herring are smoked in a manner similar to the lake herring in this country. The following description of a smoke-house in Holbek, Denmark, and the methods used therein, is from *Fiskeritidende*, No. 41, Copenhagen, October 7, 1884:

As soon as the herring are brought in from the boat, they are placed in strong brine for 3 or 4 hours, or they are left over night in a weaker brine. Some people also use the dry-salting method. The fish are then washed and strung on round, wooden sticks, three-fourths of an inch thick and 3 feet long. This stick is stuck through the gills and comes out at the mouth. According to the size, from 18 to 24 fish are strung on every stick, always in such a manner as not to touch each other. They are then hung in the open air and dried in the sunshine, if possible, and then put in the oven for smoking. The smokehouse has four ovens, built from time to time as the demands of the trade required. From 16,000 to 24,000 herring can be smoked per day. In one of three large ovens 1,600 herring can be smoked at the same time. The chimney itself should not be less than one yard square on the inside, as otherwise it is not capable of receiving the steam from the fish when they are dried in the oven. The top should be covered with a thin plate of cast iron, so that the rain can not fall on the fish. For supporting the front part of the oven it is best and cheapest to use an old iron rail; any other bar will scarcely be strong enough. The oven can easily be only half the size of one of the larger ones, but the larger it is the more profitable it will be as regards the quantity of fuel consumed. In front of the oven iron plates are hung on an iron pipe, and these plates are taken off when shavings are put on the fire. From these plates and up to the iron bar the opening is covered by a piece of linen cloth, as it is necessary to look into the oven frequently in order to see that the flames do not rise too high and burn the tails of the fish. If this should be the case, the flames must at once be quenched by moist sawdust. The fuel used is exclusively oak and beech shavings, particularly from coopers who make large barrels, as the shavings must not be too fine; beech and oak sawdust are also used, but shavings and sawdust of pine wood should never be employed, as it is apt to give to the fish a resinous flavor. The smoking process may take from 3 to 6 hours, according to the drying which the fish have undergone in the air. After the fish have been smoked they are generally allowed to hang one night to cool off, and are in the morning packed in boxes holding 80 fish each.

SMOKED SALMON.

Smoked salmon is among the choicest of fishery products, and its cure represents the highest development in fish-smoking as practiced in this country. The annual product approximates 2,800,000 pounds, which is sold at from 16 to 45 cents per pound wholesale. It is prepared principally in New York, Boston, Chicago, and Philadelphia, there being 8 or 10 smoking-houses in New York City and vicinity, 4 in Boston, 2 or 3 in Philadelphia, 2 in Chicago, and several on the Pacific coast and other points.

The great bulk of the supplies for the smoking-houses consists of salmon pickled in barrels, which come principally from Labrador, Newfoundland, and Hudson Bay, and more recently from Alaska and other Pacific coast points. The trade in pickled salmon from the east coast of the British Provinces, known to the trade as Halifax salmon, began early in the present century, developed principally between 1830 and 1840, and since has ranged between 3,000 and 10,000 barrels annually, the receipts during the last 30 years averaging 5,500 barrels, valued at about $15 per barrel. The Pacific coast salmon have been used for smoking in the Eastern States only since 1885, and the extent of their use was of little consequence prior to 1890. The favorite pickled salmon for smoking are those from Hudson Bay, with Labrador and Newfoundland ranking next in order. They range in weight from 5 to 13 pounds salted, except that some few from the Hudson Bay weigh even 20 pounds, and the wholesale price in Boston or New York during the past few years has been from $17 to $20 per barrel of 200 pounds. Practically all of the pickled salmon from the east coast of the British Provinces are smoked, the quantity going to the consumers in brine being less than 2 per cent. While not so red as the Pacific coast salmon, they are richer and finer-grained. The Pacific salmon cost on the Pacific coast usually about $9 or $10 per barrel of 200 pounds, while the cost of transportation to the Atlantic coast by rail is $3.30 and by vessel $1.20 per barrel.

The choicest salmon for smoking are those received fresh or frozen from Nova Scotia and New Brunswick, especially from the Restigouche River and vicinity. They are quite large, averaging 12 or 14 pounds each, some attaining a weight of 40 pounds or more. Some curers use fresh salmon only when the New York market is so glutted as to run the price down below 10 cents per pound dressed, the fish being then purchased, brine-salted, and kept for the smoking season. But the best class of smokers receive regular shipments from the Restigouche and vicinity and place them in cold storage, whence they are removed for smoking as the trade demands. Many years ago, when salmon were abundant in the Penobscot, Kennebec, and Connecticut rivers, they were smoked in Maine and Connecticut, but practically all New England salmon are now consumed fresh. Since the salted fish constitutes the bulk of the receipts at the smoking-houses, the methods of their treatment are first described.

As the daily needs of the trade require, the salmon are removed from the barrels, immersed in vats of fresh water for 2 or 3 hours, then washed with a bristle brush to remove incrusted salt, slime, etc., and immersed in another vat of water for 16 to 60 hours, according to the temperature of the water and the degree of saltiness of the fish. If desirable, the length of the soaking can be shortened by using warm water. In some houses they are soaked for 12 hours in running water. The fish are then water-horsed in piles, skin up except the lower layer, the piles being 2 or 3 feet high, with boards on top on which stones are placed for compressing the fish, but water-horsing is not practiced by all curers. After this pressure has been applied 4 or 5 hours the flesh of the fish is smoothed with the side of a flat knife, all ragged parts being pressed down. Each fish is then trussed with two or three thin, flat wooden sticks, so as to keep it spread out, the rough-pointed sticks being fastened transversely across the back on the skin side, the end of each stick slightly entering but not passing through the skin. If the heads have been left on, as in case of northern or Halifax salmon, a small stick or pin of hickory or other hard wood is shoved through the head at the eyes. A rope-yarn cord is next passed around this pin and about the gills, or about the uppermost of the flat stretching-sticks, in such a manner that when suspended thereby the weight is distributed proportionately and by means of which the fish may be hung from the sticks in the smokehouse. The Pacific coast salmon, which have the heads removed, are usually tied up by a cord passing through the napes or around the tail, and if very large they are sometimes cut into strips before being smoked. Some curers hang the fish up by means of five or six iron or wire hooks passing through the flesh, thus doing away with the sticks and cords above described.

The fish are permitted to drain for several hours, when they are hung in the upper part of the smokehouse, away from the heat, but not so high as to be in the warm air which accumulates at the top of the bay. Usually only two rows or tiers are smoked at a time, and in the lofty smokehouses the smoking is continued for 18 to 36 hours. About 24 hours are usually required, but on dry windy days 16 to 18 hours are sufficient, and during sultry weather 30 or more are necessary. When low smokehouses are used, in which the fish are hung within 8 or 10 feet of the fire, as at Washington, D. C., the smoking is usually completed in about 12 hours. The smoke should be even throughout and with little fire. In some houses a light fire is built under the fish as soon as placed in the smokehouse, and this is continued for 6 or 7 hours, when a shovelful of sawdust is added and the smoking continued 12 or 14 hours.

When sufficiently smoked, the fish are permitted to cool and are then packed, usually with paper or matting wrapped about them, the spreading-sticks at the back being left in. A barrel of pickled salmon yields about 180 pounds of smoked fish if the smoking be done in October, but if postponed until the following June it will yield only about 165 pounds. The average wholesale price in New York or Boston for smoked Halifax salmon is about 18 to 20 cents per pound, and for Pacific coast fish about 12 to 14 cents per pound. They will keep in good condition for 10 days or longer under favorable conditions, but are used mostly in the vicinity where cured. Smoked salmon have been shipped to New York from Nova Scotia, but although they looked well on being opened they had a tendency to mold soon after being unpacked.

The following general method of smoking salted salmon in Sweden and Germany differs from the foregoing in several particulars:

The fish are immersed for 48 hours in soft cold water, which during that time is changed at least three times. Then with a medium stiff brush and warm water each fish is well cleaned outside and inside, and by means of a cord about the tail is hung in a tub of clear cold water, where it remains for 12 hours, when it is suspended in the air for 6 hours to dry. After that it is laid on a clean table, and when well drained it is trussed or braced with three sticks along the back, as in case of Halifax salmon, and suspended for 2 hours in the smokehouse over a gentle heat, then subjected to a dense smoke for 24 to 36 hours, until it acquires a dark-red color. The cure is then complete, and after cooling the fish is ready for the market.

In preparing frozen salmon for smoking, the fish on removal from cold storage are thawed out either by immersing them in water over night or laying them on boards in a moderate temperature and turning them over every 2 or 3 hours for 8 or 10 hours, when they are usually sufficiently thawed for handling. The fish are then split down the belly from head to tail, so as to lie out flat, the viscera removed, and in some cases the head and four-fifths of the backbone. This is customary with the Pacific coast salmon, but in New York and Boston, where Nova Scotia salmon are used principally, the head and backbone generally remain. In some instances the fish are split down the back, depending on the state of their preservation.

If the fish must be handled with little expense, so as to sell at a low price, they are next placed in tight barrels or butts with about 50 pounds of No. 2 salt and from 5 to 10 pounds of granulated sugar to 200 pounds of fish. On the second day add brine made by dissolving 30 pounds of salt in 5 gallons of water. After the fifth or sixth day the fish are removed and soaked in fresh water for about 3 hours, and are then attached by five or six hooks to the smoke-sticks, dried, and smoked in the manner described for salted salmon. The product by this method sells for 20 to 30 cents per pound wholesale, but sometimes much lower. In Chicago in 1898 the writer saw salmon which had been held in cold storage for three years and then smoked after the above method and sold at 16 cents per pound, resulting, of course, in no profit because of the heavy cold-storage charges.

Usually much more care is exercised in preparing smoked salmon from fresh or frozen fish, and especially when using Nova Scotia fish. Immediately after thawing, or after removal from the ice, if fresh, the fish should be sponged dry and a mixture of equal parts of saltpeter and salt introduced into the thick portion of the flesh. This may be accomplished by making 3 or 4 cuts about 8 inches apart through the skin, but not so far as to penetrate the stomach membrane, after which the openings are closed as well as practicable by bringing the cuts together; or it may be introduced by means of a small hollow tube with a plunger to force it in as the tube is withdrawn. The fish are next split and eviscerated and carefully rubbed by hand with a composition of salt and saltpeter, 2 parts of the former to 1 part of the latter. This mixture is thoroughly spread over each fish, and at the same time wherever the surface is cut or broken the fibers are brought together, so that the fish presents a smooth, neat appearance. A curer on the Pacific coast runs a small instrument down the thick part of the flesh on each side of the backbone and thus removes about one-fourth inch of skin on each side the full length of the back, so that the saltpeter and salt may quickly permeate the flesh. The fish are next placed in hogshead butts, skin down and 3 or 4 fish to the layer, with one half inch of salt in the bottom and sprinkled over each layer of fish. Pickle of about 90° test is then added to cover the fish, and after remaining in pickle about 2 days they are removed and prepared for hanging up, by placing a wooden pin through the head and 2 or 3 flat sticks at the back to stretch the fish out in the manner already described. After passing a rope yarn about the sticks the fish are suspended in running water for 30 or 40 minutes and then hung in the open air about 6 hours to drain and be partly dried by the wind, when they are suspended in the upper part of the smokehouse, away from the heat, and subjected to a gentle smoking for about 24 hours under normal conditions. Salmon cured in this manner are known usually as Nova Scotia salmon, in contradistinction to the salted salmon from the north, generally known as Halifax salmon. 100 pounds of round fish make about 65 or 70 pounds smoked, which sell for 30 to 45 cents per pound wholesale and 60 to 75 cents per pound retail.

With a view to preventing the inner surface of salmon and similar fish from cracking, which injures its appearance and also makes it liable to mildew quickly, and to prevent the fish from falling from the smoke-sticks, and to retain the natural juices, a process has been invented by which a piece of membranous material, such as animal bladder, etc., is placed in contact with the inner surface of the fish, which by means of its natural glutinous ingredients is held there securely. Then the fish, covered on the outside by its natural skin and on the inside by the artificial coating, is placed horizontally in a frame consisting of a number of triangular transverse metal-frame standards having base ledge projections and wire screen surfaces fitted thereon, inclined in opposite directions and open at the back and bottom, for exposing the fish to the smoke.*

In smoking fresh salmon in Holland each fish is wiped clean, split down the back from the nose to the tail, the head being left on, and several incisions made inside of the abdominal cavity in the thick of the flesh, but not sufficiently deep to penetrate the skin. The blood is carefully washed out, the stomach cavity well cleaned, and the whole fish washed several times. The skin is then cut or gashed laterally nearly the length of the fish, and on each side of this cut several short ones are made, this being done to permit the salt to penetrate the flesh more readily, so that each part of the fish may become equally salted. In salting, the fish are placed on top of each other in heaps of about 6 salmon each on tables, with the flesh upward.

In making the piles, each fish is well sprinkled with fine Lisbon salt, and to prevent the stomach from forming a receptacle for the brine by sinking down, a thin, curved oak board is laid between the fish. The fish remain in piles from 2 to 4 days, when they are struck through sufficiently for smoking; but if they are not needed at once, they may be kept in ice houses or cold cellars for 2 or 3 months. Before the fish are smoked they are well washed and hung up to dry in the air, or during damp weather they are dried in the smokehouse. A fire is made from small pieces of oak wood in the center of the floor, and after this has burnt half an hour a smoke is made with oak shavings and fagots, over which ashes are scattered. For some markets the fish need not be smoked more than 12 or 11 hours, but for other markets it is necessary to smoke them 3 or 4 days.

The following method of smoking fresh salmon prevails in Germany: Each fish is first rubbed free from slime, etc., with a towel which has been dipped in salt or brackish water; then it is split down the belly, eviscerated, and thoroughly cleaned inside as well as outside. Most of the backbone is removed with a sharp knife, some being left near the tail to strengthen that part of the fish, and the flesh adjacent to the backbone is pressed flat so as to present a smooth appearance, or as though there had been no backbone. Bay leaves, from which the stems have been removed, are next spread thickly with salt on the inside of the fish and the sides brought firmly together. It is then packed in dry salt and bay leaves and a weighted board laid upon the fish. After 30 hours or so under this pressure the fish is immersed in fresh water for half an hour, the salt, etc., in the meantime being wiped off, after which it is laid out flat and trussed in the usual manner with 3 flat sticks and suspended in the air for about 6 hours. When sufficiently aired the fish is placed in the smokehouse and dried by a moderately warm smoke for 3 hours, when the smoke is increased and continued for about 36 hours or until the flesh assumes a bright red color.

A somewhat novel method* of preparing salmon for smoking was introduced in this country in 1878 by Lyman Woodruff, of Ellensburg, Oreg., by means of which it is claimed that much of the original flavor, color, and plumpness of the fish may be retained.

---

* See Letters Patent No. 577672, February 23, 1897, in favor of C. Waldemann, of Cöslin, Germany.

The inventor's description of the process is as follows:

After the fish is caught I open and clean it. I then place it in clean lime water, in which I let it remain for about 20 minutes. After removing it from the lime water I wash it clean and place it on the table, flesh side up, where I let it lie for about 10 minutes, when I wipe it dry, both inside and outside, with a clean dry cloth. For an ordinary salmon, weighing 16 pounds, I take 1 teaspoonful of finely ground black pepper and rub it well into the flesh side of the fish; next I rub in one-fourth of a teaspoonful of pulverized saltpeter in the same way, and then 1 tablespoonful of fine salt. These substances I rub in separately, rubbing each one until it disappears. The fish having been thus prepared, I sprinkle a thin layer of brown sugar over it and fold the two sides together. I let it lie in this condition for 12 hours, when I wipe the back of the fish dry and apply a coating of linseed oil to the back with a paint brush, when it is ready to be smoked. In smoking the fish I commence by creating a heavy smoke, and allow it to gradually subside in quantity until the proper volume is obtained, in order to keep off the flies.

SMOKED HALIBUT.

The preserving of halibut is effected principally by salting, but in that condition these fish are not readily marketed and smoking is applied to improve the flavor. The industry is confined almost exclusively to Gloucester, Mass., but during the past few years small quantities have been smoked at Boston and other points. Originally the smokers utilized only the surplus halibut from the fresh-fish trade, but the popularity of the article increasing, the Bank vessels began, about 1850, to salt the halibut taken by them when it was inconvenient to take them to market fresh. In 1855 the quantity of smoked halibut prepared approximated 400,000 pounds. The business reached its maximum in 1872, when about 3,000,000 pounds were prepared. Since that time the increasing scarcity of the fish and the enhanced demand for it in the fresh-fish trade have diminished the quantity. In 1880 it amounted to about 2,000,000 pounds, while in recent years it has averaged about 1,600,000 pounds, selling at about 10 cents per pound wholesale.

Although most of the halibut for smoking is received in a salted condition from Grand Bank, Western Bank, Iceland, Greenland, and more recently from Bacalieu Bank, some are received from the vessels supplying the fresh-fish market, when the market is glutted. That was the exclusive source of the supply prior to 1860, when the halibut fishery on Grand Bank and Western Bank was begun. Since the origin of the Bacalieu Bank fishery, in 1895, the smokers have received quantities of surplus gray halibut too large for the fresh-fish market.

Many of these fresh halibut are known among the trade as "seconds" or "sour halibut," the coating or membrane of the abdominal cavity becoming slightly tainted, and since the taint will quickly spread to the entire fish it is necessary that they be salted at once. When the flesh sours it puffs out, and is good for nothing except fertilizer.

The process of dressing and salting halibut is as follows: A dressing or fletching gang consists of two men, and there are four gangs to the vessel. Each being provided with a strong gaff hook having a garden spade handle, they place the halibut on a slanting cutting board on its dark side. One of the fletchers thrusts a thin knife, about 16 inches long and 1½ inches wide, into the body of the fish near the base of the dorsal fin through to the backbone, the blade being held horizontally, and cuts close to the ribs, removing a broad streak from one-half of the upper side of the fish. The fletcher on the opposite side of the table makes a cut similar to the above, separating the whole upper half of the fish from the backbone and the ribs. Two gashes are then cut in

---

* Letters Patent No. 204647, dated June 1, 1878.

the fletch, one at each end, by means of which it is removed from the cutting board. The other side of the fish is then treated likewise, making two fletches from each halibut. Formerly in fishing near Iceland, when all the fins were saved, the fletching knife was entered not so close to the fins, and when the fletches were removed the fins were cut off. During the four or five years preceding 1898 few of the fins were saved on account of their large size and fatness.

The whole fletches are at once salted in kenches in the vessel's hold, in the same manner as codfish, with the skin side down and a layer of Trapani salt over each layer of fish, 8 or 9 bushels of salt being used to each 1,000 pounds of fish. The whole fletches are supposed to hold the pickle better than if they were cut in smaller pieces, and consequently weigh more. After remaining about 15 days they are rekenched, during which time the surplus salt is shaken off. To avoid compression some fishermen place the fletches in large 400-pound boxes and pile the boxes on top of each other. On reaching port the fish are removed from the vessel's hold and placed back down, with salt, in kenches 3 feet high in the fish-house, where they may remain for a year or more without further handling. It is not unusual for smokehouses at Gloucester to have half a million pounds or more of salted halibut on hand at one time. When it is necessary to hold them over during July and August, the appearance of the fish is improved if they are kept cool, and for that purpose one of the halibut smokers at Gloucester has a small ammonia refrigerating plant, with suitable cold chambers connected, where the temperature is kept about 45° or 50° F.

The fresh halibut received at the smokehouses from the market vessels are cut in small fletches and salted in butts, back down, similar to those used for salting codfish, with about 5 bushels of Trapani salt scattered among 1,000 pounds of fish. There they remain from one to two weeks, when they are removed and salted in kenches similar to those on the vessels; or they may be scrubbed, soaked, water-horsed, and smoked at once; but this is not usually done, because of the desirability of working off the old stock. It is important that the fletches be thoroughly salted, otherwise the smoked product will be liable to spoil quickly.

When the market demand warrants their use, the fletches are removed from the kenches, washed thoroughly in fresh water with corn brooms or bristle brushes, and soaked in water for 3 or 4 hours. The water is then changed and they are again soaked for about the same length of time. This soaking is necessary in order to remove the coating of salt from the fish, and to soften its fiber so that the smoke may penetrate the flesh. On completion of the soaking they are water-horsed, skin side up, for 5 or 6 hours with weights on top. They are next placed on flakes similar to those used in curing codfish, where they are exposed to the sun's action for about 24 hours, which may extend through several days, the fish being placed in small piles and covered with flake boxes during the night or rainy weather. After drying the fletches are cut in small pieces, from 2 to 6 pieces to the fletch, with a gash in each piece where the flesh is thin and the skin appears tough.

The fletches are then strung on smooth, round, hard-wood sticks about 2 feet long and ¾ inch in diameter, or, as at Boston, small iron or steel rods 3½ feet long, the sticks passing through the splits or gashes cut in the fletches, and from 5 to 7 pieces to each stick 2 feet in length, and 8 to 12 pieces to those 3½ feet long, each piece being 2 or 3 inches from the adjacent ones to permit the smoke to pass freely among them. The sticks with the fletches attached are then passed into the smokehouse.

The principal smokehouse at Gloucester consists of a series of 10 compartments side by side, and 6 feet wide, 14 feet long, and about 20 feet high. The flooring is of lattice work, under which is a vault about 6 feet deep for generating the smoke. Within each compartment is a lane or passageway extending the length of the room, on each side of which are arranged 3 sets of parallel bars, one end of each set resting on upright poles forming the side of the lane and the other resting against the wall of the compartment. The first set of parallel bars is about 4 feet from the lattice floor, the second 3 feet above the first, and the third 3 feet above the second and an equal distance below the roof. Upon these bars are placed the ends of the sticks which hold the halibut. Each stick will carry from 5 to 7 fletches and about 20 sticks will rest on each pair of parallel bars. The capacity of each of the 10 compartments is about 120 sticks, or 600 to 800 fletches, equivalent to about 5,000 pounds of smoked halibut.

The fuel for smoking consists principally of ship carpenters' chips of oak, though recently oak edgings have been used, with sawdust to smother the flames. Some houses use sawdust exclusively. This fuel is arranged along the sides of the vault, the entire middle space being left vacant, and the fires are built and continued until the smoking is completed, usually in from 2 to 5 days. In damp weather moisture collects on the fish and the process may then require a week. During the winter it is possible to keep the fires hotter, and the smoking may be completed in 2 days. Care must be taken not to let the fires get too hot, for then the fletches may become too soft to hang on the sticks, dropping to the floor. At all times the doors are kept slightly open to permit a circulation of air to keep the halibut cool, and they will also "take the smoke" better. The entire decrease in weight by dressing and smoking is about 70 per cent, a live halibut weighing 100 pounds producing about 30 pounds of smoked fish; but as received from the kenches in the vessels 100 pounds yield about 82 pounds.

There are three principal grades of smoked halibut, namely, "heavy chunks," "medium chunks," and "strips," classification being made according to the thickness of the flesh. The napes and the thin parts of the tail do not go to the regular grocery or fish trade, being used mostly for "free lunch" at restaurants. While connoisseurs prefer the meat somewhat dark in color, yet the bulk of the trade requires it of a light straw-color. It is packed in boxes ranging in size from 1 pound to 500 pounds, the 30-pound boxes being the most popular. About 4 years ago quite a business was started in packing it in 1-pound and 2-pound "bricks," similar to boneless cod, but the trade was destroyed by persons placing smoked pollock on the market as halibut. During hot weather a small quantity of salt is sprinkled over the halibut as it is being packed in the boxes, and under ordinary conditions the product will keep for 6 or 8 months, or even longer. When intended for warm climates, smoked halibut is sometimes placed in hermetically sealed zinc boxes incased in wooden boxes, the zinc boxes having capacity for 50 pounds each. There is a little trade in smoked halibut placed in small glass bottles, with wide tops covered with cork stoppers.

The market is confined to the northern parts of the United States, none being sent south of Washington except in small quantities to Louisville and Memphis. The principal demand is from New England, New York, Chicago, and the West, the value approximating 9 cents per pound wholesale. In 1876 an effort was made to introduce smoked halibut into the European markets, and samples were sent from Gloucester to London, Liverpool, Glasgow, Cadiz, and St. Petersburg. No returns were received except from London, and English dealers expressed the opinion that it was too salt for their trade. At the Berlin Fishery Exposition in 1880, smoked halibut was exhibited by Messrs. Wm. H. Wonson & Sons, and a medal was awarded for its superior quality.

SMOKED HADDOCK OR FINNAN HADDIE.

The curing of haddock by smoking originated about the middle of the eighteenth century at Findon, Scotland, the cured product being known in the English markets as "Findon haddie," which later was modified into "Finnan haddie." Originally it was salted and dried, and afterwards soaked and placed over a smoldering fire of dried peat. But the demand soon becoming very great, it was cured in special buildings erected for the purpose, and at present large quantities are prepared at Aberdeen, Scotland, at Grimsby and Hull, England, and at other places in northern Europe in a manner similar to that employed in the New England States.

Finnan haddie was first prepared in the United States at Rockport, Mass., about 1850, but the business was soon abandoned. About ten years later Thomas McEwan, a Scotchman, began to cure it in a small way, at Portland, Me., the product being marketed principally in Canada. D. Weyer engaged in the business about 1865, and on Mr. McEwan's death in 1872 he was succeeded by John Loveitt. From time to time other firms came into the business, and as the product became better known its sale in the United States increased. In 1878 about 2,250,000 pounds of dressed haddock were smoked in Portland alone, making about 1,200,000 pounds of Finnan haddie, whereas in 1889 the five companies engaged in this business employed 48 men and utilized 3,570,000 pounds of dressed haddock, costing $71,400, which yielded 1,963,500 pounds smoked, worth $88,357 wholesale. The facilities for obtaining haddock at Boston led to the establishment of the business there in 1887, and at present the output at Boston equals that at Portland. Finnan haddie are also prepared at Eastport, Me., at New York City, and at Gloucester, Mass., where the business was established in 1893, and at St. Johns and Digby in the British Provinces. The present annual product in this country approximates 4,000,000 pounds, worth $200,000.

While the North American Provinces still receive a large part of the output, the consumption of Finnan haddie in the New England, Middle, and Central States is increasing, especially in Chicago and Cincinnati, some being sent as far as California. Formerly they were sold by the dozen, and consequently only the small fish were smoked, but at present the sales are made by weight, and haddock of all sizes are used. A singular feature in connection with the Finnan haddie trade in the winter of 1898 and 1899 was the exportation of small haddock from Boston to Digby for smoking purposes, those fish being obtainable cheaper in Boston than on the Nova Scotia coast.

To make a good product of Finnan haddie requires experience, as well as much care and attention, especially in the pickling and smoking. When received at the curing house the fish have usually been split down the belly to the vent and eviscerated, as if prepared for the fresh-fish market. They are first beheaded and washed thoroughly with a stiff brush, all the black membrane lining the abdominal cavity being removed. They are next split down to the tail and a cut made along the right side of the backbone so that they will lie out flat, and the rough edges of the backbone are removed. They are then immersed in strong salt brine, made of Liverpool, Cadiz, or Trapani salt, for 1 or 2 hours, according to the weather and the temperature, as well as to the size and condition of the fish and the particular flavor desired, the exact length of time for pickling being learned only by experience.

On removal from the brine they are fastened to the sticks from which they are suspended in the smokehouses, the napes being stretched out flat and pierced by two small iron spikes or nails fixed in the smoke-sticks. These sticks are about 1½ or 2 inches square at the end and 4 feet long, 3 fish being usually hung from each. The sticks with the fish attached are placed on frames for a few hours to allow the moisture to dry from the fish, when they are suspended in the smokehouse, which is generally like those used in smoking herring, the sticks being placed in tiers, one above another, with space between to allow the smoke to circulate. A fire of hard wood, usually oak, is started over the floor of the smoking kiln and allowed to burn from 8 to 18 hours, when sawdust is applied, smoldering the fire and producing a dense smoke, which thoroughly impregnates the fish. In smokehouses with a low ceiling the smoking can be completed in 4 or 5 hours. In some smokehouses no wood is used, the curing being effected by burning hard-wood sawdust, rock maple or beech being preferred, and the temperature is kept as high as practicable without burning the fish, which are placed high up in the bays. The time of cooking or smoking depends on the condition of the fish, temperature of the air, and the probable time to elapse before consumption, but never exceeds one night.

When the smoking is completed the fish are removed from the smokehouse and placed on racks for cooling, and when thoroughly cooled they are packed in boxes containing from 20 to 400 pounds each, but mostly 50 pound boxes, and shipped to the trade, usually by express. Only enough are cured at a time to supply the immediate demand, as it is important that they reach the retail dealers in good condition. During warm weather they will keep only a few days, but when the weather is cool they will, under ordinary conditions, keep from 10 days to 2 or even 3 weeks. If it is desirable to keep them longer they must be smoked much harder.

The season for Finnan haddie begins in October and lasts until the following April. 100 pounds of round fish yield about 55 pounds smoked, and the wholesale price ranges from $3\frac{1}{2}$ to 6 cents per pound. The choicest haddie are tender. The inside is of a light yellowish-brown or straw-color. It is alleged that some curers add saffron to the pickling brine to improve the color of the fish.

SMOKED STURGEON.

Practically all of the sturgeon flesh used in this country is smoked before going to the consumers. This is usually done in the large centers of German population, and principally in New York, Chicago, Milwaukee, Sandusky, Buffalo, and Philadelphia. The business was started in 1857 by Mr. B. K. Peebles in New York City, and reached its greatest height about 1890, the scarcity of sturgeon during recent years restricting the extent. In New York alone about 1,000,000 pounds are smoked annually, constituting over one-half of the fish smoked in that city. Along the Great Lakes the smoking of sturgeon began about 1865, these fish not being used there prior to that time. As first prepared at Sandusky and Toledo they were dressed, salted, and smoked in large strips for 8 or 10 days and sold as smoked halibut. But the smokers soon adopted methods similar to those in New York, and the business increased and was of considerable extent ten or fifteen years ago. In 1872 Mr. J. W. Milner reported to the U. S. Fish Commission that 13,800 sturgeon, averaging 50 pounds each in weight, were smoke-cured at Sandusky, and in 1880 it was found that the business had increased to 1,258,100 pounds; but the decreasing abundance of sturgeon on the lakes has resulted in a falling off in the quantity smoked. The total annual product in the United States is now about 4,000,000 pounds, worth $720,000.

Sturgeon for smoking are received from the Delaware River and other estuaries of the Atlantic coast, from the Great Lakes, and from the Columbia River. The lake sturgeon (*Acipenser rubicundus*) is the most desirable for smoking, the product selling in New York City for 24 or 26 cents per pound. Columbia River sturgeon ranks next, with an average value of 3 or 4 cents less than the Great Lakes. The Atlantic coast sturgeon (*Acipenser sturio*) shows yellow streaks when smoked, and the meat is also somewhat more coarse and red than the delicatessen trade desires, and sells in New York for 6 or 8 cents less than the smoked Great Lakes sturgeon, or about 16 to 20 cents per pound. The prices prevailing in New York are quoted, since that is the principal market in the country for smoked sturgeon and the choicest product is there prepared, but these prices are somewhat higher than those prevailing at other points, especially along the Great Lakes. In Chicago smoked lake sturgeon usually sells at 18 to 20 cents per pound wholesale.

In the early history of the industry the sturgeon intended for use after the fishing season was over were dressed and salted in butts or barrels, whence they were removed as required, being soaked out before smoking, as is now the case with Halifax salmon. At present, however, the surplus sturgeon are almost invariably kept frozen in cold storage until required. Before freezing, the fish are dressed, the heads, tails, viscera, fins, and backbones being removed. Practice differs in the various localities in regard to removing the skin. Until the last 6 or 8 years all sturgeon were skinned, and that is at present the common practice with the Atlantic coast fish. But about 1890 the practice of leaving the skin on was introduced along the Great Lakes, and is now general in that locality. Formerly the Columbia River sturgeon were skinned before being frozen, but at present nearly if not quite all of those placed in cold storage have the skin left on. Before freezing, the fish are usually cut into four pieces, about the size of the freezing pan, or in smaller pieces suitable for smoking, the former being more frequent. The methods of freezing and subsequent cold storage have already been described.

On removal from cold storage the fish are thawed by exposure to air of moderate temperature, being turned once or twice during the operation, or, better still, by immersing them in water of medium temperature. They are then treated exactly as though received fresh from the fishermen. They are cut into suitable chunks, 2 or 3 inches wide and weighing 1½ or 2 pounds, the width varying according to the thickness of the meat. These chunks are then brine-salted in barrels or hogshead butts, about 5 pounds of No. 2 salt to 100 pounds of fish being sprinkled among the chunks and sufficient brine of about 85° strength being added to cover the fish.

Some smokers, however, use no dry salt, depending entirely on very strong brine in which the fish remain from 6 to 16 hours, according to the temperature and strength of the brine as well as the size of the pieces. One very successful smoker uses dry salt without brine during the summer, and in the winter uses brine only of about 98° salinometer test in order to economize time, since it takes about 18 hours to strike the fish in dry-salting, whereas 10 hours are sufficient for striking in brine. In general dry-salting is preferred, since its tendency is to make the flesh harder and firmer. In some localities the trade requires very light, salted fish, and they remain in brine only 15 or 20 minutes, being stirred about during the immersion.

On removal from the pickle the pieces of flesh are strung on steel or iron rods about one-third of an inch in diameter and 3 feet long, the rod passing through the thin part of the chunk and 8 or 10 chunks being strung on each rod. They are at once dipped in fresh water to remove surplus salt, slime, etc. In some establishments the pieces are soused in fresh water immediately on removal from the pickle. In case the fish have been salted only 15 or 20 minutes, they are not dipped or rinsed. After drain ing for a few moments, or, better still, drying in the open air for several hours, they are suspended in the lower part of the smokehouse from 4 to 6 feet above the fire, where they are exposed to a gentle smoke with doors open anywhere from 1 to 5 hours, according to the weather and the flavor of the product desired. When the weather is sultry it requires twice as long as when it is clear. Then the doors or dampers are closed and a hot hickory, maple, or oak fire is built and the fish cooked from 1 to 2 hours, care being taken that it does not become too hot and melt or fall from the rods. On completion of the cooking process the meat is allowed to cool, either by opening the doors of the smokehouse or by removal to the open air, when it is ready for the trade.

While the foregoing are the methods in general use, yet many smokers have special processes of their own. One of the most successful smokers on the Great Lakes operates as follows: The small pieces of flesh, 1½ to 2 pounds in weight, are first rubbed with No. 2 packers' salt and put in tight barrels with salt sprinkled among them, about 20 pounds of salt in all being used to each 100 pounds of fish. In about 7 or 8 hours, when the fish are saturated with the salt, they are removed, rinsed in two waters, strung on wire hooks, and suspended from the smoke-sticks. After draining for an hour they are placed in the smokehouse in 2 or 3 rows, 5 to 7 feet above the floor, and subjected to a hard wood smoke for 7 or 8 hours at an even temperature.

One hundred pounds of dressed sturgeon yields from 63 to 70 pounds smoked, and the product usually keeps one or two weeks under ordinary conditions.

Notwithstanding the great scarcity of sturgeon and its consequent high price, the consumption of smoked sturgeon amounts to about 4,000,000 pounds annually.

It is not practicable to hold smoked sturgeon in cold storage, because of its tendency to mold, but it is canned to a small extent.

SMOKED CATFISH.

The increasing scarcity of sturgeon with the consequent high price has resulted in the smoking of channel catfish as a substitute. These are obtained chiefly from the Mississippi River, especially in the vicinity of Memphis, and they are smoked in Chicago, St. Louis, and the Middle Mississippi Valley. This industry is of very recent development, but as it furnishes a satisfactory substitute for sturgeon, which are becoming so costly, it will probably grow to considerable proportions.

Being intended as a substitute, the catfish are smoked in identically the same manner as are sturgeon. The fish as received at the smokehouse are usually beheaded and eviscerated. They are skinned and cut into small pieces, weighing about 1 or 1½ pounds each, and are pickled for 6 or 8 hours in tight barrels. This may be accom plished by rubbing the pieces with salt and placing them in the barrel either with dry salt scattered among them, or simply by placing them in the barrel with dry salt or with strong brine. On removal from the brine the pieces are rinsed by dipping in fresh water, to remove slime, surplus salt, etc.; they are then attached to the smoke-sticks and drained for an hour or so, and placed in the smokehouse, where they are smoked for 7 or 8 hours in the same manner as sturgeon are treated. 100 pounds of dressed catfish yield from 65 to 70 pounds smoked, and the product sells usually at about 15 or 16 cents per pound. The total annual product of smoked catfish in the United States probably does not exceed 50,000 pounds, and its sale is confined prin cipally to those who are willing to accept a substitute because of its being cheaper.

At several points in the Mississippi Valley the small catfish are smoked whole, like lake herring. They are split to the vent and eviscerated, the head and in some instances the skin being left on, struck with salt in tight barrels, and smoked for a few hours in the manner described for lake herring. The demand is small, the business amounting probably to 10,000 or 15,000 pounds.

SMOKED EELS.

Smoking eels is one of the industries introduced into this country by German residents, and it is carried on in New York, Philadelphia, Buffalo, Sandusky, Chicago, Milwaukee, Washington, and various minor places. The product will keep only a week or two under ordinary conditions in cool weather, and the extent of the business in each locality is generally limited to the local demand.

Generally the eels are received at the smokehouse fresh, directly from the fisheries, but some are also received frozen from cold storage. In the latter case they are thawed by immersing them in water a few hours or by exposure in the open air. Some smokers "slime" the eels with salt: that is, rub the skin with a small quantity of fine salt to remove the slime therefrom. In dressing, the fish are split from the head to the vent and the viscera removed. It is desirable to continue the splitting down to the end of the tail sufficiently deep to remove the large vein along the backbone, but sometimes this may be pulled out without splitting the fish more than an inch or two beyond the vent. Few smokers, however, give attention to this item. The eels are immersed in strong brine from 1¾ to 7½ hours, according to strength of brine, size of fish, and the desired flavor. This brine should be quite strong, about 20 pounds of Liverpool or other good salt being required for each 100 pounds of fish.

In New York the eels are usually pickled for 2 hours, while on the Great Lakes the length of the time is generally about 7 hours. On removal of the fish they are washed, bristle brushes being used by some smokers, while others simply dip the fish in water for removing the slime and surplus salt. A few smokers throw them in a tub of water and beat them with a net for several minutes to accomplish the same purpose. The eels are next strung on iron or steel rods one-third inch in diameter, the rod passing through the head of each eel, or through the throat cartilage and out the mouth, and hung in the open air a few hours for drying. But if the atmosphere be moist or the saving of time necessary they may at once be placed in the smokehouse.

In New York, where small brick ovens are used, the fish are subjected to a mild smoke for about 4 or 5 hours until they have acquired the proper color, when the fires are gradually increased and they are hot-smoked or cooked for 30 or 40 minutes. At Buffalo and some of the other Great Lakes ports, the smoking is usually at an even temperature throughout and continues for 6 or 8 hours. Mahogany or cedar sawdust is used in New York for making the smoke, while hickory or white-oak wood is used for cooking, the latter being preferred. In Washington the eels are suspended in the hogshead smokehouses over a fire made of oak and hickory wood and dried for 20 minutes, when the hogshead is covered with sacking and thus hot-smoked for 3 or 4 hours, the fires being sprinkled with water from time to time to produce a hot vapor. The smoking must be carefully attended, for if the heat becomes too great the fish will curl up out of shape. A good test to determine whether the cooking is sufficient is the ease with which the skin may be separated or peeled from the flesh where the eel has been split.

The decrease in weight by dressing and smoking is about 35 per cent. 100 pounds of round eels yielding 65 to 75 pounds smoked. In New York the product sells for about 20 cents per pound, while at Buffalo and other Great Lakes points the price is usually 14 to 15 cents per pound. When eels have been pickled 6 or 8 hours they ordinarily keep 10 or 12 days; but when the salting has been only 2 hours, as is usual at New York, they are liable to mold after 5 or 6 days. Smoked eels keep a shorter length of time than almost any other smoked fish.

Eels are sometimes skinned before being smoked, the process being the same as above described, except that less salting and smoking is required, and it is also very difficult to keep them from falling down off the rods in the smokehouse.

The trade in smoked eels in New York is probably not 30 per cent of what it was 15 or 20 years ago, but along the Great Lakes it appears to be increasing. The annual product in the entire country is probably about 150,000 pounds, worth $27,000. There is some demand for smoked eels in cans, which is met by two fish-canning establishments in New York City. The smaller eels are used for this purpose, and they are smoked somewhat more than those sold to the delicatessen trade.

The following method of smoking eels prevails to some extent in northern Europe, especially in Germany:

The head, skin, tail, and viscera are removed, and the eel is split open the entire length, and the backbone and many of the smaller bones attached to it removed. It is then laid in strong salt brine, where it remains for 6 hours, and is then wiped dry with a linen towel and is covered with the following preparation, which has been pounded in a porcelain mortar: One large anchovy, 1 ounce of fine salt, 8 ounces of powdered sugar, 1 ounce of saltpeter, and sufficient butter to make a paste of the ingredients. The eel, thoroughly cured with this preparation, is rolled up tightly in the form of a disk, beginning at the tail end, tied with a cord to hold it in position, and sewed up in a linen cloth, which covers the disk and allows the end to project. These disks are next suspended in an ordinary chimney smokehouse and subjected to a strong smoke for 5 or 6 days, then allowed to cool and become firm, when they are ready for the table.

SMOKED MACKEREL.

There is a small business in smoking both fresh and salt mackerel in New York City and a few other points on the Atlantic seaboard, the output amounting to probably 8,000 pounds of the former and 35,000 pounds of the latter. The fresh mackerel are cured in very nearly the same way as lake herring, except that usually they are not split, being prepared round. The fish are first struck in brine, in which they remain for 12 or 14 hours, then removed and opened at the vent with the point of a knife to let the pickle in the abdominal cavity escape. They are next put on smoke-sticks, drained and dried for 2 or 3 hours, and placed in the smokehouse, where they are subjected to a gentle smoke for 4 to 5 hours, until properly colored, when fires are built and the fish cooked for a couple of hours, as in case of ciscoette or lake herring.

In preparing salt mackerel for smoking, the fish are cleaned and the dark stomach membrane removed, when they are soaked in fresh water for 6 to 12 hours, or in some localities from 15 to 24 hours, according to the size and the degree of saltiness. On completion of the soaking they are washed, strung on rods or smoke-sticks, drained, and hung in the upper part of the smokehouse and subjected to a gentle smoking for 5 to 15 hours at a low temperature.

No. 2 mackerel bring about 16 cents per pound and extra large smoked mackerel 20 to 30 cents per pound, but generally it is the smaller fish that are used for this purpose. The trade in these fish is very much less than formerly, the quantity used in New York City being only about one-tenth of what it was from 1880 to 1885, but the business during that period was much greater than theretofore, resulting from the salted mackerel being received at the markets in June and July instead of a couple of months later, as formerly.

SMOKED SHAD, FLOUNDERS, LAKE TROUT, CARP, ETC.

In the Chesapeake region and at various points along the coast small quantities of shad are smoked, usually in precisely the same manner as already described for river herring or alewives. Formerly many barrels of "economy shad" salted on the Kennebec River were smoked, but the demand ceased about 1880. A superior quality of smoked shad may be made by rubbing fine salt, saltpeter, and sugar or molasses over the fresh fish, and after they are struck, smoking them a few days at an even temperature. These are far superior to those prepared from salted shad.

A few flounders are smoked each year in New York and other populous centers of the Atlantic seaboard, the quantity probably amounting to about 15,000 pounds annually. The small flounders weighing half a pound or less are used, and these are eviscerated, pickled with brine in butts for about 2 hours, strung on smoke rods, drained, and cold-smoked for 8 to 10 hours. Sometimes these fish are hot-smoked for half an hour or so after the color has been set by the cold-smoking.

Menhaden and butterfish have been smoked to more or less extent during the past few years, but few are so prepared at present.

Smoked lake trout and carp are prepared to a small extent in the manner already described for lake herring or whitefish, but little demand exists for these products.

Efforts have been made to produce marketable articles of smoked hake and pollock, but the business has never assumed any commercial importance. There seems no valid reason why smoked pollock at least should not become popular, the flesh of that species seeming well suited to this method of curing. Smoked mullet is a very choice article, but practically none is prepared for the general market.

In 1885 experiments were made by the United States Fish Commission to introduce smoked kingfish, which abound off Key West. The Fish Commission report for 1885, p. LIII, states, in substance:

These fish were prepared with much care at Gloucester, and proved to be an excellent smoked fish, being tested by many experts, some of whom pronounced them equal or even superior to smoked halibut or salmon, being free from the rather rank taste that the smoked halibut sometimes has.

Tilefish have been smoked as an experiment by several persons, but experts differ as to their qualities. The Fish Commission report for 1882, p. 247, states:

In the summer of 1879 Capt. George Friend, of Gloucester, smoked some of the tilefish, and he, as well as several others who ate them, stated that they were excellent, rivaling smoked halibut in richness and flavor. On the other hand, Mr. William H. Wonson, 3d, does not speak so highly of its fine qualities as a food-fish under the same conditions. He says that while it is certainly very good and wholesome, as well as a desirable article of food when smoked, it can not compete with the halibut, and is no better, in fact, than smoked haddock.

PRESERVATION OF FISHERY PRODUCTS BY CANNING.

DEVELOPMENT AND METHODS OF CANNING.

The various processes of canning are all directed essentially (1) to preserving foods in hermetically sealed vessels from which the atmospheric air has, so far as practicable, been driven off, and (2) to destroying by heat or otherwise such germs as may be in the food before or after it is sealed up. Heat is applied to destroy the germ within the food, and the entrance of other germs or putrefactive organisms is prevented by sealing the can.

The credit for the introduction of this method of preserving foods is shared between a Mr. Soddington, who in 1807 presented a description of his process to the English Society of Arts, under the title "A method of preserving fruits without sugar, for house or sea stores," [*] and François Appert, who in 1810 published a book giving directions for a process for which he was awarded a prize of 12,000 francs offered in the preceding year by the French Government for a method of preserving perishable alimentary substances. The methods of Soddington and of Appert were essentially the same, and as follows: Glass bottles were filled almost to the top with the food, which in some cases was partly cooked, the bottles corked loosely and placed up to their necks in tepid water, the heat being gradually raised to a temperature between 170° and 190° F., and being maintained there for a period varying from 30 to 60 minutes. The bottles were then corked securely and allowed to cool slowly in the bath. In some cases Soddington filled the bottles with boiling water before sealing, and he recommended further that the corks be covered and the bottles laid upon their sides, so that the hot liquid might swell the corks. Based on the erroneous impression that exhaustion of the air is the essential feature of preserving foods, a number of methods were soon after and have until quite recently been devised for accomplishing the result. Among these methods are the use of air pumps, introducing carbonic acid or hydrocarbon gas into the vessel containing the food, etc.; but none of them have come into general use.

This general process of preservation does not appear to have been very extensively employed until the substitution of tin cans in place of glass bottles. These seem to have been used first in 1820; and in 1823 a patent for them was issued to Pierre Antoine Angilbert.[†] Preserved fish had been placed in tin cans for many years previous, but not in the manner known at present as canning.

In "A treatise on fishing for herring, cod, and salmon, and of curing or preserving them," published in Dublin in 1800, the following method of preserving salmon is noted as being practiced in Holland:

As soon as the fish is caught they cut off the end of the snout [head] and hang it up by the tail to let the blood flow out as much as possible. A short time after they open its belly and empty it and wash it carefully. Then they boil it whole in a brine of white salt, often skimmed. Before it is quite

* Hassell: Food and its Adulterations, London, 1855, 132.
† Letheby: Chemical News (American reprint), 1869, 4, 74.

boiled they take it out of the brine and are careful not to injure the skin, after which they let it cool and drip on a hurdle. Then they expose it for a day or two to the smoke of a fire made of juniper, which must make no flame. Finally, they put it into a tin box, the sides of which must be an inch higher than the thickness of the fish, and fill up the box with fresh butter, salted and melted. When the butter is coagulated they put on the cover and solder it to the lower part of the box. Some persons eat the fish without boiling it again, but it is better when it gets a second boiling. In winter good oil of olives may be used instead of butter.

Angilbert's method was very similar to the present processes, which differ in some minor features, but are uniform in principle. A definite amount of the article to be preserved, with some liquid, is placed in a tin can, over which the cover, containing a minute hole, is soldered, and the can and contents are placed in a bath of boiling water. Through the small hole the air and steam escape from the can in boiling, and the heat also kills the bacteria. The hole is closed with a drop of solder, and the process of cooking is completed.

A number of modifications and improvements have been adopted, principally in reference to shortening the time of cooking, permitting the heated air in the can to escape, softening the bones of small fish, filling and handling the cans, etc.

While it is necessary that the fish be thoroughly cooked, yet in a majority of cases it is equally important that they remain as short a time as practicable under the action of the heat. This is facilitated by increasing the temperature of the boiling water. Formerly the cans of fish were boiled in salt water, by which a temperature of 230° F. is attainable, or in water containing chloride of calcium, or sulphide of soda, whereby 250° F. may be secured. But these agents are each prejudicial to the metal of the can and the kettle, causing them to rust or wear rapidly, and by using the maximum of heat secured by the chloride of calcium process the cans often burst, with dangerous effects to the workmen. About 1874, steam-tight cylinders were introduced, in which the cans are subjected to a very high temperature by introducing steam from adjacent boilers, thus shortening the time of exposure to heat and removing liability to burst, the outward pressure in the can being counterbalanced by the inward pressure of the steam in the cylinder. This was first applied in 1874 in canning oysters. At first steam only was used, but it was soon found that the contact of the steam with the can results, to some extent, in scorching the contents that lie next to the inner surface of the can, and the oysters or fish have a slightly burnt flavor in consequence, the can itself also exhibiting a bluish color on the inside. This was remedied by placing the cans in water, below the surface of which the superheated steam is admitted. The pressure upon the intermediate water is transmitted to the outside of the can and counterbalances the pressure from the inside until the cooking has been completed. The fish or oysters on the inside of the can are also acted upon uniformly by the heat, and neither cans nor contents injured, even if the temperature be raised to 250° F. or more. By this process, which has been generally adopted, the cooking is done in much shorter time and at greatly reduced expense.

To remove the air from the can it was formerly customary to leave a minute hole in the lid, heat the can and contents by nearly submerging the can in boiling water, and then solder the small hole. At present the cans are hermetically sealed and boiled, then punctured to permit the expanded air to escape, when they are resealed, and the process of cooking completed. An improved method has been devised, in which the air is extracted from the can by means of tubes connecting the tin with a vacuum chamber, but it is little used in canning fishery products.

. An objection to the canning of small fish is the large number of bones. Ordinarily the heat developed in the process of canning destroys the cohesion of the particles of the bones, so that they may be masticated and swallowed without inconvenience, but the bones of some small fish are not so easily softened. About 1867 it was found that by placing the fish in vinegar and subjecting them to a temperature of 170° F. for several hours, according to the size of the bones, the acid of the vinegar dissolves the lime salts contained in them. This process is somewhat costly and the vinegar is objectionable to some consumers.

In 1872 it was found that the bones could be softened without using vinegar, by successive steamings for several hours, with an intervening cooling. A patent for this process was issued May 21, 1872, to Isaac L. Stanly, of New York City, who thus describes the process with special reference to canning menhaden:

First put the fish, after being dressed and prepared, into open tin or other suitable boxes or vessels, and place the same in a steam chest, which is afterwards closed. In this condition steam the fish with steam of 212° F., or thereabout, for 5 hours, more or less, after which they are taken out of the steam chest and put on tables to cool and drain for about 5 or 6 hours. The fish are then packed in tin boxes filled with olive or other oil, and the boxes afterwards closed and the covers soldered or sealed. Said boxes containing the fish are next put into a tank or chest, which is afterwards closed, and heated by steam or otherwise to a temperature of from 217 to 220 F., or thereabout, for a period of from 2 to 5 hours, according to the size of the fish or its bones.

An objection to the use of tin cans in preserving food products is the liability of the lead in the can to affect and taint the contents, which sometimes results in lead poisoning. This danger is not great when good materials are used in making the cans, except when they are filled with such articles as shrimp, the acid in which acts upon the tin. This is remedied by introducing a lining between the inner surface of the can and its contents. This lining may be textile fabric or a coating of asphaltum cement. Silver plating has been employed, and a lining of selected corn husks has also been used to a limited extent. For an account of these methods see pp. 523 et seq.

The most satisfactory way of overcoming this liability to lead poisoning is by using cans sealed without the use of solder. There are several varieties of these cans on the market at present, the joints being made tight by introducing a gasket or washer of rubber, paper or similar material between the overlapping tin and crimping or folding the edges so as to hold them together. Some of these cans are drawn by machinery out of one solid piece of tin plate, and others have an outside soldered seam in the body only. The additional cost of these cans has confined their use to the preservation of the more costly varieties of marine products.

A great improvement in canning was the introduction, about 25 years ago, of the use of solder in the form of twisted strands cut into rings of the proper size and containing on its surface the proper quantity of flux. By twisting a strand, or by uniting two or more strands into a cord, the exterior will contain a multitude of small depressions. The wire is converted into rings by winding it on a wooden cylinder of the required diameter, a cut is made through the wire the whole length of the cylinder, and the rings are slipped off. The rings are coated with resin or other suitable flux, the depressions become filled with it and hold a sufficient quantity for soldering purposes. This quantity is regulated by the method of twisting the wire or the manner of applying the flux. If to be used for square tops, the wire is wound on a rectangular block of the required size and then cut. It is apparent that when these rings are applied to the capping of tin cans no more solder or resin need be used than is actually necessary to solder on the cap, saving a large quantity of material over the old method of using bar solder and avoiding smearing the top of the can.

In 1879 Charles C. Lane, of New Westminster, British Columbia, introduced an improvement in cans, conforming to the natural shape of the fish, to avoid cutting it into small pieces. The can is in two parts, approximating respectively the shape of the two halves of the fish divided transversely to its length, and is so fitted that it may be adjusted longitudinally to the length of the fish, one half being somewhat smaller than the other at the open end, so that it will fit into the other. The patentee thus describes his invention: the claim relates especially to the method of constructing the pieces forming the can, so as to avoid waste and reduce the amount of soldering to a minimum:

The can is made in two parts, one part being somewhat smaller than the other at the open end, so that it will fit into the other. Each of these parts is made in two pieces, the pieces being stamped out of sheet metal and soldered at the joints or meeting longitudinal edges. These halves of the ends have each stamped in them a recess or flattened portion, which admit of a certain expansion during the boiling process, and of contraction when the air is blown off and the can and contents are finally cooled. This flattened portion or recess formed in these pieces makes a flat place on the outside and facilitates packing of the cans or storing on the shelves of the dealers in good order. When lying on either side the can rests on these flat parts, and they may be placed one above another without sliding about on account of the irregularity of their general shape.

Each can is formed of four pieces, two of which are stamped out to be approximately of the form of the head and shoulders of a fish, while the other two are shaped to fit the tail half of the fish. These are got out in quantities, and the two halves of each part are united, and the longitudinal seam is soldered by the aid of the mechanical bath, which is easily accomplished when the can is empty. This enables me to do the largest portion of the work of can-making, including the long side seam or joint, mechanically, with but little hard labor, and during the idle season and before the fish commence running. When the two halves are thus made ready to receive the fish it will be seen that the surrounding transverse joint is all that is left to be done, and this renders the labor to be performed in the actual canning so slight that the fish may be put up in this manner to compete with any other form of canning.

After the separate ends are made in the manner described the fish is placed in the open end of one in the proper position, and the other half is placed over the other end of the fish. The two open ends thus come together, the smaller slipping inside the larger. At this central joint I place a V-shaped piece or strip of a soldering metal, which fits between the two parts to form the transverse joint in the final soldering, both by filling it up and by partially melting.

As the fish vary somewhat in size, it will be seen that the parts must be telescoped or pushed together until they fit the fish snugly before the transverse or central joint is soldered. This will insure a perfect fit to every fish, and will prevent it from sliding about in the can.

By making the can in this manner small sheets of tin may be used with very little waste in cutting, and at the same time cans can be made which will correspond to the size and shape of the fish. It will be manifest that either of the halves may be sealed up independently of the other by simply fitting a flat head or cover to the large, open end and soldering it in place after the can is filled.[*]

Among the devices contrived during the last fifteen years to facilitate the canning process, machines for filling the cans are numerous. But these have not been so generally adopted in the canning of fish as in preserving vegetable products, the fish in hand filled cans presenting a neater appearance than those filled by machinery. However, on account of the great saving of labor, they are used in a number of salmon canneries of the Pacific coast.

One of the first of these, invented[†] in 1880 by A. H. Moore, of Ellensburg, Oreg., is so arranged that the fish, after being cut up in suitably sized pieces, is placed in a

---

[*] Letters Patent No. 221325, November 4, 1879.   [†] Letters Patent No. 227283, May 4, 1880.

hopper and forced or fed by a plunger into a mold corresponding in size to the can. A knife then cuts off the mass, and another plunger forces the mass of fish from the mold into the can, which is then removed and sealed. The principal objection to this is that the portion of fish is not properly shaped to fit the can, being left flat on one side and great pressure has to be exerted to force the fish into the mold.

During the same year an improvement on this process was invented * by Robert D. Hume, the well-known salmon-canner of the Pacific coast. This improvement consists in the employment of a carrying belt operated by an automatic pawl and ratchet, whereby the material is carried forward into a chute, through which it is led into a shaping and compressing cylinder. In combination with this chute there is employed a pair of slim cylindrical shearing knives, arranged to rotate about a longitudinal axis, one within the other, in concentric circles, which cut the material to the exact cylindrical shape necessary to fill the can without unequal compression.

Numerous other can-filling machines have been invented, for a description of which see the following letters patent:

Patent No.	Date.	In favor of—	Patent No.	Date.	In favor of—
259142	June 13, 1882	Wm. West, Keene, Canada.	356122	Jan. 18, 1887	John B. Hodapp, Mankato, Minn.
262575	Aug. 15, 1882	Augustine Crosby, Benton, Me.	358498	Mar. 1, 1887	H. R. Stickney, Portland, Me.
291799	Jan. 8, 1884	J. Stevens, Woodstown, N. J.	360541	Apr. 5, 1887	Do.
297540	Apr. 29, 1884	Geo. Ackermann, Cincinnati, Ohio	361177	Apr. 12, 1887	G. L. Merrill, Syracuse, N. Y.
301897	July 15, 1884	Edmund Jordan, Brooklyn, N. Y.	372876	Nov. 8, 1887	H. R. Stickney, Portland, Me.
304063	Aug. 26, 1884	Volney Barker, Portland, Me.	373306	Nov. 15, 1887	D. D. Rauney, Lewistown, Ill.
306658	Oct. 14, 1884	J. Stevens, Woodstown, N. J.			

The preservation of foods by the canning process has now become one of the world's great industries, it being applied to alimentary substances of almost every description, and the product is of enormous extent. It was early used in the preservation of sardines, lobsters, etc. In 1824 John Moir & Son, of Aberdeen, Scotland, began the canning of salmon, game, and other meats.

Canning was first employed on the American continent by Charles Mitchell, at Halifax, Nova Scotia, in 1841, and in 1842 by U. S. Treat at Eastport. Me. The canning of oysters was commenced about 1844; Pacific coast salmon were canned first in 1866, and the preparation of sardines in this country dates from 1875. Yet at present the total value of the domestic output of these three products as canned approximates $15,000,000 annually.

There are five general classes of canned marine products—viz, (1) plain-boiled, steamed, or otherwise cooked; (2) preserved in oil; (3) prepared with vinegar, sauces, spices, jellies, etc.; (4) cooked with vegetables, etc., and (5) preserved by some other process, but placed in cans for convenience of marketing.

The first class includes salmon, mackerel, herring, menhaden, cod, halibut, smelt, oysters, clams, lobsters, crabs, shrimp, green turtle, etc.

Sardines almost exclusively make up the second class.

The third class includes various forms of herring prepared as "brook trout," "ocean trout," etc., mackerel, eels, sturgeon, oysters, lobsters, crabs, etc.

The fourth class comprises fish chowder, clam chowder, codfish balls, green-turtle stew, terrapin stew, and deviled crabs.

The fifth class is made up of (a) smoked herring, halibut, haddock, carp, pickerel, lake trout, salmon, eels, sturgeon, etc., and (b) brine-salted mackerel, cod, and caviar.

* Letters Patent No. 233149, dated October 19, 1880.

CANNING SALMON.

The canning of salmon appears to have originated at Aberdeen, Scotland, in 1821; but prior to the establishment of salmon canneries in the United States, in 1864, the application of the process to this fish was very limited. During the last 30 years this industry has been confined to the western coast of the North American continent and to certain Asiatic countries bordering the Pacific coast. It has become one of the great fishery industries of the world, the annual output exceeding $10,000,000 in value, over 99 per cent being prepared on the American continent.

On the western coast of the United States the industry was begun in 1864 by Messrs. Hapgood, Hume & Co., at Washington, on the Sacramento River. A member of this firm had been engaged in canning lobsters in New Brunswick, on the shore of the Bay of Chaleur, and methods somewhat similar were applied to the canning of salmon. The machinery and appliances were very crude as compared with modern devices. The fish, cut into transverse sections of suitable lengths, were placed in the cans and the cover attached, with ventholes open. The cans were then nearly submerged in fresh water contained in large round-bottomed iron kettles and boiled for an hour, after which they were removed and the vent closed. They were next placed without arrangement in an iron bath kettle containing salt water heated to a temperature generally from 228° to 230° F. After an hour's bath the cans were removed and placed in a tank of cold water. When cooled they were wiped off, the ends painted with red lead, the sides labeled, and the cans packed in the cases. No process was employed for testing for leaks, and consequently about one half of the product of the first year spoiled.* Much difficulty was experienced in placing the canned salmon on the San Francisco market, but eventually the entire pack was sent in separate lots to Australia, where it netted $16 per case to the shippers.†

The trade gradually increased from year to year with the improved transportation facilities and the development of markets for the product. In 1866 the first Columbia River cannery was established at Eagle Cliff, about 40 miles above Astoria. In 1874 canning was begun in British Columbia, and in 1882 Alaska began to make a showing. The total pack on the west coast of North America in 1892 was 1,323,000 cases of 48 1-pound cans each, approximating in value $6,549,000; and in 1895 it was 2,175,986 cases, worth $10,081,907 at first hands.

During the first years of the trade South America and Australia furnished the consumers of the canned salmon, but as the output increased an English market was sought. The latter did not at first take kindly to the American product, but after persistent efforts on the part of some of the most extensive London wholesale dealers the article became better known and the people of Great Britain soon became the principal consumers, sometimes using 500,000 cases in a single year.

Several species of salmon are utilized in the canneries of the west coast, the principal ones being chinook or quinnat salmon (*Oncorhynchus tschawytscha*), blueback salmon or redfish (*O. nerka*), silver salmon (*O. kisutch*), steelhead (*Salmo gairdneri*), dog salmon (*O. keta*), and humpback (*O. gorbuscha*).

---

* Hume's Salmon of the Pacific Coast, p. 8.
† Report U. S. Fish Commission, 1888, pp. 167, 168.

The following summary shows the number of cases of each species canned on the Pacific coast during the four years ending in 1895:

States.	Chinook.	Blue-back.	Silver.	Steel head.	Dog.	Hump-back.	Total.
1892—Washington	134, 253	19, 441	28, 708	26, 945	29, 411		238, 758
Oregon	237, 684	51, 106	60, 293	45, 403			394, 486
California	14, 334		1, 550				15, 884
Total	386, 271	70, 547	90, 551	72, 348	29, 411		649, 128
1893—Washington	129, 078	55, 237	31, 707	25, 663	23, 480	17, 530	282, 695
Oregon	176, 024	23, 074	62, 913	39, 563	9, 230		310, 804
California	26, 436		500				26, 936
Total	331, 538	78, 311	95, 120	65, 226	32, 710	17, 530	620, 435
1894—Washington	156, 549	53, 717	32, 118	23, 209	33, 952	9, 049	308, 594
Oregon	216, 507	25, 523	100, 087	38, 829	3, 162		384, 108
California	31, 663		500				32, 163
Total	404, 719	79, 240	132, 705	62, 038	37, 114	9, 049	724, 865
1895—Washington	157, 187	70, 304	81, 957	18, 985	48, 686	23, 633	400, 752
Oregon	316, 284	12, 854	138, 981	30, 693	27, 027		525, 839
California	28, 635		400				29, 035
Total	502, 106	83, 158	221, 338	49, 678	75, 713	23, 633	955, 626

NOTE.—468,970 cases of salmon were packed in Alaska in 1892 and 619,379 cases in 1895, making the total pack for the United States in 1892 1,118,098 cases, and in 1895 1,575,005 cases.

The extent of the salmon-canning industry of the Pacific States in 1895 is shown in the following table:

States.	Persons employed.	No. of canneries.	Value.	Cash capital.	Total investment.	Salmon utilized.		Salmon canned.	
						Lbs.	Value.	Cases.	Value.
California	198	4	$62, 000	$64, 000	$126, 000	1, 906, 525	$52, 591	29, 035	$128, 632
Oregon	1, 960	26	719, 225	942, 500	1, 661, 725	35, 299, 241	1, 184, 529	525, 839	2, 456, 698
Washington	1, 146	17	374, 650	601, 000	975, 650	27, 441, 724	731, 522	400, 752	1, 638, 938
Total	3, 304	47	1, 155, 875	1, 607, 500	2, 763, 375	64, 647, 490	1, 968, 642	955, 626	4, 224, 268

In Japan about 200,000 cans of salmon (*Oncorhynchus perryii* and *O. haberi*), amounting in value to 24,000 yen ($12,000), are prepared annually, principally at Hokkaido and by the Sanmitara and Fujino Company.

The following description of the methods of salmon canning is from Mr. W. A. Wilcox's "Notes on the Fisheries of the Pacific coast," (U. S. Fish Commission Report for 1896, pp. 583–587):

As at present conducted there is but slight difference in the manner of preparing canned salmon in any of the canneries. As a rule the factories are located adjacent to or very near the fishing-grounds, so that at the most but a few hours elapse from the time the fish are freely swimming until they are caught, delivered at the cannery, dressed, canned, cooked, and packed, thus insuring a perfectly fresh product, old or stale fish never being met with at a salmon cannery. The neatness and cleanliness of the canneries is one of the first things noticed by visitors during the packing season.

The notes here presented on the methods of salmon canning were taken in 1896 at a cannery on the Columbia River, and with few exceptions represent the canneries of the entire west coast.

The buildings connected with a salmon cannery are always built at the water's edge or partly over the water, so that vessels or boats may come alongside and deliver their fish and supplies or receive the packed products. As a rule they are large, roomy, one-story frame structures, the business of receiving, cooking, and packing of salmon all being in the one large, high, and well-lighted room. The lofts are used for the storage of empty packing-cases, empty cans, nets, etc., and in some instances large rooms are there used for the manufacture of cans. Adjacent to the cannery are the rude quarters in which the Chinese employees live and near by is usually the home of the superintendent.

Chinese have a monopoly in the canning of salmon, but never engage in their capture. Before the season opens contracts are made with some large Chinese firm of San Francisco or Portland to do the work so far as relates to receiving raw products and turning the same over canned, packed, and ready for shipment.

As a rule the fish are bought from the fishermen at so much apiece or per pound, a price for the season having previously been agreed on; but in some cases the fishermen are hired by the month, with or without board, the fishing boats and nets in that event being furnished by the cannery.

Contracts with the Chinese usually call for the packing of at least a certain number of cases, of 48 pounds each, at prices ranging from 30 to 40 cents a case for 1-pound cans, and higher for half-pound oval or other special cans.

A working gang of from 30 to 75 Chinese, in charge of a native expert foreman, is sent to the cannery in advance of the opening of the season. The men are constantly under the orders of the Chinese foreman, and he in turn is under the supervision of the superintendent. The foreman divides up the duties, assigning a gang for each part of the work from the time the fish are landed until they are cased for shipment. These gangs follow their particular part of the work all through the season, only in exceptional cases being called to any branch except their own. The receiving and dressing gang, being the first to begin, are the first to finish their labor, while the packers are the last to begin and end the work of the day. If fish are plentiful, all of the men work from about 7 a. m. to 6 p. m., with only a stop for the midday meal. If salmon are scarce, the men may have but a few hours' work.

On the completion of the work of any gang the men must before leaving thoroughly clean their section. In doing so a hose is used, with abundance of water, brooms, and scrubbing brushes, and when the day's work is over the interior, platforms, and wharves are left scrupulously clean and ready for the work of the following day.

As the fishermen arrive their catch is thrown out on the wharf, where it is received by the Chinese and carried inside the cannery and thrown into boxes on the scales. Having been weighed, a receipt is given to the fishermen, and the fish begin their journey through the cannery, that only ends after they have been canned, cooked, packed, and cased ready for shipment.

From the weighing scales the fish are thrown out on the floor and receive their first washing from a stream of water that is played on them from a hose, the fish being turned over with a pitchfork, as may be necessary, to thoroughly remove all gurry and dirt. In some instances, where fish are received faster than they can be immediately handled, they are kept cool and fresh by having, as needed, a fine spray of ice-cold water thrown over them from an overhead revolving pipe. The first gang receives the fish on the dressing tables, which are near the door. Here the first of the work begins, and to follow it through from its entrance to its exit, canned and cased, is an interesting sight to thousands of visitors during the packing season. The first operator seizes fish after fish, and with a few quick strokes of a large butcher knife severs head, fins, and tail. The next man opens the fish, removes the viscera, and scrapes the carcass inside and out. Through an opening in the floor all offal and waste are at once thrown into the river and quickly consumed by schools of scavenger fish or the large number of gulls that hover in the vicinity waiting for their food. At some of the canneries near Astoria receptacles for waste are provided by those interested in oil and fertilizer factories.

The fish is then shoved along to the man standing by the side of the header and cleaner for the next washing, and at the same time is scraped with a knife that removes the scales. The fish is then passed along into a second tank of clear water, where it receives its final washing and cleaning and is made ready for cutting in proper-sized pieces. A series of semicircular knife-blades is attached to a roller, the blades being equal distances apart, corresponding with the size or depth of cans to be filled. One end of the roller is hinged, to the other end a handle is attached. The knives are raised by means of the handle, the fish is placed under them, and with one quick, sharp blow the fish is entirely cut up into lengths suitable for canning. For 1-pound tall cans, 7 knives are attached to the roller; for 1-pound flat cans, 13 knives; for ½-pound cans, 17 knives. The fish are now in suitable lengths, but must be sliced into sizes proper to enter the cans. This is quickly performed, and the pieces are passed on to the filling gang.

Several men stand at one or both sides of the filling table, each supplied with small scales adjusted to the weight of the cans to be filled. In some canneries cans are filled by machinery, but this is usually done by hand. As soon as filled the can is placed on the scale. If it shows full or over weight it is passed on, no fish being removed; but if short weight, the can is put one side to receive enough to make up the deficiency.

From the filler the can passes to a man who places it on a swiftly revolving plate having a closely fitting cap, and a small but strong stream of water is made to play against the revolving can, removing all grease and dirt. A small scrap of flat tin is put on top of its contents, and the prepared top is fitted on. In order to keep the top in place pending soldering, the can next goes to a man who places it on a revolving plate, while, by means of a lever operated with his foot, the top piece is forced down and securely held at the same time the edges are being smoothly crimped. The sealing of the top is ingeniously accomplished. A brick furnace has on its top a long V-shaped trough that is kept filled with molten solder. At the upper end of the solder trough and a few inches higher is a similar one filled with muriatic acid. From the crimping machine the cans drop in an almost continuous stream into the trough with acid. A heavy endless chain passes along just over the troughs, and by the friction from the chain the cans are carried on first through the acid, which touches only the outer edge of the can and prepares it to receive the solder through which it next passes. The top of each can has a small hole punched in it to permit the escape of air as the can becomes heated in its passage through the trough of molten solder. Here may be noticed the utility of the small piece of tin before mentioned as being put in the can, without which the hole is liable to be clogged up with fish, and, the vent being closed, the cans are liable to be badly strained, made leaky, and have the entire top blown out.

The endless chain, having rolled the cans through the solder, drops them on an inclined plane some 30 or 40 feet in length. Shortly before reaching the bottom of this incline the cans are met with a shower bath of cold water from perforated pipes alongside of the incline. The bath is simply to cool them so that they can be instantly handled. The cans are at once placed on iron trays, known as coolers; they consist of an iron frame 35 inches square, 3 inches deep, with slatted iron bottom. One cooler holds 86 one-pound flat cans, or 160 one-pound tall cans. The small venthole on top of the can is next sealed. The cooler filled with cans is then attached to an overhead rail track and transferred to a large wooden vat filled with water, having a temperature of 212° F. The cans remain in this bath only a few moments, which is ample time to test them for leaks. If a can is not perfectly air-tight, this fact is at once made evident by small air bubbles rising from the can while in the bath, and it is at once removed for repairs, another can being substituted. When the test is completed, the cooler is placed on a small iron-framed car having a capacity of 8 coolers, one on top of another. The cans are now ready for the cooking of the salmon in large retorts.

Retorts are made of wood or iron, but are usually of boiler iron, have a round shape, and are about 13 feet long and 5 feet in diameter. A steam pipe extends along near the bottom. This is perforated for the escape of steam, which passes through a small amount of water with which the pipe is covered. On an iron track just over the pipes the loaded cars are run. Retorts usually have an opening or door at only one end, but in the cannery now being described there was an opening at each end and two retorts were used, the few feet separating them being connected by a track by which cars of coolers having passed through the first retort can pass on into the second. Each retort has a capacity of 4 cars, or 3,200 one-pound tall cans or 2,952 one-pound flat cans. Cans of salmon remain in the first retort under a steam temperature of 230° for one hour. They are then run out, vented, and at once resealed. As the top of each can is perforated with a small, sharp-pointed iron, the heated air or steam is expelled, and before its place can be taken with cold air the vent is closed by a drop of solder, and the can may be said to be free of air and air-tight. The cans are now ready for another cooking in the second retort. Here the temperature is 240°, in which one-pound tall cans remain 1 hour and flat cans 1½ hours. Retorts are under a steam pressure of 7 to 10 pounds to the square inch.

On removing the cans from the retorts they have a stream of cold water thrown on them, by which they are cooled and cleaned. They are now finally tested for leaks or imperfections by tapping each can on the top with a small piece of iron, an experienced ear quickly detecting by sound any imperfection. Imperfect cans are replaced by others, and the cans pass on to be lacquered, labeled, and packed in boxes, each holding 48 one-pound cans or 24 two-pound cans. They are then ready for a distribution that reaches almost every portion of the civilized globe.

The cannery at which these notes were taken was provided with electric lights and ample steam power; the rooms were well ventilated and lighted; its walls were white with paint or whitewash. It is located on the Columbia River with the Cascade Range of mountains towering from 1,500 to 2,500 feet just at its back. From these mountains the icy-cold and very pure water used at the cannery is brought.

Much attention is given to variety and styles in cans and labels, which yearly show improvement in style and design. Sixteen varieties of cans were used in the pack of 1895. Brands also receive much consideration, and in many cases have a high value on account of the enviable reputation of the goods previously packed under them.

To supply the annual demand for 60,000,000 to 80,000,000 tin cans in which to pack the salmon caught in the waters of the United States on the Pacific Coast is of itself a large business. Many packing firms make a part or all of the cans they need. This is more for the purpose of keeping desirable men employed between seasons than for any saving in expenses for cans. The bulk of the cans are turned out from factories at San Francisco. In April, 1893, the Pacific Can Company of San Francisco, by opening a branch factory at Astoria, filled a long-felt want of the packers in Oregon and Washington. For ten months in the year this Astoria branch gives employment to 80 persons, of whom 25 are females and 55 males. Chinese are not employed in this establishment. The weekly pay roll averages $750, or $30,000 a year. The plant represents an investment of $80,000.

Of late each year finds some new shape or size of can used in salmon packing, there being 16 varieties of cans for salmon manufactured by this company in 1895. The total number of cans turned out in 1895 amounted to 18,500,000, of which 2,000,000 had the key attachment, on which the royalty is 75 cents per 1,000 cans. A large quantity of the various cans is kept on hand, from which the canneries of the Columbia and coast rivers and those of Puget Sound are supplied as needed. The tin plate used amounted to 45,000 boxes, of which 10,000 were of American manufacture and 35,000 were imported. The average price of plate during 1895, including an import duty of $1.20 a box, was $4.10 a box, averaging 100 pounds with 112 sheets each. American-made plate is 50 cents a box lower at Chicago than that imported, but 50 cents higher by the time it reaches Astoria, owing to the difference in freight—by sailing vessel from Europe of $7 a ton, or by rail from the East of $11 a ton.

Commander Jefferson F. Moser, U. S. Navy, commanding the U. S. Fish Commission steamer *Albatross*, has given a description of the processes of canning salmon in Alaska, in the Fish Commission Bulletin for 1898, pages 22–34.

CANNING OYSTERS.

Preserving oysters by cooking and hermetically sealing them originated in the United States in 1844, in the establishment of Thomas Kensett, Baltimore, Md., but the trade appears to have been developed by A. Field & Co. of that city. From the beginning of the industry up to the present time it has been prosecuted mainly at Baltimore, probably not 3 per cent of the total product since 1844 having been prepared outside of that city. In addition to Baltimore, oysters are canned at one or two other Chesapeake ports and at Apalachicola, Fla., Biloxi and Bay St. Louis, Miss., and Morgan City, La. The term "cove" is sometimes applied to oysters prepared in this manner, and the higher prices prevailing in the fresh-oyster trade usually require that the medium and small-size oysters be used.

Originally in the canning business the raw oysters were opened by hand, but about 1858 Lew McMurry, of Baltimore, began scalding them, this process causing the shells to open and allowing the oysters to be removed with far less labor than would be required if they had not been heated. Steaming originated about 1860, the oysters being placed in baskets holding about 3 pecks each, and these to the number of about 300 were placed in a large box and there steamed. In 1862 Henry Evans, of Baltimore, devised the present method of using, in combination with the steam box, large cars, in which the oysters are placed. The sides and bottoms of these cars are made of iron bars, so that the steam may readily pass among the oysters, partly

---

* See Letters Patent No. 35511, dated June 10, 1862.

cooking them and causing the shells to open freely. The track on which the cars run is laid from the wharf to and into the steam box, and thence to the shucking room.

The details of the present process of steaming and canning follow, this description applying especially to the business at Baltimore.

The oysters when taken from the vessels are placed in cars of iron framework, 6 or 8 feet long, with capacity for about 20 bushels. These cars are run on a light iron track, which is laid from the wharf through a steam-tight chest or box, to the shucking shed. This steam chest is a rectangular oak box, 15 to 20 feet long, lined with sheet iron, fitted with appliances for turning on steam to any desired pressure, and with a door at either end which shuts closely and is so packed with felt or some other material as to make the joint between the door and box as nearly steamtight as practicable. When a car is filled with oysters in the shell it is run into the steam chest and there left for 15 minutes, with the doors closed and steam admitted. The chest is then opened and the car run into the shucking room, its place in the chest being immediately occupied by another car. By having a sufficient number of cars the laborers may be constantly employed, loading and unloading in succession as they are steamed and emptied.

In the shucking sheds the cars are surrounded by the shuckers, sometimes to the number of several hundred, each provided with a knife and a can arranged so as to hook to the upper bar of the iron framework of the car. The steaming causes the oyster shells to open more or less widely, and the meat is readily removed.

The opened oysters are then washed thoroughly in cold water and transferred to the "fillers' table," and the cans, when filled, after being weighed individually, are taken to the soldering table and there "capped"; that is, hermetically sealed. From the "cappers" they are placed in a cylindrical crate or basket and lowered into a large cylindrical kettle, called the "process kettle" or "retort," which is partly filled with water, where they are again steamed to such a degree as to destroy all germs of fermentation. After this they are placed, crate and all, in a vat of cold water, this serving the double purpose of arresting the operation of cooking by cooling them and of testing for leaks. When sufficiently cool to be handled the cans are transferred to another department, labeled, and packed in boxes for shipment.

The shuckers usually work in gangs of 6 or 8 persons, comprising sometimes whole families of men, women, and children. Those in Baltimore number about 4,000, ranging in ages from 12 to 60 years, and are mostly women and children, the work being light and peculiarly adapted to them. They are mainly of foreign parentage. Few scenes are more interesting than those observed on a visit to the shucking room of any one of the large canning houses. At one end the cars of steaming hot oysters are received, and as these are arranged in long rows covering the length of the room the shuckers, numbering 600 or more in some establishments, surround the cars and with rapidly working knives hastily and skillfully remove the yet steaming oysters. These employees are extremely industrious, and hundreds of small dwelling houses have been purchased in Baltimore with money obtained by the women and children at work in the oyster houses. The shucking is done in a cup known legally as the "oyster-gallon cup," which holds 9 pints, wine measure. The shuckers are paid at the rate of 6 cents per "cup," averaging about 65 cents per day, the total wages paid those in Baltimore amounting to about $80,000 annually.

About 800 other persons are engaged in the Baltimore canneries, of whom about three fifths are men. These employees are paid from $5 to $25 per week, their total wages amounting to about $90,000 annually. The largest item of expense in the canning houses, aside from the cost of the oysters, is the purchase of tin, labels, etc., this amounting to about $315,000 annually. The incidental expenses of the Baltimore canneries amount to about $25,000. The total cost of handling a bushel of oysters in the canning houses is about 29 cents.

The cost of the oysters for the canning trade at Baltimore has averaged during recent years about 55 cents per bushel. Each bushel yields about 50 ounces of "solid meats." These are packed in 1-pound and 2-pound cans and cans of miscellaneous sizes, most of the latter being a trifle larger than the 1-pound cans, which contain about 5 ounces of solid meats, the 2-pound cans containing 10 ounces each. The price received during the last seven or eight years for the 1-pound and 2-pound cans has averaged about 75 cents and $1.40, respectively, per dozen.

The following summary shows the extent of the oyster canning at Baltimore during the most recent year for which detailed returns are available:

Extent of Baltimore oyster canning in 1891-92.

Capital invested, employees, etc.		Expenses.		Products.		
Establishments, etc.	Number, value, etc.	Items.	Amount.	Cans.	Ounces.	Value.
Number of establishments................	20	Cost of oysters........	$1,291,600	5-ounce cans, 9,388,650 ...	46,943,250	$764,450
Value of property...	$1,255,000	Wages paid shuckers.	73,680	10-ounce cans, 4,644,822....	46,438,220	725,515
Cash capital........	$1,170,000	Wages paid to others.	97,500	Miscellaneous cans	22,635,000	366,545
Persons employed ...	4,848	Tin cans, labels, etc ..	320,000			
Oysters received, bushels.............	2,396,763	Incidentals	25,000			
		Total	1,717,780	Total	116,016,470	1,856,510

The oysters on the Gulf of Mexico are large and less firm in structure than those of Chesapeake Bay; consequently, when canned in the ordinary manner, they tend to disintegrate and have a somewhat slimy appearance. In 1880 a process was invented by Mr. J. T. Maybury by which the texture of the oysters may be hardened so that they may be boiled without disintegrating.* This process is as follows:

To 10 gallons of pure water add one-half gallon of good commercial vinegar and 1½ gills of a saturated aqueous solution of salicylic acid, to which mixture sufficient common salt is added to impart the requisite salty flavor to the oysters. The mixture is boiled a few minutes and poured over the oysters in the cans, which are at once sealed and placed in a steam bath, the temperature of which is 202° F. This temperature is gradually raised to 240° and maintained at that degree for about 45 minutes. The cans are then vented, resealed, and steamed as before for about 30 minutes, when they are ready to be labeled and packed.

The acids serve to coagulate the fatty portion of the oyster and thereby render its body more dense and firm. The acids are harmless and the quantity is so small that they are not perceptible to the taste.

The term canning is frequently applied incorrectly to a much less permanent method of preserving oysters, viz, placing them in hermetically sealed tin cans or small wooden kegs, which contain from 25 to 200 oysters, without cooking them. By this method the germs within the can or keg and its contents are not destroyed, and the oysters will keep for only a few days, and even then the cans are usually shipped in iced boxes. Twenty years ago a very large part of the oyster trade was carried on

---

* See Letters Patent No. 230561, dated July 27, 1880.

in this way, but the extra cost of packages and the increased facilities for handling opened oysters in larger packages have almost entirely done away with the shipment in cans. It is yet practiced along the Gulf of Mexico, especially on the Texas coast, and at a few points along the Atlantic coast and in the interior of the United States.

CANNING SOFT CLAMS.

The soft clams (*Mya arenaria*) of the New England coast have been canned in quantities of greater or less extent for thirty years or more, principally by establishments in Maine as a minor part of their output. There are also a few canneries of which the principal output consists of clams. The first clam cannery in this country was established at Pine Point, Me., 8 miles west of Portland, by Messrs. Burnham & Morrill.

The process of canning in general use is as follows: First the siphon or "snout" is cut off, the thin skin or film covering removed, and the clams cleaned in the same manner as when prepared for the table. They are then placed in tin cans, holding from 6½ to 12 ounces, after which the cans are nearly filled with the liquid, diluted with either fresh, salt, or sea water, and the covers soldered on. The cans are next placed in crates and the contents are cooked in a tank of boiling water, the time of boiling depending on the freshness of the clams, usually continuing about 2 hours when the clams are fresh, and a trifle longer if they have been out of water several days. The cans are at once vented and again sealed and boiled about 1½ hours, when they are ready for labeling and boxing.

In some establishments a lining of white paper is placed on the bottom, around the sides, and at the top of the can, to prevent the contact of the contents with the tin, which sometimes results in the clams turning dark.

The product of canned clams in the United States at present amounts to about 40,000 cases annually, valued at $2.50 to $3.50 per case.

CANNING MACKEREL.

There are three distinctive varieties of canned mackerel prepared in this country, differing entirely in the methods of preparation and alike only in that they are sealed in tin cans: (1) fresh mackerel cooked in the manner usually applied to salmon—the oldest and most extensively used process; (2) broiled mackerel; (3) brine-salted mackerel, which are placed in tin cans simply for convenience in marketing.

MACKEREL PLAIN-CANNED.

The preparation of the first variety of canned mackerel was begun in this country in 1843 by Messrs. Treat, Noble & Holliday, of Eastport, Me., and was carried on by them incidentally with the canning of lobsters. During several years following 1843 the output was very small, averaging about 5,000 1-pound cans per year, the public being somewhat slow to fully appreciate the qualities of the product, but from that date to the present time the business has been continued on the Maine and Massachusetts coasts in connection with lobster and other canneries, and the extent of the product has fluctuated from year to year according to the abundance of mackerel on the coast.

In canning fresh mackerel it is quite essential that the fish be prepared as soon as practicable after being removed from the water. In dressing them the heads, tails, fins, scales, and viscera are removed, the fish cleansed, and soaked for a short time in strong brine to acquire a salty or "corned" flavor. They are next placed and sealed in tin cans which are at once immersed in boiling water until their contents are thoroughly cooked. The cans are then "vented," resealed, cooled, and prepared for the market. The cans are usually 4½ inches in height and 3 inches in diameter, and hold about 1 pound of fish; 2-pound and 3-pound cans are also used to some extent.

The price runs from $1 to $1.30 per dozen 1-pound cans, and the product from 5,000 to 3,000,000 cans annually. In 1879 the output in New England was about 900,000 1-pound cans: in 1880, 1,342,668 pounds, worth $150,449, and in 1881, 2,864,000 pounds.

The decreasing abundance of mackerel has resulted in considerable falling off in the output since 1885. The product in 1892 was about 1,000 cases, and in 1898 very few were put up, probably not over 200 cases, prepared at Provincetown and Boston.

BROILED MACKEREL.

Canning "broiled" mackerel was begun in the spring of 1880 by Eastport sardine canners. Two methods are practiced. The first, which originated with Mr. Julius Wolff, of the Eagle Preserved Fish Company, is almost identical with that used for sardines. The fish are dressed, washed, assorted according to size, dried, fried in oil, and sealed in tins with vinegar and spices.

The second method, which is in more general use, originated with Mr. Henry Sellman, of the American Sardine Company. The small mackerel are split down the belly, the heads and tails being removed, and then cleansed and put in strong brine for half an hour or so. When sufficient salt has been absorbed they are rinsed, spread on wire trays, and placed in a steam box, where they are steamed for a few minutes. The fish, yet on the trays, are baked in the rotary oven described on page 527, and are then packed with mustard, tomato sauce, or spiced vinegar, in oval-shaped tin boxes holding from 1 to 3 pounds each. The cans are sealed and subjected to a hot-water bath, after which they are vented, cooled, and labeled. These mackerel are far superior to any of the brands of herring, and the demand has always been greater than the supply. From 10,000 to 15,000 cases were prepared in 1898, about equally divided into mustards, tomatoes, and spiced vinegars. They are sometimes placed in fancy glass receptacles and are sold at high prices.

CANNING SALT MACKEREL.

The demand for small packages of salt mackerel led, about 1875, to the preparation of them in tin cans. In 1879 the equivalent of about 280,000 5-pound cans was prepared in Gloucester and Boston, about equally divided between the two places. In other parts of the country about 80,000 cans were prepared during the same year. These sold wholesale at about $5 per dozen 5 pound cans.

In preparing this article, the commercial, brine-salted mackerel, usually of the better grades, are used, No. 2 being the most popular size. These are washed and scraped, to give them a neat appearance, and the heads and tails are cut off, and if large the fish are sometimes cut transversely in two pieces. In packing, a small quantity of fine salt is sprinkled in the bottom of the can and the fish are carefully arranged flesh side up, except the top layer, which is placed flesh side down. A small quantity of salt is then sprinkled over the fish and the top is soldered on. A puncture about one-third inch in diameter is then made at the side of the can, through which the can is completely filled with salt brine. A tin button is soldered over this puncture and the can is cleaned and labeled for market. A barrel of mackerel will "mess" about 180 pounds, which will fill 3 cases each containing 1 dozen 5-pound cans. This work is usually done at the salting houses, the necessary cans being purchased of the can-makers, and a solderer is employed by the hour.

In addition to the 5-pound cans, 3-pound, 1 pound, and 10 pound cans have been used. The 5 pound cans are of two shapes, each of which is about 3½ inches deep; the first is round and 9 inches in diameter, the second is oval and 6½ inches wide and 9½ inches long. The scarcity of mackerel during the past few years has resulted in very few being put up in this way. When mackerel were abundant, as 12 or 15 years ago, a single firm used to ship 1,500 to 2,500 cases of canned salt mackerel each week, but probably not more than 1,200 cases were prepared in 1898.

CANNING LOBSTERS.

On the American continent lobsters were first canned in 1842, at Eastport, Me., by Messrs. Treat, Noble & Holliday. As is frequently the case in the establishment of new industries, the capital was limited, the appliances were crude, and the quality of the product could not always be depended upon. In 1843 the firm secured the services of Mr. Charles Mitchell, who had engaged in canning during the preceding year at Halifax, Nova Scotia, and who had ten years' experience in the same business in Scotland. Mr. Mitchell improved the processes according to methods employed in his native country, and no further difficulty was experienced in preparing a desirable grade of goods. Letters patent were applied for, but the matter was not pressed and the claim was not granted. During the few years following the origin of the business the 1-pound cans of lobsters sold at 5 cents each, and 3½ pounds. live weight, were required to make a 1-pound can. No lobsters under 2 pounds in weight were used.

In 1880 there were 23 lobster canneries on the United States coast, and the output amounted to 2,049,806 cans, worth $238,000; in 1889 it was 1,170,561 cans, worth $142,613, and in 1892 it equaled 1,235,160 cans, valued at $195,941. Since that year the output has been very small, the product during 1897 and 1898 on the whole of the United States coast not exceeding 20,000 cans annually. While there appears to be no accurate data as to the output of canned lobsters prior to 1880, yet it is well known to have been of much greater extent than at present.

Prior to 1870 the industry was confined almost exclusively to the United States, but the growing scarcity and the increased price of lobsters on the Maine coast soon resulted in the establishment of numerous canneries in the British Provinces by New England capitalists. The number of these canneries has greatly increased, and they now furnish nearly the whole supply of this product. The quantity of canned lobsters produced in the Dominion of Canada during the year 1892 amounted to 12,524,498 1 pound cans, valued at $1,758,425, and in Newfoundland 1,258,308 cans, valued at $176,083, making, with the 1,235,160 cans produced in the United States, an aggregate of 15,017,966 1-pound cans, worth $2,130,449. From 1870 to 1893, inclusive, the total product of canned lobsters in the Dominion of Canada was 254,106,936 1-pound cans, with a total value approximating $38,116,000.

It is cheaper to can lobsters in the British Provinces than in the United States. In this country lobsters, when obtainable, do not cost much more than in the Provinces, but wages are at least 50 per cent higher here than in Canada, where the men receive about $1 per day and the women and children about 50 cents per day. In addition to this, there is a duty on tin in the United States, while that article is free of duty in the Provinces.

The process of canning lobsters on the coast of Maine and in the British North American Provinces is as follows:

The lobsters are first boiled in a large vat or kettle about 20 minutes, after which they are heaped on large tables, usually with the backs up, care being taken to have the bodies more or less straightened out. The boiling is frequently done in the afternoon, in order that the lobsters may have sufficient time to cool during the night. The next morning certain men, designated as "breakers," break off the claws and tails from the bodies, throwing the latter with the refuse, for the reason that, though the carapax contains some good meat, it is difficult to extract and clean it. The sweetbreads, however, are generally saved. The claws are then split by the "crackers," using a small hatchet or cleaver, which opens them so that the meat can be readily taken out. Formerly the tail was split in a similar manner to the claws, but at present the meat is punched out from the tail by means of a small "thole" pin, or other suitable pointed implement. The meat is next thoroughly washed in water, the tin cans filled and weighed to insure uniformity, and then covered and cleaned, after which they go to the sealers, who solder the covers down. Next comes the bathing, the most difficult part of the process. The cans are immersed in boiling water for about an hour, when they are taken out and "vented," a small hole being punched in the cover to release the air, after which they are sealed again and boiled for 2 hours longer. They are afterwards allowed to cool, tested to insure their being tight, and then scoured, painted, and labeled. If the cans are boiled in a retort, say at a steam pressure of 15 pounds to the square inch, which is equivalent to 250° F., the time of boiling is reduced about one-half.*

The preservation of "shell lobsters" was originated as early as 1840 in Christiania, Norway, by Jacob March. In that year he took out a patent in his native country for putting them up in such a manner as to exhibit the red color of fresh-boiled lobsters. It appears that he dipped them in boiling salt water until they reached this color, and then made an incision in the soft part under the tail, thus releasing the water within them, and then placed them in hermetically sealed vessels. The process was never employed extensively and was abandoned within a few years.

The canning of shell lobsters in this country was begun in 1879, at Southwest Harbor, Me. This product is the outgrowth of a demand in the British market for whole lobsters for garnishing purposes. Finding difficulty in obtaining lobsters, as commonly prepared for the trade, sufficiently fresh for this purpose, the London agent for one of the leading packing establishments in Maine suggested the idea of meeting this demand, and satisfactory results were obtained after many experiments. The lobsters, 12 and 14 inches in length and of good condition, are selected from the general stock and boiled, the tail bent under the body, and without being removed from the shell are packed in long cylindrical cans suitable for this purpose. The method of boiling is similar to that ordinarily used in canning lobsters, the only difference being

---

*See The Fishery Industries of the United States, sec. v, vol. 2.

that they are boiled a little longer in order that the heat may thoroughly penetrate the shell and preserve the meat.

"Deviled lobsters" in half-pound cans have been prepared at several canneries, the article originating in 1871.*

Twenty years ago London was the principal distributing market of the world for canned lobsters, England, France, and Germany receiving about three-fourths of the entire product of the world, which at that time amounted to about 400,000 cases of 48 1-pound cans. The consumption in the United States, then comparatively small, has steadily increased; so that while the total pack has decreased to about 250,000 cases a year, the United States consumes about 100,000, or 40 per cent.

CANNING SHRIMP.

The shrimp-canning industry was established in this country in 1875 by Messrs. G. W. Dunbar & Sons, at New Orleans, La. Their factory was occupied mainly with the canning of various fruits, and utilized shrimp only during the months in which they are most abundant. As the product became better known the extent of the business increased and canned shrimp is now, next to oysters, the most valuable of the marine products canned on the Gulf of Mexico.

The quality of the product during the first year of the business was unsatisfactory, as the direct contact of the shrimp with the tin caused, during the process of cooking and thereafter, a precipitation of black or dark matter which discolored the shrimp and detracted from their flavor and richness, and the liquid in the shrimp constituted a medium for diffusing the coloring matter throughout the can, so that all portions of the contents were equally affected and discolored. This precipitation or coloring is believed to be caused by the action of sulphur contained in the shrimp on the metal of the can. After much experimenting, Messrs. Dunbar overcame the trouble by interposing a lining between the meat and the can, which protected the tin from the action of any acids contained in the shrimp. They also place the shrimp in the lined can while they are in a dry or moist condition and devoid of free liquid, and seal the can without adding any liquid to its contents. The lining consists of any textile fabric in the form of a cylindrical bag or sack, the diameter of which, when filled, permits it to fit snugly within the can. The use of this lining was protected by Letters Patent 178916, dated June 20, 1876, and Reissue 9957, dated December 6, 1881.

The process of canning shrimp at that time, according to the specifications accompanying the letters patent, was as follows:

The shells having been removed from the shrimp in the usual manner, the fish is thrown into salt water of about 6 , and there remains for an hour, more or less, and from thence to kettles filled with water and brought to a boiling heat, after which they are placed on drippers and cooled and thoroughly rinsed with fresh cold water, from which, so soon as thoroughly dripped and in a moist condition and without the addition of salted or otherwise prepared liquid, they are placed in the sack, the same having been previously arranged in the can. So soon as the sack is filled, the mouth thereof being properly secured, the lid or head is placed in position on the can and immediately sealed. The cans are then subjected to a steam bath or placed in kettles containing boiling water and boiled for two hours at the highest temperature attainable, and which completes the process.

The present method differs somewhat from the foregoing in a few minor particulars. As soon as practicable after being landed from the luggers, the shrimp are

---

* The Fishery Industries of the United States, sec. v, vol. 2.

boiled in salt water for 30 minutes, and separated from the shells by girls, who use only their fingers for this purpose. The shrimp are then dumped on a small platform, and the shell particles, tentacles, etc., are picked off, after which they are passed through a blower on an endless platform, where the remaining dust and other foreign matters are removed. They are next placed in 1-pound cans containing 10 ounces each, the lining having been inserted. When the cans are filled, the lining covers are adjusted and the lids sealed on and the cans placed in a bath of boiling water for 2 hours. On removal from this bath they are ready for labeling.

Messrs. G. W. Dunbar's Sons (successors to G. W. Dunbar & Sons) now have the only shrimp cannery in Louisiana, though for a few years prior to 1891 a factory owned by Messrs. A. Booth & Co., of Chicago, was operated at Morgan City. At Biloxi, Miss., are three canneries which devote considerable attention to preparing shrimp, the first of these having been established about 1880.

In 1897 the output of canned shrimp in Louisiana and Mississippi was 1,395,168 1, 1½, and 2 pound cans, which sold for $156,190.

In 1879 a shrimp cannery was established at Galveston, Tex., by Messrs. Pecor Brothers. To prevent discoloration of the shrimp, instead of placing a lining of some textile fabric between the fish and the can, as in the Dunbar process, this firm lined the can with a coating of asphaltum cement, which was permitted to dry thoroughly. Strips of paper were then cut, and, after being coated with a hot solution of paraffin, were placed within the can so as to fully cover its interior surface.* The can was filled with shrimp, and the subsequent treatment was substantially as hereinbefore described. In 1880 this cannery was reported as having put up 75,000 1-pound cans of shrimp, worth $13,000, but a large decrease in abundance of shrimp in Galveston Bay necessitated its closing down a few years later.

Another method of lining the inside of tin cans to prevent the direct contact of the shrimp with the metal was devised by Louis Lenglet, of St. Louis, Mo. This consists in providing a lining of corn husks, covering the inside annular body of the can as well as the top and bottom. It is claimed that corn husks have the advantage of requiring no previous treatment with acids to neutralize or destroy any peculiar odor or flavor of the material, and that such lining keeps its shape well, is sufficiently thin and flexible, and not expensive.

Shrimp are also canned in Japan by a process similar to that employed in this country. Specimens of the Japanese product were exhibited at the World's Fair, Chicago, 1893.

CANNING CRABS.

The canning of crabs originated in this country about 1878, with Mr. James McMenamin, then located at Norfolk, Va. He first attempted to follow the methods used in canning lobsters, but not meeting with satisfactory results he started out on original lines, the product being first placed on the market in the fall of 1878. On account of the greater abundance of crabs in that vicinity, Mr. McMenamin moved to Hampton, Va., in April of 1879, and began operations on a large scale. In that year another cannery was established at Hampton by Mr. T. T. Bryce. In 1879 these two canneries used 6,000,000 crabs, producing 84,000 2-pound cans, worth $16,800. One

---

* Letters Patent No. 226547, dated April 6, 1880.

or two other canneries have been established since, the principal one being at Biloxi, Miss. The present annual output in the United States is valued at about $45,000.

The season for crab canning in the Chesapeake begins in April and continues until October, except that sometimes the work is discontinued during June and July on account of the large number of crabs with spawn, in which condition they are not suitable for canning.

The crabs are placed in open slat-work cars, usually of a size sufficient for holding about 250 dozen, and are rolled into a steam box, where they are cooked 20 or 30 minutes, when they become red. The cars are then rolled out of the steam chest and the crabs passed to the "strippers," who remove the outer shells, viscera, and small claws. The crabs next pass to the "pickers," consisting principally of women and children, who remove the meat from the bodies and the claws, crushing the latter with the handle of the knife employed in the work. Different operatives are employed for picking the bodies and the claws. The pickers generally receive from 2 to 3 cents per pound for this work, and the most skilled among them prepare 40 or 50 pounds per day, but the average quantity is about one-half of this.

After being weighed the meat is placed into cans of two sizes, 1 pound and 2 pounds, about 12 crabs being required to each 1-pound can. The cans are sealed, boiled for half an hour, and then vented. They are at once resealed and boiled for a second time (making the third time that the meat has been cooked) for about 2 hours. The length of the second boiling may be shortened by increasing the possible temperature of the water, which is usually done by adding chloride of calcium thereto.

When the cans have been properly cleaned and labeled, they are packed in cases holding 48 1-pound or 24 2-pound cans, and sold wholesale at about $8 and $6 per case, respectively. A package of shells usually goes with each case of cans, four shells being allowed for each 1-pound can.*

It is stated that in 1891, 3,838 barrels of crabs, worth $6,141, were canned in Virginia, yielding 1,095 cases of 1-pound cans and 2,880 cases of 2-pound cans, worth $7,884 and $16,128, respectively. In 1890, 6,365 barrels, or 2,386,256 crabs, worth $5,090, were used, producing 1,277 cases of 1-pound cans and 5,472 2-pound cans, worth $9,194 and $30,613, respectively. In 1897 the output was 1,992 cases of 1-pound cans and 3,898 cases of 2-pound cans, worth $14,177 and $22,061, respectively.

When intended for nearby markets, and for consumption within 4 or 5 days, the crabs are not usually canned, but the meat, after being removed from the shell, as in case of canning, is placed in 10-gallon tins, a piece of ice placed in each tin to keep the flesh cool, and the tins placed in refrigerators. When orders are received, the ice is removed and the meat repacked in tin or wooden buckets of suitable size, containing ½ gallon, 5 gallons, 8 gallons, and 10 gallons, and the buckets placed in a small barrel in much the same way that ice cream is packed for shipment, ice being placed all about the can or bucket. Sometimes a small lump of ice is placed in the bucket, and some dealers also use an antiseptic, such as boracic acid. The crab meat is sold at $1 to $1.50 per gallon, wholesale, and with each 10-gallon bucket half a bushel of empty shells are sent without extra charge. This trade is carried on at Annapolis, Crisfield, Hampton, Norfolk, and other ports on the Chesapeake Bay.

---

*Fishery Industries of the United States, sec. v, vol. 2, pp. 646-647.

SARDINES.

The most valuable of the marine products canned in oil is the preparation of small fishes of the herring family, in the form known generally as sardines. This industry was established at Nantes, France, about the year 1834, and it was introduced in the United States about 1875. It has reached its greatest development in Brittany, the most costly brands on the market being canned on that coast. Sardines are now prepared in France, Spain, Portugal, Italy, Norway, United States, Brazil, Mexico, etc., but the industry has three principal geographical centers: (1) the Mediterranean coasts, (2) the Bay of Biscay and the Atlantic coasts of Spain, and (3) the coast of Maine. In each one of these regions methods are employed quite at variance with those used in the others.

The preparation of sardines began in the United States about 1875 and has gradually increased in extent, though it was confined within comparatively narrow limits until 1880, when 13 new canneries were established, there being only 5 operated previously. In 1886 there were 45 canneries, and since that year the value of the output has averaged about $2,000,000 annually, all prepared on the coast of Maine, and nearly all in the counties of Washington and Hancock. On the Maine coast 37 factories were engaged in canning sardines in 1889, 46 in 1892, and 60 in 1898. A few sardines are prepared also on the Pacific coast of the United States.

SARDINE CANNING IN MAINE.

The herring used in the sardine industry in Maine are from 5 to 10 inches in length, and are taken from the middle of April to the middle of December, by means of weirs, and to some extent in gill nets and seines. The present method of preparing these fish as sardines is as follows:

On reaching the factory the fish are at once distributed along the middle of the cutting tables where they are quickly decapitated and eviscerated. It was formerly customary to remove the tails, but this is no longer practiced. As each fish is dressed it is thrown into a cutting box placed under the edge of the table. They are next washed by being dipped with a scoop net into a washing tank, through which water is constantly running, and are immediately immersed in strong brine from 15 minutes to 1 hour, the length of time depending upon the size and fatness of the fish, their freshness, and the condition of the weather. In cold weather, owing to their firmness, they must be salted longer than in the summer. As soon as they are sufficiently "struck" the herring are removed from the brine and allowed to drain in baskets. They are afterwards carried to the flaking room and arranged upon flakes, which are wooden frames about 3 feet long and 22 inches wide, filled in with wood or galvanized wire stretched across and separated by 1 or 2 inches, so as to give a free circulation of air and to touch the fish at only a few points, in order that evaporation may go on from all parts of the body. Each flake holds about 110 fish, placed in rows with the tails in the same direction, so that when fixed in the drying room, with the anterior part lowest, the moisture will more readily drip from them.

Next comes the drying process, which is of much importance, and great care must be observed that no decomposition occurs before it is completed. Originally the fish were dried in the open air by action of the sun, as is the present practice in France, but owing to the moist atmospheric conditions about Eastport, even those factories that make use of this method whenever possible are obliged to supplement it by artificial means. The factories are, one after another, discarding entirely the sun-drying process, occasionally using a drying room, but more frequently adopting a patented drying apparatus similar to a baker's rotary oven. Drying rooms are usually located on the top floor of the cannery, with movable racks for holding the flakes obliquely, each rack containing 40 or 50 flakes, placed about 3 inches apart and directly over each other. The room is supplied with a constant current of warm, dry air, brought from stoves or furnaces in the lower part of the building by means of large pipes, and which finally escapes through ventilators in the roof.

The oven that has been quite generally adopted in the sardine canneries was introduced by Henry Sellman in 1880,* and is similar to an ordinary rotary oven of large size, and serves not only to dry but at the same time to cook the fish. It is about 15 feet square and 18 feet high, and contains 6 or 8 skeleton iron frames attached to arms extending from a cylinder and which remain in a horizontal position while revolving in the oven, like the cars of a Ferris wheel. In these ovens the fish are subjected to a temperature of about 250° F. for 10 to 25 minutes, according to the size of the fish, but the time required for drying varies. In open air or in a drying-room it may take the greater part of a day; in a furnace heated drying apparatus from 2 to 7 hours, and in an oven only a few minutes, as before stated. The use of ovens is becoming more popular, and most of the canneries are now fitted with them.

When fish are oven-dried they need no further cooking, but are at once cooled and packed in cans. In other cases they are placed in shallow wire baskets or other proper receptacles and immersed in oil of suitable quality and heated to a temperature of about 220° F. This is for the purpose of frying and expelling from the fish all moisture remaining in them after the drying process. Cotton-seed oil is used mostly, and it is placed in a pan to the depth of about 2 inches and the fish immersed in it from 1 to 3 minutes. This oil can be used only a short time, since water and gluten from the fish pass into it and injure its flavor. For this reason the pan must be cleaned frequently and the oil renewed. The oil is boiled either by direct furnace heat or by the passing of steam through coils of pipe in the frying tank. The latter was introduced in 1884 and has many advantages over the old method of direct furnace heat. About half of the factories which fry their fish do so by means of steam, and, as is the case with other improved methods, the number is increasing.

Mr. R. E. Earll states on page 178 of U. S. Fish Commission Bulletin for 1887:

It is said that the fish which have been fried have a better flavor, and, having absorbed more oil, keep longer than those baked in an oven. It is claimed, however, by those using ovens, that by the baking process very much depends upon the skill of the baker, and that at its best it may produce results equal if not superior to those of the old system. It appears that the first fish fried in a given quantity of oil are better than the best baked fish, but that, as it is necessary, in order to keep the expenses within reasonable limits, to use the same oil for frying successively a great many pans of fish, the fluid soon becomes filled with scales and small particles of fish, which burn on the bottom and impart to the product a bitter and unpleasant taste. In baking, on the other hand, when it is properly done, the fish are all of a quality equally good.

Instead of the ordinary methods of cooking, some factories employ an endless belt, 200 feet long, which runs in a wooden case 100 feet long. At one end of this case is a revolving fan which forces a blast of hot air over the fish that have been spread on

---

* Letters Patent No. 223682, dated January 20, 1880.

the belt at the other end of the case. After passing along the belt once the fish go into the bath of boiling oil and are treated in the usual manner. With this apparatus the several flakers required by the old method are represented by one woman, who spreads the fish on the belt, and a man who turns a crank which moves the machinery.[*]

After leaving the frying-pan or the oven and draining and cooling, the fish are assorted according to size, and those of like size are placed in tin cans or boxes previously filled with oil, or, in some cases, mustard sauce or vinegar with spices. Up to within a few years, although other oils were used in the frying, the sardines were packed in olive oil, either alone or mixed with inferior kinds, but at present its use has been almost entirely superseded by cotton seed and nut oils. This change is accounted for by the facts that the heavy duties make olive oil very expensive, that it will not keep as well as cotton-seed oil, and that the latter can be made exceedingly palatable. It is claimed by some canners that even at the same price cotton-seed oil is more desirable for the Maine sardines, because the fish imparts its distinctive herring flavor to olive oil much more readily than to cotton seed, the latter covering it up somewhat. It is stated that at some of the canneries even tallow oil and herring oil have been used. Peanut oil, which is sometimes used, is said to be even better than cotton-seed oil. The oil is usually flavored to the taste by adding lemon, sugar, and various spices. The olive oil used in France for sardines is said to be often largely adulterated with American cotton-seed oil, as well as with palm and other oils.

In packing the fish, those of the most desirable size are packed with a dozen in each can; the number is never smaller than seven or eight. The smaller fish are generally packed in oil in "quarter-cans," which are 4½ inches long, 3 inches wide, 1 inch deep, and contain from 9 to 12 herring measuring from 3½ to 4 inches when dressed. The larger fish, measuring from 8 to 9½ inches in length when whole, or from 4 to 4½ inches dressed, are occasionally placed in oil, but more frequently are put in mustard, spices and vinegar, tomato sauce, or other condiments, in "half cans," holding from 10 to 16 fish. These cans are 4½ inches long, 3½ wide, and 2 inches deep. Occasionally "three-quarter" cans of oil sardines, or in tomato sauce, and "half" or "three-quarter" cans in spices are put up, and in rare instances small fish are put up in mustard or spices in quarter-cans.

When the cans have been filled with fish the covers are sealed on, and the filled cans are then ready for bathing, and are placed in boiling water, where they are allowed to remain from 1½ to 2 hours, according to the size of the cans. Fish prepared with spices must be boiled longer than those prepared entirely with oil. The time of boiling may be considerably reduced by introducing a proper quantity of chloride of lime or other chemicals into the water, by which the temperature may be raised to upward of 250° in the open air. Some canneries accomplish this by using a closed bath. By either of these methods the time can be reduced to about 30 minutes.

Formerly, after the bath, the cans were punctured, to allow the inclosed air to escape, and the puncture was thereupon closed with solder. In this process, when the cans are punctured the escaping air carries a portion of the oil with it, so that when the cans are opened the fish are found to be only partly covered with oil, and consequently not in a state of perfect preservation. If the can has been properly sealed, the top and bottom become level or horizontal when boiled the second time. The fact that it does not expand into a level position is sufficient evidence that there is a defect

[*] Bull. U. S. Fish Commission for 1890, p. 117.

in the soldering, and it is at once resoldered, punctured in two or more places, and placed in hot oil until it is again filled with oil, when the punctures are soldered.

In 1884 it was discovered * that the process of venting could be avoided by making the can with concave top and bottom. The depression of the middle part causes the air in the cans to collect about the edges of the top, and the heat of the soldering tool heats the air and causes it to expand and escape in front of it as it passes along the edge, so that when the soldering is completed the air will be sufficiently expelled. Venting is no longer practiced in preparing the ordinary quarter size, but it is generally adhered to in the treatment of the half and three-quarter cans.† Some of the factories partly immerse the half and three-quarter cans in boiling oil, driving out the air and rendering venting unnecessary.

In the specifications forming a part of the Letters Patent No. 223682, dated January 20, 1880, issued to the inventor of the rotary oven for baking, the following account is given of the methods of preparing sardines on the Maine coast, and the improvements effected by using that oven:

> After the fish are landed they are subjected to the process of decapitation and disentrailment and salting for a suitable period. They are then washed clean and placed in shallow baskets to drain, after which they are separately spread on lath or other suitable frames for drying to a certain extent. After the fish have been sufficiently dried by exposure to the atmosphere or to an artificial current of warm, dry air they are placed in shallow wire baskets, or any other suitable receptacle, and immersed in oil, suitable in quality and heated to a certain degree, for the purpose of frying and expelling from the fish any parts of water which remain in them after the drying process. They are then allowed to drain, and are packed in tin cans. This mode of drying by natural or artificial currents of air and frying the fish in oil is, for reasons hereinafter stated, very deleterious to the quality of the article of fish to be put up, and the invention herein set forth tends to do away with and overcome the former objectionable method. The fish used for the purpose indicated are of a very tender and delicate nature. They do not admit of much handling, and, owing to this delicacy of nature, are subject to very rapid decomposition, as they should be salted but very slightly.
>
> The process of drying the fish, either in open air or by an artificial current of warm dry air, takes so much time, that decomposition of the fish to a greater or less extent is unavoidable, as 3 to 24 hours are consumed in drying the fish sufficiently by the modes indicated.
>
> In frying the fish in oil, as now practiced, the quality of the oil in which quantities of fish are fried is rapidly deteriorated by the water from the fish, which is not evaporated, and from the gluten from the fish passing into it. A large percentage of the fish is also lost by breaking during the process of frying in oil. In our improved process the fish, after landing, are decapitated, disentrailed, salted, and washed. They are then spread on wire netting or other frames made of suitable metal and of any suitable size. They are then subjected to a process of steaming by live steam, which is injected from a steam-boiler into an upright chamber of suitable size, lined with sheet metal, and provided with narrow internal flanges or shelves, upon which rest the wire frames which hold the fish. The steam passes through the closed box and escapes through an opening in the side or end opposite to where it is introduced. A door opening outward is also provided for obvious reasons.
>
> The time consumed in this process is from 10 to 20 minutes, according to the power of the steam employed, and may be performed within 2 hours after the fish are first landed. This steaming process has the effect of evaporating the water from the fish in a much more thorough manner than by the old process. It has also the effect to prepare the fish for the subsequent baking process, and by killing any germs in them preventing rapid decomposition, keeping them sweet, and retaining their natural flavor. After the steaming process the fish (which remain on the same frames on which they were steamed) are subjected to the baking heat of a revolving reel oven, operated by steam or any other power, until they are fully cooked or baked. They are then taken from the revolving reel oven, cooled a certain time, and packed in tin cans, which are supplied with fine oil, mustard, sauces, spices,

---

* See Letters Patent No. 289710, dated June 3, 1884.

† See the Fishery Industries of the United States, section v, vol. 1, pp. 511–518, and Bulletin U. S Fish Commission, 1886, pp. 177–179.

or vinegar, as desired. The cans are then soldered and subjected to the action of a bath of boiling water for a certain period, for the purpose of expelling all air from the cans by the usual process.

The essence of the whole mode of procedure consists in preserving the fish against decomposition by steaming and baking, as set forth, thus preventing breaking of the skin, curling and breaking of the body, and thus evaporating from the fish all water, and then, while in this baked condition, subjecting them to the preservative process of canning similar to that practiced with sardines, inclosing in tin cans with oil, mustard, spices, etc.

An appliance recently devised for testing the cans before they are filled is thus described by Mr. Ansley Hall:

It consists of a cylindrical tank about 5 feet in length and 1 foot in diameter, fixed in an upright position at the end of a table. The tank is filled with water to within about 18 inches of the top by means of a pipe leading from the boiler of the engine. Air is forced through another pipe into the space above the water by the air pump which supplies air for oxygenizing the flame of the kerosene-oil stoves used in soldering. The pressure of air, which requires to be about 12 pounds, and the quantity of water are regulated by steam and water gauges. On the table, a few feet from the tank, is a tin pan or tray, in the center of which is a rubber pad, a little larger than a sardine can. A pipe fitted with a valve leads from the tank and passes up through the pad from the under side of the table. The can when tested is placed bottom upward over the nozzle of the pipe and held in position by pressure applied with a lever worked by the foot. The operator then turns a thumb piece on the pipe, which opens the valve and lets a small stream of water into the inverted can. If it is not perfectly tight, the leak is immediately disclosed by the fine jet of water which passes through it. The water, after being used, escapes by a waste pipe in the tray. One advantage of this method is that it shows which class of solderers has done the poor work, whether the seamers or can-makers, and the defective cans are returned to them for the leaks to be mended, after which they are again tested in a similar manner. If any cans are imperfect after coming from the bath, the fault is known to lie with the sealers. An improvement is contemplated by arranging the valve to open with the lever when the pressure is applied, and thus avoid the movement of the hand in turning the thumb piece. The apparatus costs about $15, and is operated by one person.

On the Maine coast many sardines are put up in mustard and in spices, usually with a quantity of the best quality of vinegar. While these are considered equal to the sardines in oil, they are usually sold at a lower price. In 1892, 154,051 cases of sardines in mustard were prepared, the value of which was $457,445; and 10,020 cases of sardines in spices, worth $32,425, were prepared.

Nearly every year a small quantity of sardines are put up in tomato sauce, but these do not keep very well and the demand for them is small. In 1889 the Maine canneries prepared 279 cases of them, worth $762.

The following shows in detail the sardine output of Maine in 1889 and 1892:

Description.	1889.		1892.	
	Cases.	Value.	Cases.	Value.
Sardines in oil:				
Quarters	261,940	$1,013,877	396,428	$1,455,245
Halves	9,881	56,716	6,614	31,870
Three-quarters	1,025	4,100		
Sardines in mustard:				
Quarters	4,127	20,635	5,031	21,582
Three-quarters	158,069	553,212	149,020	435,863
Sardines in spices:				
Quarters	1,062	5,310	543	2,145
Three-quarters	5,609	21,034	5,705	18,011
One pound	74	277		
Two pound	10	26	710	1,643
Three pound			1,042	3,126
Odd sizes	36	126	2,000	7,500
Sardines in tomato sauce:				
One pound	256	704		
Two pound	23	58		
Total		1,676,105		1,976,985

From Mr. Hall's excellent report on the "Herring Industry of the Passamaquoddy Region" is extracted the following tabular statement, showing the cost per case of quarter-oil sardines in 1895, the statement being prepared on a basis of seven cases for convenience in manipulating some of the items.

Statement of the cost per case of quarter-oil sardines in 1895.

Material:	**Labor— Continued.**
Tin plate for 7 cases, at $3.40 per box............ $3.43	Trucking 7 cases, at 1 cent per case.............. .07
Decorating 35 sheets of tin plate................. .58	
Oil for 7 cases, at 30 cents per gallon........... 2.10	Labor for 7 cases 6.43
Solder for 7 cases, at 25 cents per case 1.75	Labor for 1 case .... .92
Fuel for soldering, soldering coppers, and acid... .21	
Shooks and nails for 7 cases..................... .53	**Expenses of shipping and selling:**
Fish, at $3.11 per hogshead 1.19	Freight on 7 cases at 10 cents per case...... .70
Coal, wood, sawdust, and salt .... .12	Commission on 7 cases, at 5 per cent. .... .87
Waste of material, 1 per cent .... .10	Discount of 1 per cent for cash payment........ .17
	Fire and marine insurance...................... .06
Material for 7 cases......................... 9.92	
Material for 1 case.......................... 1.42	Expenses on 7 cases 1.80
	Expenses on 1 case .... .25
Labor:	
Cutting, rimming, and bending tin .... .20	Total cost of 7 cases........................ 18.15
Cutting two-thirds of 1 box of tin on dies ...... .14	Total cost of 1 case......................... 2.59
Seaming cans for 7 cases, at 5 cents per case.... .35	
Making cans for 7 cases, at 12 cents per case .... .84	**Summary of the cost per case:**
Sealing cans for 7 cases, at 30 cents per case.... 2.10	Material 1.42
Cutting and flaking fish for 7 cases, at 10 cents per case...... .70	Labor .... .92
Packing 7 cases, at 10 cents per case............ .70	Cost at cannery............................. 2.34
Making 7 cases, at 1 cent per case ...... .07	Expenses of shipping and selling.............. .25
General labor on 7 cases, at 18 cents per case 1.26	Total cost per case......................... 2.59

Mr. Hall further states:

An estimate similar to the above, made in 1886, showed the cost of quarter-oils at that time to be $4 per case at the factory. The material then cost $2.83 and the labor $1.17, whereas the material now costs $1.42 and the labor 92 cents, a total of $2.34 per case. The cost per case was therefore $1.66 or 41½ per cent less in 1895 than in 1886. In the estimate for 1886 the fish were reckoned at $6 per hogshead, but the average for that year was about $9; hence it is probable that the actual difference in the cost of production was even greater than these figures show. It will be noticed that the reduction in cost since 1886 has been more largely in material than in labor, the cost of material in 1895 being nearly 50 per cent less than in 1886, while that of labor was only 21½ per cent less. Of the total difference, 85 per cent is in material and 15 per cent in labor. Prior to 1886 the cost of manufacturing sardines was somewhat greater than it has been at any time since, but it was probably never more than about $7 per case. The price of the products has therefore fallen much more rapidly than the cost of production, and consequently the profits have been constantly diminishing. It was not until after 1880 that the cheapening of the cost of the products became an imperative necessity.

In the strong competition between the various manufacturers the quality of the goods has in a measure been sacrificed to the interest of producing large quantities. When the industry was first established, it was the ambition of the packers to make the quality of the domestic product equal, if possible, to that of the sardines imported from France and other countries, and thus secure at least a part of the trade which was then wholly supplied by the foreign manufacturers. It was also hoped that when the supply should exceed the demand of the home market the surplus stock might be exported. To this end, therefore, the best quality of material was used, and the greatest care was exercised in the methods of preparation, and for a few years the quality of sardines put up at Eastport, while somewhat inferior to the best, was equal to that of the average brands imported. Had these efforts been continued until the present time, it seems not improbable that a still higher standard of excellence would have been attained. Attention was, however, soon directed toward reducing the cost of the products. One of the most important changes made was that of substituting cotton-seed and nut oils of various kinds for olive oil. This practice began to some extent before 1880, but did not become general until after that date. The cheaper oils were first introduced for trying the fish, but in a short time they were also used for packing them in the cans. Changes have also been made in the methods of preparing sardines in order to render the performance of the work more rapid and thus increase the capacity of the canneries at a reduced ratio of cost.

There appears to be some doubt in the minds of the packers whether or not the herring (*Clupea harrugus*) which is used for sardines on the coast of Maine is susceptible of being so prepared that it will be equal in quality to the best imported sardines. It may be quite safely asserted that the character of this species does not offer any insurmountable barrier. The sardine (*Clupanodon pilchardus*) used in France, which is the young of the pilchard, the English sprat (*Clupea sprattus*), and the California sardine (*Clupanodon cæruleus*) all belong to the same family of fishes as the herring, and it is probable that any superiority which one may have over another, when packed in oil, depends more on the quality of the oil and the method of treatment than on the natural characteristics of the species. That the experiment is a hopeful one as to its effect on prices is indicated by the fact that in 1895 considerable quantities of goods were so improved in quality that they were sold for at least 50 cents more per case than the best average brands. This improvement consisted mainly in frying the fish and in the exercise of greater care in their preparation. There is no doubt that their value might have been still further enhanced by the use of either olive oil or olive oil blended with other oils of a delicate flavor.

For further information regarding the preparation of sardines in Maine, and the present condition of that industry reference is made to Mr. Hall's above-mentioned report (The Herring Industry of the Passamaquoddy Region, Maine, by Ansley Hall, United States Fish Commission Report for 1896, pp. 413–487).

SARDINES ON THE PACIFIC COAST.

Within the past few years sardines of choice quality have been prepared at San Pedro, Cal. The following account of the species utilized and of the industry is from a recent report (Bulletin for 1894, pp. 227–230) of the United States Fish Commission:

The California sardine (*Clupea sagax*) is very closely related to the sardine of Europe (*C. pilchardus*), from which it chiefly differs in having no teeth and less strongly serrated scales on the belly. It attains a length of nearly a foot. It is found along the entire Pacific coast of the United States. The fish is, however, most constant in appearance and most abundant on the southern part of the coast, and it is doubtful if it exists in sufficient numbers to maintain a regular fishery north of San Francisco. Even at that place the supply is uncertain. While there have been periods of years in which the sardines were found in San Francisco Bay in large quantities, and for a considerable time in each season, for the past five years they have been very scarce.

The distribution of the anchovy (*Stolephorus ringens*) is similar to that of the sardine. It occurs in abundance along the entire coast, and is often found in enormous quantities in Puget Sound, San Francisco Bay, and elsewhere. It reaches a maximum size of about 7 inches. In most places it is known as the anchovy, but in Puget Sound, according to Swan, it is called "sardine."

The natural advantages which the west coast possesses for the canning of sardines and other similar fish are unusually good, and are superior in some respects to those of the east coast. At least the two fishes named, the sardine and the anchovy, suitable for canning as "sardines," occur in large quantities, the first named very closely resembling and being an excellent substitute for the sardine of southern Europe. The dry atmosphere and other climatic conditions of the southern coast of California are very favorable for the preparation of a good grade of canned fish. The culture of the olive supplies a native oil of superior quality, which is essential in the canning of the best goods. Another item of importance to canners in this connection is the abundance of cheap labor. The chief desideratum in the establishment of a factory for the canning of sardines (and other similar species) is a regular supply of fish during a certain period. This is thought to be of greater importance than an abundance of fish at uncertain or irregular intervals.

While the sardine ranges along the whole western coast of the United States, and is at times very abundant even as far north as Puget Sound, it is doubtful if in Washington or Oregon a supply sufficiently large and regular exists to warrant the machinery, etc. This matter has already received the consideration of some salmon canners; but the general canning of sardines by salmon packers is not anticipated as long as the supply of salmon lasts.

Personal observation and inquiry, the testimony of fishermen and dealers, and the studies of ichthyologists afford ground for the belief that the successful operation of a sardine cannery can not be expected any farther north than San Francisco, and the history of the industry at that place seems to indicate that the northern limit of satisfactory work is even farther south. South of San Francisco, the prospects of profitable business appear to be in direct relation to the latitude; the more southern the location of the cannery the more constant and abundant the supply of fish.

In 1889, a canning factory was established in San Francisco which continued in operation until August, 1893. During the five years in which the cannery was run the yearly pack was from 5,000 to 15,000 cases. The canned fish consisted chiefly of anchovies in oil in ½-pound cans and large sardines in 1-pound and 2-pound round cans. The fish consumed at the factory were caught in San Francisco Bay with haul seines. In the earlier years sardines small enough for use in quarter-pound cans were obtained, but during the last two years of the cannery's existence no sardines of size suitable for "quarter oils" could be had. This was the chief reason for closing the works.

In this region sardines are found throughout the year. They "show" at the surface at times, and thus permit the use of the purse seines. They sometimes go in immense schools. Single hauls of several tons are often made, and 10 tons have on several occasions been taken at a single set of the seine, such a catch being obtained about May 1, 1891. In December, 1893, several very large bodies of sardines were observed, and a haul of 10 tons of small-sized fish was taken. From January to June the fish appear to gradually increase in numbers. Some schools are made up of fish of uniform size, while in others they are mixed. The smallest fish caught are 4 inches long, the largest 12 inches, the average 7 inches.

The condition of the fish in regard to fatness varies considerably with the season. Mr. J. H. Lapham, the president of the fish company operating the cannery, states that in December, 1893, when the canning began, the smaller fish were poor, while the larger ones were fat. In January and February conditions were about the same. In March the smaller fish began to improve, continued to grow fatter through April and May, and in June sardines in excellent condition, suitable for "quarter oils," were taken. In May 4 or 5 tons of large fish that were poor were seined on one occasion. The factory is under the superintendence of an experienced fish canner from Maine. It is a large two-story structure, with a salting house attached. The plant is worth about $10,000.

The principal processes to which sardines are subjected before emerging as the canned product are as follows: When the fish are unloaded from the vessel they are received into a large, airy room, where the cutting and washing are done, and then transferred to the second floor by means of an elevator. There they are next arranged on latticed trays (32 inches square) and dried. If the weather is fair and the atmosphere dry, the drying is done in the open air, occupying, as a rule, about 2½ hours. On rainy days, or when the air is especially humid, drying is accomplished inside the building by means of steam, which requires about 10 hours.

After drying, the fish are placed in wire baskets (22 inches long, 18 inches wide, 3 inches deep) and immersed in boiling oil for 2 to 6 minutes, depending on their size. The oil is contained in a shallow sink, into which the wire baskets fit and are lowered and raised by means of long wire handles. The boiling of the oil is done by means of a steam pipe entering at the side and running under the sink. After draining and thoroughly cooling, the fish go to the packers, thence to the sealers, thence to the bathmen, and, after cooling and testing for leaks, to the boxing room.

The cutting of the fish is done by men and girls, the average number of whom employed is 25. They are paid by the basket or the bucket of cut fish, and by working steadily earn about 25 cents an hour. The flakers number 12 to 14, and are the same girls who pack the fish in the cans. Ten men act as sealers and can-makers, and 10 others are employed in the remaining branches of the work.

The sizes and grades of canned sardines placed on the market from this cannery, and the wholesale prices received, are as follows: quarter oils, 100 cans to a case, $6.50 to $8.50 per case, according to the quality of the oil; half oils, 50 cans to a case, $5.60 per case; 2-pound oval cans, with mustard spices, and tomato sauce, $2.25 per dozen cans.

MENHADEN AS SARDINES.

Some years prior to the establishment of the Maine sardine industry the extensive market in the United States for sardines led to numerous attempts to obtain an article that would compete with the foreign product. Among other species the immature menhaden was tried with considerable success. The American Sardine Company prepared this species quite extensively from 1872 until the development of the sardine industry at Eastport. In 1871 the company located a factory near Port Monmouth, N. J., and for nearly a year experimented with various processes with a view to removing or softening the numerous bones in the menhaden without the use of vinegar or other acids. They adopted a process, described in Letters Patent No. 127115, dated May 21, 1872, which consisted in successive steaming, combined with an intervening cooling, which softened the bones so that they might be eaten without inconvenience.

Their treatment of the fish was as follows: When landed, which must be very soon after they are removed from the water, the fish are cleaned, scaled, and dressed, and slightly salted in hogsheads. After remaining in salt a few hours, depending on the temperature and the size of the fish, they are placed in cooking cans, which are a little larger than the market cans, and put in a steam chest, where they are subjected to a temperature of 220° F., or thereabouts, for 2 or 3 hours. On removal they are placed on tables to cool and drain for 5 or 6 hours, when they are packed in tin cans suitable for market, and the cans are then filled with oil, after which the covers are soldered on. The cans and their contents are heated in a steam chest for a length of time depending on the size of the fish, then vented, when the cans are ready for labeling and boxing.

In 1873 the American Sardine Company prepared about 30,000 dozen cans in the manner above described, and several other factories were also engaged in this business in the same locality.* During the last fifteen years, however, menhaden have not been used for the preparation of sardines in this country, herring having been found much more suitable for this purpose.

FOREIGN SARDINES.

The importation of sardines into this country was begun about 1836. By 1858 it had reached a value of $250,000 annually, and from that year to 1898 it aggregated $29,867,457. The great bulk of these came from France, with much smaller quantities from Portugal, England, and Norway.

The general method of preparing these fish in France is as follows: On arrival of the fish at the factory they are placed on tables or platforms and lightly sprinkled with salt, just enough salt being used to prevent the fish from becoming slimy. The heads, tails, and intestines are removed, and the fish are immersed in weak brine for 1¼ to 2 hours, when they are thrown into small baskets and rinsed in clear water. Next they are placed on small gridirons, and again rinsed and laid aside to dry. The drying is best done in the open air, but when the weather prevents it is done in a specially prepared oven. As soon as sufficiently dry the fish are cooked in oil and then packed with olive oil in tin boxes and cooked and vented, as in case of the Maine sardines.

In France sardines in oil are sometimes mixed with truffles. They are also prepared with tomatoes and sent in small quantities to the New York market, but the chief export in this form is to Mexico. Sardines preserved in butter are quite good, but the butter is generally of inferior quality, and it is necessary to remove it before serving the sardines, and the box must be heated to melt the butter, so that each fish may be removed entire. Sardines preserved in vinegar require to be washed before serving. The addition of oil renders the fish more palatable, though the sardine retains the taste of the vinegar and its flavor is partly destroyed.

Boneless sardines (put up at Concarneau and Davorneney, France) are quite popular in the United States markets, but they are not prepared in this country. The method of preparing them is only a slight modification of the general process. When the sardines are about half dried in the sun (if dried in an oven they can be

---

* Report of U. S. Fish Commission, 1877, pp. 137-138.

boned only with difficulty and loss), the backbone is broken with a pair of pincers near the root of the tail; then by evenly and tightly squeezing it with the fingers it is loosened along the entire length. By this manipulation the whole bone system is loosened, and, commencing at the neck, by the use of a pair of pincers the backbone and the connecting bones can be readily removed.

The present unsatisfactory condition of the Maine sardine industry suggests the advisability of preparing the fish after the manner of the Norwegian smoked sardines. The fish used in Norway is the sprat (*Clupea sprattus*), which is very much like the small Maine herring (*Clupea harengus*); but among the sprat the seines catch many herring (*Clupea harengus*), which are treated in the same manner as the sprat. The industry is prosecuted along the southern and southwestern coasts of Norway and is centered at Stavanger. When the Norwegians began canning these small fish they copied the French methods and put up a product somewhat like the Maine sardine on the market at the present time. But the fish lacked the delicate flavor necessary for competition in Europe with the products of France and Portugal, and in order to cover up the herring flavor they tried smoking them. The quality of the article proved exceedingly satisfactory, and the output now amounts to several hundred thousand cases annually. The method of preparation is as follows:

The fresh fish, as soon as practicable after capture, are put into strong brine for 4 or 5 hours, and then strung on a small iron rod, drained and dried in the open air. They are next placed in the smokehouse, where they are dried for a few minutes by lightly warmed smoke, the temperature not exceeding 100° F. Then the fires are increased and the fish are hot-smoked, being cooked without breaking the skins, so as to hold all the juices. The completion of the smoking is determined by pressing the flesh of the fish, which should separate into flakes. The cutters then remove the heads and tails and pass the fish to the packers. The latter assort and put them in the cans with olive oil, after which the cans are soldered. In steaming, a large number of sardine cans are placed at one time in the cooking tank, where they remain from 20 to 30 minutes, and are then removed and the air-hole soldered. They are next again steamed for a few minutes, cleaned and labeled. Before selling, the cans must remain for 2 or 3 months in storage, so that the oil may have sufficient time to penetrate through the flesh.

As a result of his investigations into the French methods of preparing sardines in oil and his subsequent experience with the preservation of pilchards in Cornwall, Mr. C. E. Fryer, inspector of fisheries for England and Wales, makes the following suggestions for the preparation of sprats (*Clupea sprattus*), which in many particulars resemble the small herring of Maine:

The fish should be landed in as fresh a state as possible, spread on the floor, and sprinkled with salt. They should then, without delay, be beheaded and gutted (all bruised fish being rejected), thoroughly washed, and immediately placed carefully in vats, with a thin layer of coarse British salt between each layer of fish. Here they should remain for 1 or 2 hours, after which they should be taken out, again washed, and ranged in specially prepared wire baskets (*grilles*) to dry.

One great object to be aimed at is to handle the fish as little as possible, and to pass them through the preliminary stages with all speed. With this object the baskets into which the fish are thrown by the "gutter" should be of a size to be easily handled, and should be constructed of open wicker-work, so that the fish in them can be effectually washed by merely plunging the baskets into an open tank plentifully supplied with fresh water. The wire drying baskets are so contrived that the fish will not need to be touched by the hand again after they are once placed in them till they are ready to be packed in the tins. As the wire receptacles are filled with fish they are ranged in the sun, or under shelter in a dry atmosphere in wet weather, and in a free current of air, till the fish are thoroughly dry. The *grille* is then taken to the cooking stove. This consists of a series of shallow pans (each large enough to hold a *grille* full of fish) containing boiling olive oil, in which the fish are cooked. This will take from 2 to 3 minutes, according to the size.

After standing a minute or two, to allow the superfluous oil to drain off, the *grille* is hung up to cool, when the fish are ready to be tinned. In the bottom of the tin a piece of bay leaf and a clove head or allspice (pimento) seed are placed, 2 or 3 hands being specially told off to prepare the tins and pass them on to the packers, who carefully but firmly place the fish in the tins in layers, with their tails right and left alternately. The tins are next passed to the oil fillers, who fill them up with cold olive oil. After standing sufficiently long to enable the oil to settle down into all the interstices, and filling up, if necessary, the tins reach the hands of the tinmen or solderers, who fasten down the lids. This operation requires the greatest care, and is the only one (except the analogous one of making the boxes) which calls for the services of skilled workmen. All the other operations of working and packing, etc., require neatness and dispatch, but need no technical skill; but the smallest air-hole left by the solderer in the joint of a tin will spoil it.

After closing down, the tins are collected in a crate and lowered in a large boiler, where they are kept boiling for 2 or 3 hours, according to size. This operation serves a triple purpose—it completely cooks and softens the fish, it expels any remaining air from the tins, and it proves whether or not they are hermetically sealed. On emerging from the boiler, all the tins are bulged, but as they cool they naturally contract, the top and bottom of the tin becoming slightly concave. Any tins, however, which have been imperfectly soldered remain bulged and are spoiled. A rub in sawdust will cleanse the tins, when cool, and they then are ticketed (unless made of decorated tin plate) and packed in wooden cases ready for the market.

Having thus described in general outline the method of preserving fish in tins *à la* sardine, I may perhaps usefully refer to two points of detail which it would be well to observe in the arrangement of any factory established for its adoption.

The buildings should be so arranged that the fish can find their way directly from the hands of those who perform one stage in the process into the hands of those who complete the next stage. When necessary, an arrangement of flues from the cooking range and boiler (and even from the soldering room) may be made to utilize the waste heat to assist the drying process. The tables on which the *grilles* are allowed to drain after cooking, and those at which the operation of "oiling" the boxes is carried on, should be covered with tin plate and fitted with gutters and collectors for saving waste oil, which is marketable. The oil should be stored on the floor above and conveyed to the "oiling" tables through a series of pipes with taps, so that the supply may be under immediate control. Only olive oil of the best quality should be used. Oil of a second quality may be used for working purposes. Olive oil adulterated with cotton-seed oil, or even the latter alone, is often used in preparing French "sardines," but for the best brands the best olive oil only is employed. This is the most costly item in the whole process of manufacture. In cooking the fish, care should be taken to renew the oil before it becomes thick or discolored.

For soldering the tins the only special apparatus required consists of an ingenious but simple turn-table revolving on a pivot and furnished at the top with a "cage," into which the tin fits closely while the top is being soldered. A footplate at the bottom enables the workman to rotate the table at will while, with the soldering iron in one hand and a thin stick of solder in the other, he rapidly closes the "joint" between the body of the box and the lid. As already stated, this operation is the crucial one in the whole process of preparation. In order to keep a check on different workmanship, it is usual to pay the tinmen so much for every 100 boxes "made" and "soldered down," and to deduct so much for every tin that remains bulged after boiling. As a means of identification, each workman marks the tins he makes and the lids he solders down with a special mark, and it is easy for the foreman, when examining and counting the tins, to check the number turned out by each workman and to trace to its author every flaw that leads to the rejection of a tin.

Other methods of making and closing tins are being introduced, and in this and various other details the process of preparing fish *à la* sardine is open to modification. The system above described, however, is that commonly adopted in France, and was successfully applied by me to the preparation of pilchards in Cornwall. In some French sardine factories the fish are baked in hot ovens, instead of being boiled in oil. Occasionally, again, the fish, whether baked or boiled in oil, are soldered down as soon as packed, without the addition of oil in the tins. Sometimes the fish are not subjected to any preliminary cooking, but are packed as soon as dry and thoroughly cooked by prolonging the operation of boiling in the tins. How far sprats can be treated in this and other various ways can only be determined by actual experience. The exact length of time during which they must be subjected to the several operations of salting, cooking, and boiling, and the proper proportions of spice, etc., will depend on the size of the fish, the size of the tins in which they are packed, and other considerations which must also be determined by careful experiment. That sprats can, however, be preserved in tins à la sardine is proved by the fact that at least one factory of the kind already exists on the southeast coast of England, and a ready market can no doubt be found for a largely increased supply abroad, if not at home, and more particularly in India and in our southern colonies, where supplies of fish are scarce. But owing to the shortness of the sprat season no curing establishment could probably afford to be dependent solely on the supplies of this one fish. During a great part of the year the tinmen would no doubt find continuous employment in making the tins in anticipation of the curing season; but it would be found economical to keep the other hands at work in the tinning of other kinds of fish in their season. In Scotland herrings, hake (in slices), cod, ling, and other kinds of fish, besides crabs and lobsters, would no doubt readily lend themselves to modifications of the mode of cure above described. The tinning of vegetables also serves in Cornwall and in France to keep the works going at times when fish are scarce.

Considerable quantities of young herrings are, I believe, taken at certain times in the garvie or sprat nets. This admixture of the two species has the effect of reducing the value of the catch under ordinary circumstances, but there is every probability that young herrings would make a valuable article of food if preserved à la sardine; and as each fish has to be individually handled in the process of cure it would probably not be difficult to distinguish the herrings from the sprats and "tin" them separately. On the other hand, it could be easily ascertained by experiment whether for the purpose of preparation in tins any such separation would be necessary.

It will be understood that there are various circumstances under which the application to sprats of the French system of preserving sardines must be attended with disadvantage. In the first place, the sardine season in France is in the summer months, when the fish can be readily dried without artificial heat. In Cornwall the pilchard harvest takes place later than that of sardines in France, and toward the end of the season the occurrence of rainy or damp weather is a great drawback. The sprat season is later than either, and the provision of artificial means of drying the fish will become more necessary. On the other hand, the heat of a French or Cornish summer is a disadvantage as compared with the comparative coolness of the weather at the time of the sprat harvest, while the sprat has the additional point in its favor that it is less delicate, and will stand carriage and handling better than the sardine. The bones of the sprat, however, are much harder than those of the small immature sardines generally preserved in France. The bones of the pilchard (which is an adult sardine) are much harder than those of its French relative, and those of the sprat are probably harder still. This is one of several points which must be taken into consideration in any proposal to place tinned sprats into competition with tinned sardines. The greater cheapness of sprats will, no doubt, be a question of some importance in determining the issue of such competition. (Fifth Report of Fishery Board for Scotland, pp. 218-221.)

CANNING EELS.

At several of the canneries on the Atlantic coast small quantities of eels are prepared each year. The extent of this branch of the canning trade has been limited on account of the small demand for the product and the scarcity of eels in those localities in which the process has been tried. For this purpose the salt water eels from the Gulf of Maine are used and especially those from Washington County, Me., and Barnstable County, Mass., and small or medium sized ones are selected. After the head, skin, and viscera are removed, the eels are cut in suitable lengths and placed on wire trays and cooked in a steam retort, or, in some cases, fried in an oven for 20 or 30 minutes. They are next placed in cans, either plain with a small amount of jellies to hold them firmly together or with a sauce made of vinegar and spices. The cans are either tall round, large oval, or similar in shape to those in which sardines are packed. Canned eels are prepared principally at Eastport and Camden, Me., and

New York City. With a cannery located on some portion of the United States coast where eels are abundant and the demand for other purposes limited, as at the mouths of some of the rivers emptying into the Chesapeake Bay, it seems probable that an important and profitable business could easily be developed.

There is a small output in New York City of smoked eels in cans. These fish are eviscerated and smoked in the usual manner, with head and skin on (see page 504), after which they are cut into 6 or 8 inch lengths, or slightly less than the height of the can, and these pieces placed close together in the cans, the interstices being filled with diluted cotton-seed oil suitably flavored with vinegar, cloves, etc.

MISCELLANEOUS CANNING.

HERRING.

Owing to the scarcity of mackerel on the United States coast, and the consequent high cost of canned mackerel, herring are frequently used as a substitute therefor, it going on the market under the brand of "herring mackerel," "blueback mackerel," etc. The method of preparation differs in no particular from that applied to the mackerel. The principal factories for their preparation are on the Maine coast, and the product amounts to about 20,000 cases annually. Herring are also put up in spices, in mustard sauce, and in tomato sauce, the output approximating 12,000 cases annually, and the process of canning is substantially the same as that applied to mackerel. These fish are usually branded "brook trout."

MENHADEN.

At several canneries on the Maine coast menhaden have been canned and placed on the market in 1-pound cans as "ocean trout," "herring mackerel," "blueback mackerel," etc., and have met with ready sale at about 65 cents per dozen cans. In 1889 378,272 cans of menhaden were prepared in Maine, but since then these fish have been so scarce on that coast that comparatively few are canned.

SMELT.

The canning of smelt was first begun late in the fall of 1879 in Boston. They were thoroughly cooked in butter and packed in 1-pound cans, 5 dozen cans in a case.* This business has been abandoned, and at the present time no smelt are canned in this country. In 1885, when the pack of oil sardines was smaller than usual, owing to the scarcity of small fish suitable for quarter cans, experiments were made in the canning of smelt as a substitute for herring in the manufacture of sardines, but they were found to be dry and hard, and deficient in flavor, and efforts in this line were soon discontinued.†

SMOKED STURGEON.

In the canning of smoked sturgeon the fresh fish are cut into pieces adapted to the size of the can for which they are intended and placed in a wire drum, the cross-section of which is equal to the cross-section of the can. This drum is so arranged that one side or head enters the receptacle, and by means of a spring or clasp is pressed into the drum, thus slightly compressing the contents. While it is subjected to the action of the smoke, and as the fish becomes more and more compact, the movable head will gradually press it against the fixed head, so that the contents take the shape of a disk with comparatively flat sides. The drum is so suspended that it may be turned or rotated from time to time, so that the juice that settles to the bottom is brought to the top and compelled to flow through the mass again, thus retaining it in the flesh. On completion of the smoking the disks of fish are removed from the drum and placed in cans with a small quantity of cotton seed oil, and the cans are hermetically sealed. On account of the scarcity and the consequent high price of sturgeon during recent years, comparatively small quantities are canned. The product is very palatable and will keep for a year or two under favorable conditions.

HALIBUT.

The generally brisk demand for fresh and smoked halibut has prevented many attempts in New England to preserve them by canning. On the Pacific coast, both in Alaska and in the State of Washington, and at Klawak, Prince of Wales Island, fresh halibut have been canned, but in no great quantities. There is no doubt but that the fish is suitable to be thus preserved.

SPANISH MACKEREL.

In 1879 the owner of an oyster and clam cannery at Ocracoke Inlet, North Carolina, purchased small quantities of Spanish mackerel and put up a few hundred 2-pound cans. Shortly afterwards, at the suggestion of Professor Baird, experiments were made in canning Spanish mackerel at Cherrystone, Va., to ascertain their relative value as compared with other kinds of canned fish. The reports of the canneries were that they are no better than fish of ordinary grades, and that as a canned fish they are inferior to the common mackerel (*Scomber scombrus*).*

GREEN TURTLE.

The canning of green turtle (*Chelonia mydas*) in this country was first begun in 1869 on the coast of Texas at the beef packeries located on Aransas Bay. When these canneries were closed, about 10 years afterwards, a small factory was established at Fulton, on the same bay, for preparing turtle meat in tin cans for market. This cannery was in operation up to 1896, using annually about 1,000 turtles, weighing 250,000 pounds, and preparing about 40,000 two-pound cans of turtle meat and 800 two and three-pound cans of "turtle soup." On account of the increasing scarcity of green turtle on the Texas coast, the cannery has not been in operation since 1896.

Small quantities of green turtle meat are incidentally canned at times at various other places. Each cannery uses methods peculiar to itself, so that it is scarcely practicable to describe any general method of preparing this product. To prepare it so that it will keep a suitable length of time, requires close attention and the greatest cleanliness.

GIANT SCALLOPS.

The Bulletin of the U. S. Fish Commission for 1889 contains the following account of experiments in canning the giant scallop (*Pecten magellanicus*):

About 1876 the Castine Packing Company undertook to put scallops on the market in a canned condition, as is now so commonly done with clams in many localities on the coast of Maine. It is said that the company was unable to properly preserve the thick, solid meats, and the effort was abortive. Six years ago, however, the attempt was renewed and was in a measure successful. It was found that by previously frying the meats they could be canned without difficulty, but the method was considered too costly and was not put to much practical use.†

---

* Report U. S. Fish Commission, 1880, p. 111; 1881, pp. 221–227.
† Bulletin of the U. S. Fish Commission for 1889, p. 320.

CODFISH BALLS, ETC.

In 1878 the preparation of codfish balls was begun by a Boston canner. This product consists of codfish and potatoes cooked with beef tallow, with the addition of a small quantity of saltpeter, the whole being hermetically sealed in tin cans. The usual method of preparation is as follows: For 100 pounds of salt codfish, 125 pounds of potatoes, 10 pounds of raw onions, and 13 pounds of pure beef tallow are required The fish are soaked in tepid water to remove the salt and then reduced to a pulp: the potatoes are boiled, skinned, and mashed; and these ingredients are warmed and mixed thoroughly with the chopped raw onions and beef tallow, adding 6 pounds of saltpeter and 6 ounces of pepper or other suitable flavoring condiments. While the ingredients are being mixed they are chopped as fine as practicable by machinery. The warm mixture is then placed in 1, 2, or 3 pound tin cans and sealed. The cans and contents are boiled at a very high temperature for 2 or 3 hours. On cooling and labeling the product is ready for market. In 1879, 11,000 cases, equivalent to 264,000 two-pound cans, were prepared by Boston canners, the value of which was $38,500. The present annual output is somewhat less, owing to the increased popularity of boneless codfish.

Among numerous other fishery products preserved in cans are clam chowder, fish chowder, Finnan haddie, smoked lake trout, smoked pike, smoked carp, caviar, etc.

The following summary shows some of the varieties of canned fishery products on the New York market and the average wholesale price in 1898:

Designation.	Price per dozen.	Designation.	Price per dozen.
Carp, silver, smoked, one-pound cans	$2.50	Lobsters:	
Caviar:		One-pound cans, tall	$2.70
Quarter-pound cans	1.85	Half-pound cans, flat	1.70
Half-pound cans	3.15	One-pound cans, flat	3.15
One pound cans	5.10	Pickled, one-pound cans	2.20
Two-pound cans	9.20	Pickled, two-pound cans	3.45
One-eighth-kilo, cans	2.10	Mackerel:	
Quarter-kilo, cans	3.00	Genuine, one-pound cans	1.15
Half-kilo, cans	7.00	"Herring mackerel," one-pound cans	.70
Clams:		Soused, three-quarter-pound cans	1.10
Eastern, soft-shell, one-pound cans	.90	Mustard, three-quarter pound cans	1.05
Eastern, soft-shell, two-pound cans	1.45	Oysters, standard, one-pound cans	.70
Little-neck, hard-shell, one-pound cans	1.00	Oysters, standard, two-pound cans	1.30
Little-neck, hard-shell, two-pound cans	1.90	Pickered, smoked, one-pound cans	2.50
Clam chowder, two-pound cans	1.50	Salmon:	
Clam chowder, three-pound cans	2.00	Columbia River, half-pound cans, flat	.80
Clam juice, two-pound cans	1.35	Columbia River, one-pound cans, flat	1.45
Codfish balls, two-pound cans	1.30	Columbia River, one-pound cans, tall	1.40
Crab meat, one-pound cans	1.75	Alaska, red, one-pound cans	1.15
Crab meat, two-pound cans	2.75	Alaska, medium, one-pound cans	1.10
Crabs, deviled, one-pound cans	2.25	Alaska, pink, one-pound cans	.90
Crab soup, imported, one-pound cans	3.50	Alaska, sockeye, one-pound cans, tall	1.15
Crab soup, imported, two pound cans	6.50	Alaska, sockeye, one-pound cans, flat	1.20
Eels, in jelly, one pound cans	1.60	Sardels:	
Eels, in jelly, two-pound cans	2.62	Half-pound cans	2.00
Eels, in jelly, five-pound cans	6.25	One-pound cans	3.75
Eels, Hamburg, one-pound cans	2.25	Shrimp, one-pound cans	1.75
Eels, Hamburg, two-pound cans	3.20	Sturgeon:	
Eels, Hamburg, five-pound cans	7.50	Pickled, one-pound cans	1.90
Eel soup, imported, one-pound cans	3.25	Smoked, one-pound cans	2.20
Eel soup, imported, two-pound cans	6.00	Terrapin meat, one-pound cans	8.00
Finnan haddie, one pound cans	1.05		
Finnan haddie, Scotch, one-pound cans	1.75		Price per 100 tins.
Green-turtle meat, one-pound cans	10.00	Sardines:	
Green turtle soup, one-pound cans	8.00	Oils, domestic standard, quarters	$2.85
Herring:		Oils, domestic extra, quarters	3.25
Domestic, fresh, one-pound cans	.95	Oils, imported, quarters	6.25
Domestic, bloater, one pound cans	1.10	Oils, choice, quarters	9.00
Scotch, fresh, one-pound cans	1.40	Oils, domestic standard, halves	4.70
Scotch, kippered, one-pound cans	1.50	Mustards, quarters, small	2.75
Scotch, in tomato sauce, one-pound cans	1.40	Mustards, quarters, extra large	3.00
Yarmouth bloaters	2.25	Mustards, three-quarters	5.25
Bismark, three-pound cans	2.75	Spiced, quarters	3.35
Kieler sprotten, 1½-pound cans	6.00	Spiced, three quarters	5.20
Kieler sprotten, three-pound cans	10.00	In tomato sauce, quarters	3.25
Lake trout, smoked, one-pound cans	2.50	Norway smoked, quarters	8.50

PREPARATION OF FISH EGGS FOR FOOD.

The roes or eggs of fish are among the most valuable of the miscellaneous food products of the fisheries. The most important are the roes of the sturgeon, mullet, herring, shad, whitefish, cod, and haddock. Some of these eggs are sold to the consumers while fresh, especially the eggs of shad, river herring, whitefish, and haddock. In pickling sea herring the roes are usually left in the fish and no special treatment is applied to them. The eggs of the sturgeon, mullet, and of a few other species are nearly always removed from the fish and separately prepared, and it is to the treatment of these that the present chapter more particularly relates. Sturgeon eggs are salted in brine and sold under the name caviar, the domestic product approximating 300,000 pounds annually, worth $225,000. Mullet roes are dry-salted or pickled in brine all along the United States coast from North Carolina to Florida. The Indians of the Northwest coast dry considerable quantities of roe from various species of fish, the product being stored for winter use, when it is pounded between two stones, immersed in water, and beaten with wooden spoons into a creamy consistency, or it is boiled with sorrel and different dried berries and molded in wooden frames into cakes about 12 inches square and 1 inch thick.

CAVIAR.

Caviar is made from the eggs of sturgeon or similar species of fish, which are suitably salted and held in tight packages in brine. It is the most costly food product obtained in the United States fisheries, and while highly relished by many persons, a liking for it must usually be acquired. For many years the manufacture of caviar was monopolized by the Russians, most of it being prepared on the Volga River and Caspian Sea, where large quantities are even now annually put up, the trade centering at Astrakhan. The product in Russia amounts to about 8,000,000 pounds annually, and it is in great demand in Europe, especially in those countries bordering the eastern half of the Mediterranean.

The abundance of sturgeon in the United States led to the preparation of caviar on the Hudson River about 1850, and three years later on the Delaware River. It was prepared at Sandusky, Ohio, first in 1855, and soon afterwards its manufacture was begun at other points on the Great Lakes and the various rivers on the Atlantic coast; in 1885 its preparation extended to the Columbia River on the Pacific coast, and subsequently to Lake of the Woods. An acquaintance with its peculiar process of manufacture became of considerable value, sums ranging from $100 to $500 being frequently paid for instructions in the secret method. At present, on account of the high price at which the article sells—from 50 to 90 cents per pound—every locality in America in which sturgeons abound is vigorously fished, and on the Delaware River female sturgeons with ovaries in suitable condition sell ordinarily for $10 each, and as high as $60 worth of products has been made from one fish.

The best caviar made in the United States is from the eggs of the lake sturgeon (*Acipenser rubicundus*), these being larger than those of the common species (*A. sturio*). The latter is the sole source of caviar produced on the Atlantic coast, the short-nosed species (*A. brevirostris*) not being found in sufficient quantities for this purpose. The lake caviar sold in 1898 for about 80 cents per pound, whereas the Delaware product sold for 60 cents, and the Southern Atlantic for 50 cents per pound.

The caviar prepared on the Pacific coast is from the *A. transmontanus*, and sells usually for 40 cents per pound. During the past three or four years the eggs of the shovel-nose sturgeon (*A. scaphirhynchus*) have been used to a small extent for making caviar, most of this product coming from the Mississippi River, especially in the vicinity of Memphis. This caviar is not choice and usually sells for about 30 cents per pound, or half that of Delaware caviar.

The increasing scarcity of sturgeon and the high price of caviar have led to many attempts at finding a substitute for sturgeon eggs, but so far with very little success. The eggs of horseshoe crabs (*Limulus polyphemus*) have been used, but they are small, and become hard and tasteless when salted. Garfish eggs have also been tried, and while of good size, they are without flavor and have a disagreeable and even repulsive odor. The most successful substitute yet found is shad eggs, which have been prepared in identically the same manner as those of sturgeon and mixed with the latter. The resulting product sells for a lower price than caviar made entirely from sturgeon eggs, but the decreased value is more than counterbalanced by the increased quantity.

The product of caviar in the United States amounted in 1898 to about 2,800 kegs of 125 to 160 pounds each. Of these, 100 kegs were of the large-grain variety from the Great Lakes, Lake of the Woods, Lake Winnipeg, etc.; 100 kegs from Columbia and Fraser rivers; 200 kegs of small grain, from the South Atlantic coast, and the remaining 2,400 kegs, of the medium-grain variety, from the Delaware, south coast of Long Island, and other waters of the Middle States. About 500 kegs were consumed in this country, the remaining 2,300 being exported to Europe.

Small quantities of caviar are imported into this country annually, the supplies coming from the Volga and the Elbe. The wholesale price in New York varies from 80 cents to $4 per pound, depending on the quality of the grain and the extent to which it has been salted. The higher-priced varieties are very lightly salted and must be kept at a low temperature.

The equipment for making caviar is simple and inexpensive, consisting, in addition to the floats, slaughter-house, etc., necessary for handling sturgeon meat, of several sieves with wire meshes, a few large-sized buckets, tubs, and a number of tight kegs for holding the product. The first step in the process is to remove the roe from the sturgeon, which should be done as soon as practicable after the fish is caught. The sturgeon is turned on its back or side, a gash is cut from the neck to the vent, and the eggs are removed. Care must be taken to avoid bringing the eggs in contact with fresh water, since it softens them and breaks the shells. The quantity of roe removed from each fish varies considerably. The Delaware sturgeon yield from 6 to 12 gallons, including the investing membranes of the ovaries and the supporting tissues, the latter being only a very small part of the organ, so that there is but little waste from this source. The Columbia River sturgeon yield nearly as much as the Delaware sturgeon, but those from the Great Lakes average only 2 or 3 gallons.

The masses of eggs and membranous tissue are at once placed upon a wire sieve, the meshes of which are just large enough for the eggs to pass through as the masses are rubbed back and forth by the workman's hand. A tub, can, or box is placed under the sieve to receive the eggs as they pass through. It is convenient, where much caviar is prepared, to fit the sieve over a zinc-lined trough, about 18 inches deep, 2 feet wide, and 4 feet long, with its bottom sloping to one end, where an outlet is arranged. As the eggs are gently pressed by the hand and worked back and forth across the meshes, they become separated from the membranous tissue and from each other and fall into the receptacle, whence they are removed and placed in clean half-barrel tubs.

In the tubs the eggs are at once mixed with a compound of Lüneburg salt, 100 pounds of eggs requiring 13 pounds of Lüneburg salt, to which is added 1 pound of "preservaline," a proprietary composition of certain antiseptics, such as boracic acid, salicylic acid, etc. The Lüneburg salt costs about $4 per cask of 300 pounds, and the preservaline costs about $19 per 100 pounds. The preservaline has been in use for about twelve years, and since it gives satisfaction and costs only a trifle compared with the value of the caviar, no disposition exists to experiment with a substitute. The mixing of the eggs with the salt is accomplished by gently stirring the mass by hand for a few minutes. The immediate effect of the salt is to cut the slime or glutinous coating from the eggs and to dry the mass, but very soon its strong affinity for moisture causes it to extract the watery constituents of the eggs, and in 10 or 15 minutes a very copious brine is formed, and upon its surface a frothy substance collects. This is skimmed off and the eggs placed in sieves of a finer mesh, about 8 or 10 pounds of eggs in each sieve. For convenience in draining, these sieves may be placed on a sloping plank, with strips nailed on each edge to elevate them, or in some houses they are placed over an opening in the floor. The draining must be thorough, and requires from 12 to 20 hours.

The process is now complete and the caviar is at once placed in small, clean oaken kegs, which have been thoroughly steamed, with capacity for holding from 125 to 160 pounds. The kegs cost about $1 each, being made of red or white oak or Norway or Oregon pine. The Delaware kegs hold about 125 or 130 pounds, those used on the Great Lakes about 160 pounds, and those on the Columbia River about 145 or 150 pounds. The kegs when filled with caviar should be kept in a cool but not freezing temperature and be allowed to stand for a considerable time, in order that the gas may escape. During this time the caviar settles several inches and the keg should be again filled before being headed up.

Experience is essential to the preparation of a high grade of caviar, as the extent of the salting, draining, etc., depends on the condition of the eggs, the temperature, and the state of the weather. It is usually customary to keep light and dark varieties of roe separate, since mixing the two gives a speckled appearance to the product. In storage the caviar should be held at a low temperature, 38° to 40° F. being found most satisfactory, and under favorable conditions it may be kept for several years.

The principal market for caviar in kegs is New York City. There are numerous buyers at various fishery points on the coast and in the interior, who collect the output of the smaller manufacturers and ship it to New York dealers, who export the greater part of it to Germany. Occasionally a manufacturer on the Delaware may ship his product to Germany direct, but more frequently it passes through the hands of New York dealers.

Caviar is packed in kegs for the wholesale market, and is never handled in any other form of package by the original producer, but a keg holding more than a customer usually desires, the large dealers prepare it in hermetically sealed tin cans for retail trade. When prepared in this manner it is sometimes subjected to a process different from that employed for packing it in kegs.

In 1875 Max Ams, an extensive dealer in fishery products in New York, devised and patented * a process, which is as follows:

After the eggs have been sieved and salted in the usual manner, except that preservaline is not generally used, they are placed in tin cans, which are immediately soldered and then exposed to water in a gentle heat, which is very gradually increased to not less than 140° nor more than 200° F. The can is then vented and immediately reclosed to retain the caviar in an air-tight package. By this process the salt mixed with the eggs will be combined with the extraneous matter sufficiently to protect it against decomposition and to constitute a protective covering for the eggs. If the temperature be less than 140°, this effect would not be obtained and decomposition would probably ensue; and if the heat exceeds 200°, the essential oils would evaporate and the eggs be left dry, brittle, and tasteless.

The usual size of cans for the retail trade in this country is ½ pound, ¼ pound, 1 pound, and 2 pounds. The price received for ½ pound cans is about $1.85 per dozen, and for 1-pound cans $5.10 per dozen. Other sizes sell at proportionate rates.

A very choice product of caviar, which, however, seems to be little known in this country, is the freshly salted eggs. The fresh eggs on removal from the fish are at once mixed with a small quantity of salt and served in that condition within 2 or 3 hours. This makes a delightful dish, quite superior to the usual caviar of commerce. In order to obtain the article in Moscow and St. Petersburg, the living sturgeon are transported from the Volga in tank cars, so that the eggs may be had perfectly fresh.

A special method of preparing caviar was patented in this country in 1851,† which does not appear, however, to have ever been employed to any great extent. This process is as follows:

The roe, being removed from the fish, are squeezed gently by hand in order to remove the individual ova from the membranes by which they are covered. Sprinkle a small quantity of fine salt in a clean tub and place in the tub a layer of ova and a layer of salt, to the extent of 100 pounds of roe and about 5 pounds of salt. When it has remained about 6 hours, pour 6 quarts of strong brine-pickle over the mixture. After 12 hours a like quantity of pickle is again poured over. In from 30 to 50 hours, according to the state of the weather, the ova will rise or float on the pickle, while certain refuse matter will settle to the bottom of the tub, the extraneous matter being separated from the ova by a process similar to fermentation. The ova are then spread about half an inch thick on sheets, and are exposed to the air from 20 to 40 hours, being turned over in the sheets in the meantime 4 or 5 times a day. When dry, mix with it about 2 ounces of black pepper and 3 pints of oil extracted from the liver or milt of the male sturgeon, the purpose of the oil being to restore to the roe the sturgeon flavor removed by the salting process. Let it stand for 10 or 12 days and then pack in kegs for market.

RUSSIAN METHODS OF PREPARING CAVIAR.

Large quantities of caviar are manufactured in Russia, especially in the vicinity of the Caspian Sea, not only from the eggs of sturgeon but of various other species. Of the sturgeon caviar two kinds are prepared, (1) fresh or grained, and (2) hard or pressed caviar; the former is more valuable than the latter, selling at Astrakhan from $21 to $25 per pood (36.112 pounds), while pressed caviar sells at $15 to $17 per pood. The method of preparing each kind is as follows:‡

In preparing by either method the roe of the sturgeon is spread on a net stretched on a wooden frame and with narrow meshes forming a sieve. The grains are passed through the meshes by slightly pressing the whole mass with the hand till nothing remains on the sieve but the cellular tissue, the

* Letters Patent No. 169668, November 9, 1875.   † See Letters Patent No 7895, January 7, 1851.
‡ See Notice sur les Pêcheries et la Chasse aux Phoques dans la Mer Blanche, l'Ocean Glacial et la Mer Caspienne. Par Alexandre Schultz, St.-Pétersbourg, 1873. Also Rapport sur les Expositions Internationales de Pêche, par J.-L. Soubeiran. Paris. 1871.

fat, and the muscle, the grains falling into a wooden receptacle placed underneath. If grained caviar is to be made, the roe is sprinkled with very clean and fine salt, and the whole mass is stirred with a wooden fork having eight or ten prongs. The quantity of salt required varies, according to the season, from 6 to 15 pounds per 100 pounds of roe; more salt being required in warm than in cold weather. It is desirable that as little salt be used as is absolutely necessary for preserving the caviar. The roe mixed with salt first presents the appearance of dough when stirred, but when each grain has been impregnated with salt the whole mass swells, and in stirring a slight noise is perceptible like that produced by stirring grains of corn. This noise is a sign that the process is completed.' The caviar is packed in casks made of linden wood, as this imparts no disagreeable flavor to the contents.

For manufacturing pressed caviar a tub half filled with brine is placed under the sieve, the strength of the brine varying with the temperature and the season. To impregnate the grains evenly with brine the whole mass is stirred with a wooden fork, always turning it from the same side. This is continued for 10 minutes in summer and about half that long in winter. Then the roe is removed with fine sieves and, after the brine has drained therefrom, it is put in receptacles made of the bark of the linden, 3 poods (108 pounds) to each sack, each of which is placed under compression to remove all the brine from the roe and to transform it to a solid mass, remaining under compression for about 6 days. During the pressing many grains are crushed and a portion of their contents flows out with the brine, the loss in weight amounting to about 30 per cent. The pressed caviar is then removed from the sacks and packed in casks containing usually 30 poods (1,080 pounds) each, the inside of which is covered with "napkin linen," this being the reason why the caviar is frequently called *caviar à la serviette* (napkin caviar). The finest quality of pressed caviar, that which has been least salted and pressed, is packed in straight linen bags of cylindrical shape, and is called *caviar à sac* (sack caviar). Caviar is also shipped in hermetically sealed tin cans.

Mr. Schultz states:

The fatness of the roe depends on the quality of the fish and the season when it is caught. The fattest is that made from the roe of sturgeon caught in the Caspian between July 8 and August 15. This roe is left only a few hours in the brine and then taken out and packed, without being pressed, in casks holding from 5 to 10 poods (180 to 360 pounds) each. If the fish has been dead so long that the roes are somewhat spoiled, the roes and ovaries are placed in the brine until they are thoroughly impregnated with salt and then pressed and packed in large casks containing about 1,000 pounds. This is sold at a very low price, from 5 to 8 cents per pound, wholesale.

The choicest caviar in the Russian trade is from the roe of. the bélouga (*Acipenser huso*), the eggs being large and of good appearance; but for the bulk of caviar the roe of the common sturgeon (*A. guldenstädtii*) and of the sévriouga (*A. stellatus*) is used. Choice caviar is made from the eggs of the sterlad (*A. ruthenus*), which, however, does not enter into commerce, being used by the fishermen and their neighbors.

The eggs of the bream (*Abramis brama*), of the perch (*Lucioperca sandra*), and of the "vobla" or chub (*Leuciscus rutilus*) are also used for making a form of caviar, which finds a market principally in Constantinople and Greece. Merchants from Greece visit the fishing establishments near Astrakhan, purchase the fresh eggs, and have the caviar prepared under their own supervision in a manner quite similar to the salting of mullet roes in the United States. The roe bags with the eggs therein are carefully removed and mixed with dry salt in bulk. After sufficient salting the mass is placed between boards weighted down by heavy stones, and after remaining thus for a month is shipped in casks. In the retail trade it is usually cut into disk-like slices and is much sought after in Greece. From 500,000 to 700,000 pounds of the caviar from perch eggs are prepared every year in Kuban. During recent years the Greek Islanders have prepared large quantities of roes from the above named species of fish.

Day states[*] that the roe of carp (*Cyprinus carpio*) is made into caviar by Jews in Italy and Eastern Europe, as by their regulations they may not eat caviar made of sturgeon, that fish being destitute of scales.

---

[*] The Fishes of Great Britain and Ireland, by Francis Day, vol. II, p. 162.

The fishermen of the Dardanelles prepare a kind of cheese from the roe of several species of fish by drying it in the air and then pressing it. By dipping it in melted wax, a crust is formed over it which prevents its being affected by the air. Inside this crust the roe undergoes a sort of fermentation, giving it so piquant a flavor that one can eat but little of it at a time. It is said to taste like a mixture of fine sardines, caviar, and old cheese. Before it is eaten, the crust of wax is taken off, and if it has become moldy, which frequently happens, it is soaked in strong vinegar.*

In Germany a form of caviar is made from the eggs of the pike, in the following manner. The fresh eggs on removal from the fish are rinsed in cold water and rubbed through a coarse sieve to separate them from the membranous tissues enveloping them. On completion of this, they are rinsed two or three times and are placed in a finer-meshed sieve to drain. Next, they are well mixed with fine salt and flavoring ingredients, there being added to each 100 pounds of eggs about 3½ pounds of fine salt. 2½ ounces of citric acid, and a small quantity of lemon oil. After being thoroughly mixed with these ingredients the eggs are put in a cool place, and after remaining undisturbed for eight days the jars or tubs containing them are tightly sealed.

MULLET ROES.

Mullet roes are considered great delicacies in nearly all countries in which this fish abounds, and large quantities are prepared along the southern coast of the United States and in countries bordering the Mediterranean. At maturity, which occurs in September and October, the roe of the Southern Atlantic mullet is from 5 to 8 inches long and 1½ to 2½ inches in diameter. These are saved in nearly all the mullet fisheries of the United States, and are sold either fresh or dried, about 300,000 pounds, worth $20,000, being salted annually on the west coast of Florida.

Dried mullet roes are prepared along the southern coast from North Carolina to Florida, inclusive, in a manner quite similar to the drying of mullet. The roes utilized are from the matured females which have not begun spawning, for as the spawning time approaches the eggs soften and burst the surrounding membranes or roe bags, when they are useless for salting or drying. Nothing but firm roes should be used for salting, and soft roes, roes from fish which have been caught some hours, as well as roes from roe bags half emptied should, if used at all, be salted separately.

In removing them care should be taken to avoid breaking the roe bags or injuring or bruising the eggs, but they should be free from portions of the surrounding viscera. If the tubs in which the roes are gathered have holes through which the water can run off, some salt is saved and a better product is secured, the water making the roes soft and less liable to keep. The roes still in the roe bags are then placed in boxes or barrels with salt sprinkled among them, or in some cases they are placed in brine, where they remain for ten or twelve hours, but the former method is preferred. An excess of salting must be avoided, since it causes the egg-sacs to break and the eggs are ruined on exposure to the sun and to pressure, or they become dark and brittle. If properly treated a good article can be made of roes that have become somewhat soft by salting it immediately on removal from the fish, by using more salt than for the firm roe, or by resalting it. Medium grain salt is preferred for salting. Coarse salt should be avoided, since it is liable to become imbedded in the roe membrane and give

---

* Norsk Fiskeritidende, vol. v, No. 2, Bergen, April, 1886.

it a burnt look. About 1 peck of Turks Island or Liverpool salt to 160 pounds of eggs usually gives the best results.

On removal from the salt the roes are spread out on boards and exposed to the sun for about one week, being taken in at night to prevent the moisture and dews from falling on them, and every morning they are turned over to thoroughly aid in drying them. Care must be taken to prevent them from becoming wet after the drying has begun, and upon the first indication of a rain they should be placed under shelter. Sometimes after one day's exposure other boards are laid on top of the roes so as to slightly compress them. When properly cured, they are 4 to 8 inches long, 2 to 4 inches wide, and one-half to two-thirds of an inch thick, and vary in color from a yellowish brown to dark red, according to the freshness of the roe, carefulness of handling, degree of saltness, and length of drying. The roes are then sent *en masse* to market in baskets, boxes, or the like, and sold from 40 to 60 cents per dozen, according to the size and carefulness in curing.

In the West Indies and in many countries bordering on the Mediterranean Sea, mullet roes are prepared by methods similar to those employed in this country. In Greece almost the same process is used, except that when dried the roes are generally dipped in melted beeswax. Those obtained from Tunis are very highly esteemed, about 150,000 being sold in Italy each year at about 20 cents each.

Mr. Day states:

In Italy, the hard roe of mullet is converted into cakes termed bolarge or bolargo, which are prepared by washing and sprinkling with salt and pressing between two boards. This may be smoked or sun-dried and is considered a good appetizer to promote thirst. But in India the same article is somewhat similarly treated and considered excellent for curries.

Readers of Pepys will recall the eating of bolargo in England, as the gossip says:

Sir W. Penn came out in his shirt on to his leads and there we stayed talking and singing and eating bolargo, bread and butter till twelve at night, it being moonshine, and so to bed very nearly fuddled.

SALTED SHAD ROES.

A small quantity of shad roes are brine-salted in North Carolina, Virginia, and Maryland each year, these being so prepared only when the state of the fresh-fish market or the transportation facilities makes it necessary to pickle the female fish. The roes are removed from the fish in dressing the latter, care being taken not to cut or injure the roe bags. As soon as practicable thereafter they are washed by stirring them with the hands in tubs of water, and are then placed in tubs of strong brine with dry Liverpool salt sprinkled among them and at the top. Every 12 hours during the ensuing 5 days the roes are gently stirred to separate them from each other and to have them uniformly salted. The sixth day they are removed from the pickle, drained, and placed in suitable packages, with dry Turks Island salt sprinkled at the bottom, through the roes, and at the top. The package should then be filled with strong brine made of Liverpool salt. A variety of packages are used, the most convenient being 20-pound kits, which when filled with salted roes sell usually for about $2 each.

EGGS OF COD, HADDOCK, ETC.

It is somewhat remarkable that the roes of cod, haddock, and other ground fish are not more extensively used for commercial purposes in this country. A large trade exists in cod roes in Norway, the eggs being salted and shipped to France to be used as a bait in the sardine fishery. About fifteen years ago a small trade developed in exporting cod roes from this country for use in the same fisheries. The price received was about $2.75 per barrel net, and the price usually paid in France is about 50 francs per barrel. In 1879–80, 3,200 barrels of cod and pollock roe were salted at Gloucester and shipped to France via New York, but on account of discriminative duties these shipments were soon abandoned. An attempt has been made to introduce cod roe as a bait in the Eastport sardine fishery, but without success.

The only roe now saved in the New England fisheries is that of the haddock, which is brought ashore fresh, especially by the shore vessels, the proceeds from the sale usually going to the cook or to the crew. It is taken principally in the spring, from 35 to 75 pounds being secured for each 1,000 pounds of dressed fish. It is sold at prices ranging from 25 cents to $2.25 per bucket of 25 pounds, and the annual product is about 600,000 pounds, for which the fishermen receive $14,000. It is purchased by consumers while fresh, and does not receive any special method of preservation.

The possibilities for utilizing a part of the roe now wasted in our New England fisheries furnish sufficient reason for incorporating herein the following description of the methods of making cod caviar in Norway:

For the preparation of cod caviar the Norwegians use the whole ovaries of the cod which are salted in barrels, and mostly in the Lofoden winter fisheries. The roe must be salted whole without injuring or breaking the enveloping membrane, and must not be salted too much, just sufficient to impart a nice orange-red color. When the salted ovaries are removed from the barrels they are first thoroughly washed several times in fresh water, and then hung on wires or ropes in the open air, but protected from too strong sunshine. After they have dried about 24 hours they are taken down for smoking. For this purpose they are hung in the same way in the smoking-house on sticks or rods or put on frames covered with old nets or wirework and cold smoked for two or three days, or until they become of a dark-brown color. After smoking, the enveloping membrane or skin of each roe sack is torn and removed, and the eggs packed in good, tight barrels, which are then tightly closed and placed in a normally cool place for a month or six weeks. At the end of that period the eggs begin to ferment somewhat, which may be detected by the swelling of the barrel. It is well not to wait too long, but to examine the barrels every week or so, and as soon as fermentation has begun a sufficient quantity of salt should be put into the roe, to prevent the product from spoiling.

By the fermentation the roe receives a slight acid flavor and a taste resembling that of fermenting beer or wine, and this fermentation must be stopped by adding salt at a definite point, which is to be learned by experience only. The salt used to stop the fermentation must be of the very best quality, and if the roe seems to be dry a little good French olive oil is added to moisten the product. After the roe has been thoroughly mixed with the salt it is put in 1-pound glass bottles that are sealed with cork stoppers.

FOOD EXTRACTS OF MARINE PRODUCTS.

Various methods have been introduced for preparing extracts of the alimentary principles of marine products, especially of those that are otherwise wasted. In Norway and other countries of northern Europe a number of preparations in the nature of pastes or extracts are made from fish. A well known instance is the fish meal of Norway, which is composed of the flesh of fish reduced to powder, in which all of the nourishing materials are concentrated and condensed, with the addition of certain other substances. It is claimed that it contains 4 times as much nutritious matter as beef, and 16 times as much as milk or rye bread. On the coast of Cochin China, large quantities of a fish paste are prepared from the shrimp and small fish inhabiting the inshore waters. It is stated that this sauce is brought to perfection by being buried in the earth for several years. About $500,000 worth is consumed in the French provinces alone.

EXTRACTS OF FISH.

Following the idea of Baron von Liebig in preparing the well-known article of commerce known as "extract of meat," several attempts have been made to prepare a similar article from fish. In the case of meats, the substances soluble in water are extracted from the tissues, and the albumens are then coagulated by the aid of heat or by the addition of dilute acid. The fluid remaining after the coagulated albumen has been skimmed off consists of the extractives and the salts soluble in water, and this is evaporated down to a semifluid condition, in which it is placed on the market.

In 1876, Stephen L. Goodale, for many years secretary of the Maine Board of Agriculture, introduced a method of preparing a food extract from fish, especially applicable in connection with the use of menhaden for oil and fertilizer. His process, as improved in 1880 and covered by Letters Patent No. 248586, dated October 25, 1881, was as follows:

Clean the fish and boil for a short time to coagulate the albumen contained in the muscle juices. Separate the liquid from the solid matter by drainage and pressure and allow the liquid to stand in a suitable vessel until any oil which may have passed over in the liquid has risen to the surface and been removed. The liquid is then aerated at the highest practicable temperature, either by introducing a current of heated air or of heated steam, or by ebullition with free access of air, when a substance causing turbidity is precipitated, the complete precipitation being ascertained by examining samples taken out from time to time in a glass tube or heater. If gelatine be present in the liquid, which is the case if skins and bones are not excluded in cleaning the fish, the precipitate will be finer and slower in falling than if muscular flesh alone were used. The liquor will also attain a somewhat darker color, resembling that of light wine, and be reduced in bulk by the further concentration incident to the means used to effect precipitation. When the precipitation is completed the precipitate should be removed from the liquid either by drawing off or by filtration in any convenient manner. The clear liquid thus obtained is evaporated, as is customary in making meat extracts, the evaporation to be continued until the desired consistency has been reached, which is usually about that of honey. The product may be put up in cans, bottles, or other closed vessels.

The inventor states that a barrel of menhaden yields about 3 pounds of the extract, that the article compares favorably with Liebig's extract of beef and retains its flavor under any ordinary condition of temperature or climate. While it has never been prepared for the general market, it seems not improbable that it might have a considerable patronage if properly introduced.

A somewhat similar process was invented [*] in 1882 by Carl Adolph Sahlström, of Jönköping, Sweden, for producing a nutritious extract from the flesh of the shark, whale, seal, and other sea animals. This process was as follows:

The raw material is cut up into as small pieces as possible by mechanical means and is placed in a vat provided with stirring apparatus. A quantity of clean water, free from lime, is boiled and cooled down to from 6 to 15 C., and to this is added so much dissolved hypermanganic alkali as will impart to the water a light-red color (say from 1 to 10 grams for every 100 liters of water) and from 20 to 100 grams of water of ammonia. Sufficient of this liquor is added to the finely cut raw material to give thereto the consistence of thin gruel, and the stirring apparatus is then set to work. After a period of from 10 to 30 minutes the mass is removed from the vat and is placed in a centrifugal apparatus for the purpose of separating the liquor, which carries the fat with it. The inner part of the centrifugal apparatus is preferably covered with cloth. When all the fluid is separated the mass is again soaked in fresh liquor and passed through the centrifugal apparatus, and this is done as often as may be necessary to remove all the fat. All the fluid obtained is mixed together and left to stand in a deep tank for a period varying according to the temperature and until complete separation takes place. The fat and oil rise to the top of the liquid and are removed for further treatment. The oil is separated for special treatment. The solid mass remaining in the centrifugal machine is also reserved for further treatment. The fluid thus obtained, free from any particle of fat, is then mixed with 1 to 10 grams of common salt to each 100 liters of the fluid, is boiled as quickly as possible until the albumen coagulates, and is then filtered. The clear fluid is evaporated in vacuo or otherwise till it attains the consistence of treacle. It is then poured into a shallow vessel, which can be heated by steam. From 0.1 to 8 per cent of sugar is then added, for the purpose of preserving the extract and of imparting a taste thereto similar to that of Liebig's extract of meat. The extract is heated to a temperature of 100° C., and kept constantly stirred until the desired consistence is attained. Vegetables or extracts thereof, or any other flavoring matter, or flour or other material for imparting a higher nutritive power or to give solidity, may be added at pleasure.

A factory was established at Aberdeen, Scotland, in 1885, under the superintendence of Sahlström, in which quantities of the extract were prepared from whale flesh. It was reported that the product possessed no flavor of the crude flesh whatever, and was quite similar to that prepared from ox flesh. It does not appear that anything is done in this line at present.

In a discussion of extracts of fish, published in 1885, Prof. William Stirling, of Owens College and Victoria University, Manchester, states:

The Normal Company, under the superintendence of Mr. Sahlström, has recently established a factory in Aberdeen, and has manufactured large quantities of a similar extract from whale flesh. This extract presents all the characters of an extract made from the flesh of the ox. Such an extract forms an excellent basis for a soup, having all the flavor of an extract of ox flesh. But extracts of fish can be made in a similar way, the product being, as far as sensible characters are concerned, indistinguishable from that of ox flesh. These are points of difference depending on the slightly different chemical composition of fish and flesh; for, even in the same animal, there is a difference in the chemical composition of individual muscles. Such fish extracts have no flavor of fish whatever, and possess all the aromatic flavor of meat extract, and I understand that they can be made much more cheaply than extract of meat. At a certain point in the process of extraction all the fishy flavor disappears. As a general rule, these extracts are made by boiling a watery extract of the fish muscles, after acidulation and precipitation of the proteids or albumins, in an open vessel with a double jacket, so that steam can be admitted between the layers of the jacket, and thus keep up

Letters Patent No. 353822, dated December 7, 1886.

obullition. Such extracts will keep for a very long time, and they are available for all the purposes for which meat extract is available. The question has still to be tested dietetically whether such extracts are in any way superior to those of meat. In any case they are quite equal to meat extracts in stimulating and restorative properties.

Such extracts, however, can also be made from other marine animals, e. g., crabs and shellfish generally. In these cases the extract is so made that it retains the flavor of the crab or shellfish. Thus there may be manufactured on the spot a large amount of extract which undoubtedly has a commercial and dietetic value. In a properly adjusted dietary, however, mere stimulants and restoratives are not sufficient, but there must be a proper amount and adjustment of the proteids (albumins) carbohydrates (such as starches and sugars), fats, and mineral salts. The question arises, then, Can not a cheap and useful food be made so as to combine these substances in proper proportion? The whole order of the legume tribe, represented by peas, beans, and lentils, have a high dietetic value, and this fact was made use of by the Germans in the manufacture of the famous "Erbswurst," or "iron ration," which played so prominent a part in the dietary of the Prussian soldiers during the Franco-German war.

As a matter of fact, in most soups what one obtains is really the extractives and salts and some flavoring materials. The substances in meat which give rise to the sensations of flavor and sapidity are really most important from a physiological point of view, for they excite powerfully the secretion of the digestive juices, and this greatly aids the process of digestion. Hence the value of mixing even highly nutritious food with sapid articles. Every one is familiar with the fact that tasteless articles very soon pall on one's palate, and how nauseating they become after a time. (Fourth Annual Report of the Fishery Board for Scotland, pp. 257, 258.)

With a view to producing a more digestible and nutritious as well as a more economical article of food than the dried cod of commerce, L. M. Haskins, of Boston, Mass., introduced in 1881 a combination of fish flesh, bone, and salt, ground together and desiccated. His process of manufacture was as follows:

The edible composition consists of fish bone and fish flesh ground together with common salt in a mill or between grinding rolls, so as to be reduced to a powdered state and thoroughly mixed or combined. Sixty pounds of the flesh, 20 pounds of the bone, and salt sufficient to give the mixture the requisite savor and preserve it from decay under ordinary circumstances are found to afford in a ground state an excellent edible composition. The proportions of the ingredients of the composition may, however, be varied, as occasion may require, to produce a palatable and suitable article of food.*

The inventor claims that this composition, by containing the alkaline and gelatinous properties of the bone in a powdered state, is not only readily digestible, and, from a sanitary point of view, better as an article of food than salted fish without any osseous additions, but that it can be manufactured and sold at a cheaper rate comparatively. It is well known that wheat or other flour without the admixture of the bran is not so digestible or beneficial as food as it is with a due amount of the bran, the latter containing the constituents necessary to the formation of bone. So this composition, by containing osseous elements in a finely reduced state, is rendered thereby not only easier digested but better as a food, especially when suitably cooked.

The following method of preparing fish meal was introduced by F. B. Nichols and Cathcart Thomson, of Halifax, Nova Scotia, and patented May 1, 1883:

The fish are headed and split and a portion of the backbone is removed in the same manner as for making the ordinary dry-salted fish. The pieces are then washed and all bloody portions removed. Very little salt should, it is said, be used in curing, as heavy salting makes an inferior meal, even when the excess is removed by water previous to drying. For some qualities of meal it is preferred to dry without salt. In this state the fish would soon spoil and very rapid drying must be resorted to in order to save them. The immediate application of currents of hot air would accomplish this, but would render the skin so friable as to defeat the after process and in other respects injure it for making meal, and open-air drying would not be speedy enough to keep the fish from tainting. In order to obviate these difficulties the fish-drying house and apparatus of the patent granted this inventor

---

* Letters Patent No. 241357, dated May 10, 1881.

December 6, 1881, No. 260382, is employed. The drying must be more thorough than for ordinary dried fish in order to make the fish hard and crisp. The hard-dried fish are made small enough to be fed into the hopper of a mill to be coarsely ground. Almost any kind of grinding mill may be used, provided it is not too sharp and is set high for coarse grinding for the first run. This run should be bolted through sieves having about 144 meshes to the square inch. About 75 per cent of it should pass through the bolt. The remainder, which is too coarse to pass, consists of the bones and the skin with considerable fish flesh adhering to it. In order to utilize this it is reground with the mill set closer and again passed through the bolt. If on examination much fish adheres to the skin it should be subjected to another grinding with a still closer set of the mill and again passed through the bolt. The residue from this, consisting principally of skin, bones, and scales, should not amount to more than 10 per cent of the weight of the dried fish and may be utilized as manure. The product of the last grindings contains considerable of the white portion of the skin, with fragments of bone and enough of the black skin to give a coarse, dirty appearance to the meal. In order to remedy this it should be again ground in a sharper and closer set mill to reduce it to a fine meal, and this, being passed through a bolt having about 400 meshes to the square inch, gives a fine product and contains the most nourishing portion of the fish. The last product can be either used alone or incorporated with the first by uniform mixing.

The inventors say:

We are aware that fish meal has been previously made; but in all previous processes, so far as we are aware, the fish used have been so salt as to require soaking the meal to remove the excess of salt before cooking, and the skin, fins, tail, and larger bones removed before grinding. We propose to use fish dried with little or no salt, and to grind them without removing either skin, bones, or other refuse contained in fins or tail, and to separate them by bolting.

In Europe "pastes" are made of anchovies, bloaters, shrimp, etc., the output being considerable. The following is one of the methods used in preparing anchovy paste. For each gallon of fish take 1 pound of salt, ½ pound of saltpeter, 1 ounce of sal prunella (saltpeter deprived of water of crystallization by heat) and a few grains of cochineal, and pound the whole well together in a mortar. In a stone jar place a layer of the ingredients, then a layer of fish, and so on until the jar is filled, press them hard down and cover up carefully, and let them remain for six months, when the paste is ready for use.

Somewhat similar to the above are the very delicious sardine butter (*Sardellenbutter*), crab butter (*Krebsbutter*), and crawfish butter prepared in Europe. These sell very high, 60 or 75 cents being the usual price for a 2-ounce bottle. Mrs. M. von Eisenhardt furnishes the following process for making crawfish butter:

Remove the meat from 100 boiled crawfish, dry the shells, put them with one pound of butter into a mortar and pound them fine. Then place in a saucepan over a fire and stir 5 minutes, add 2 quarts boiling water and cook for 5 minutes. It should then be strained through a napkin into cold water, and as soon as cold and firm remove it from the water and stir it in a saucepan over the fire for a few minutes, when it is ready for use. It should be placed in small glass jars and stored in a cool place.

In Japan lean pieces of fresh flatfish, eels, shark, etc., are freed from the bones, pounded in a stone mortar, and at the same time mixed with a certain quantity of salt, flour, sweet wine, white of an egg, and sacchariferous algæ (*Laminaria*), until the mixture assumes a paste-like consistency. This mixture is molded into various shapes, such as semicylindric, on a curved wooden plate; hollow cylindric, around a bamboo stick; discoid, on a circular plate, etc. These are heated over a charcoal fire, and then steamed and baked. The product may be kept from 3 to 20 days, according to the amount of the desiccation and the season of the year.

The secret of preparing several choice forms of fishery products has become lost. The method of preparing the *garum sociorum* of the Romans, a kind of fish sauce, is now unknown. Athenaeus and several other ancient writers speak of it in most glowing terms, and Pliny, who states that it is an extract from the entrails of certain fish that had undergone the process of fermentation, further says:

> The Greeks, in former times, prepared "garum" from the fish called by that name. The best "garum" comes now from Carthage, in Spain (Carthagena), and is called "garum sociorum." You can scarcely buy two boxes (each containing about 10 pounds) for a thousand pieces of money. No fluid, except scented waters, sells for so high a price, and it is in great demand by all classes of society. The fishermen of Mauritania, Betica, and Cartega prepare it from mackerel fresh from the ocean, which alone are fit for this purpose. The "garum" from Klozomene, Pompeii, and Liptes is also highly praised; and the prepared fish from Antipolis, Thurium, and Dalmatia are no less to be recommended. (Pliny, Hist. Nat., XXXI, 8.)

EXTRACTS OF CLAMS AND OYSTERS.

It is generally conceded that clams, both hard (*Venus mercenaria*) and soft (*Mya arenaria*), form one of the most nourishing and easily assimilated of all foods, especially when the hard indigestible portions are eliminated. For this reason there are many preparations of these marine products on the market, possessing excellent medicinal and restorative qualities, making them almost invaluable for invalids or convalescents. In making these preparations the solid matter is usually separated from the liquid and the latter reduced in bulk by evaporation. The extract thus obtained is rich in nutriment, is easily assimilated, and is a valuable tonic for people of weak or impaired digestive organs, and also as an article of food either alone or combined with water, milk, etc. The juices of oysters and other mollusks are also used at times for preparing similar articles, but they do not possess the nutritive qualities of clam extracts.

The first of these proprietary compounds was introduced by Butler G. Noble in 1867,[*] the extract being prepared in the following manner:

> The clams are removed from the shells, rinsed so as to remove grit or sand, cut into small pieces; a small quantity of fresh water is added and the whole boiled for about an hour. The free liquor is then poured off, the fibrous mass subjected to pressure, and the liquid obtained by this pressure is subjected to a process of evaporation at a temperature not exceeding 190° F., and as much lower as is practicable, until it is reduced to a thick paste, which is further reduced to a state of dryness in proper drying chambers. During the process of making, salt, pepper, and other condiments may be added if desired. This extract, which can be made into cakes of any size or reduced to powder, is readily soluble in water and contains the essential elements of nutrition and flavor peculiar to the clam. It is recommended that it be used in the making of soup, in flavoring, or for a variety of other purposes in cookery.

A patent[†] was granted to the same inventor for a similar process of drying the juice or natural liquor of oysters, which in the shucking-houses is generally drained off and thrown away. This waste material was to be reduced to comparative dryness by any of the means of evaporation, and then pressed into cakes or any other desired form. It is stated that a 2-ounce tablet may contain the nutritive ingredients of 4 quarts of fresh oyster juice and produce, with the addition of boiling water, 4 quarts of strong oyster soup, retaining the natural flavor of the oyster, to which may be added some freshly cooked oysters for verisimilitude. It does not appear that either of these processes is now used to any commercial extent.

---

[*] Letters Patent No. 66616, dated July 9, 1867.
[†] Letters Patent No. 66732, dated July 16, 1867.

In 1875 Charles Alden introduced a process [*] of preserving desiccated oysters, clams, etc., for food in combination with vegetable or other alimentary matter. His process is as follows:

The clams or other shellfish are taken from the shell and the natural liquor separated from the meat by straining through a sieve, or by any other convenient means. The body or meat is then desiccated by evaporation to a dry condition, so that it can be pulverized or granulated by crushing or grinding. The liquor, after separating the body or meat of the fish, is strained, to separate impurities, and sufficient bread crumbs or other farinaceous or alimentary material to absorb the whole of it added to the same, after which the mass is desiccated by evaporation in the same manner as the meat of the clams, and, when dry, is pulverized and granulated and added to the desiccated meat. Salt and other desired seasoning substances may be added to the compound, and the whole, after being thoroughly mixed, is put in suitable packages for use.

By this process it is claimed that all the natural elements of oysters or clams are preserved in suitable condition for use in making soups, chowder, fritters, and for other culinary purposes.

Letters patent [†] issued in 1877 to H. W. Buttles, of New York City, cover a process differing little from Alden's, and consisting in crushing the flesh of shellfish and desiccating it with the juice, then combining the residuum with salt and certain farinaceous substances. The method is thus described by the patentee:

In preparing the clams for desiccation according to this process the meat is reduced to a pulp in its own juice by passing the meat and juice of the freshly-opened clams through a mill constructed on the principle of the "beating engine" used by paper-makers in the preparation of paper pulp from rags, the clam meat, flowing in its juices, being caused to pass between a revolving cylinder armed with knives arranged parallel to its axis, and stationary knives fixed below it, the two sets of knives being so approximated as to readily cut that which passes between them, the pulp being made to circulate in a suitable channel from the knives back again to the opposite side thereof by means of the revolution of the cutting cylinder. Or the clams may be crushed and thus reduced to pulp by means of a wheel revolving in a circular trough, or by means of any of the improved forms of meat-chopping machines known to the art.

Having reduced the clam meat to a pulp in its own juice by any suitable means, substantially as described, it is next desiccated, either by subjecting it to strong currents of moderately-heated air upon revolving cylinders or disks, as in the process for desiccating eggs patented by Lamont, Quick, and others, or by exposing it to a moderate heat in suitable vessels placed in a receiver wherein a vacuum more or less perfect has been produced. In either case the clam pulp must not be subjected to a temperature so high that the albumin in the pulp shall be cooked or in the least coagulated and hardened while desiccating.

The clam meat thus desiccated in its own juices is prepared for market and use by reducing the resultant hard brittle mass to an impalpable powder, and then admixing it with common salt, finely powdered, and with a proper proportion of pure and unadulterated, cooked and uncooked, pulverent farinaceous substance, derived either from cereals, such as wheat flour, or from roots, such as potato starch, a proportion of about 60 per cent of clam, 32 per cent of farinaceous material, and 8 per cent of salt, yielding an excellent product. Or the clam pulp, prepared substantially as above described, may be admixed with bread crumbs, cracker dust, or other farinaceous preparation, before desiccation, in sufficient proportion to form a paste or dough, and the resultant hard, dry compound be reduced to a powder for use.

Another process of desiccating clams and other shellfish was introduced [‡] in 1890 by S. G. Van Gilder, of Philadelphia:

In carrying out this method the clams, oysters, or other shellfish are first removed from their shells and separated from their natural liquor by a draining process, accompanied by slight compression, if necessary, to expel all the liquor. After this separation, the meaty portion of the fish is

* Letters Patent No. 168703, dated October 11, 1875.
† Letters Patent No. 191024, dated May 22, 1877.
‡ Letters Patent No. 440519, dated November 11, 1890.

reduced to a finely comminuted, pulpy mass by chopping, grinding, or in any other suitable manner, in which state it is mixed with a portion of the natural liquor previously separated from the fish, and then subjected to a boiling heat—say 212 . After this cooking process the solid matter is again separated from the liquid and the latter combined with what remains of the raw liquor. Then the combined juices are subjected to a boiling and skimming process to remove all superfluous matter and concentrate and refine the liquor. This boiling and skimming process serves to eliminate objectionable matters floating in the liquor and concentrate and cook the juices, so that the resultant product will be more refined and will keep in a prime condition for any length of time. To this refined and concentrated liquor is then added a suitable quantity of some farinaceous substance, such as flour, meal, cracker dust, bread crumbs, etc., after which the whole is subjected to a boiling temperature, which will serve to cook the same and thicken and coagulate the albuminous and starchy matters contained therein. Then this coagulated mass is thoroughly mixed with the pulpy mass, and the whole subjected to a moderate degree of heat to evaporate all the moisture from it, and thereby desiccate it. The heat for the purpose of desiccation may be applied by steam, hot air, the vacuum process, or otherwise, in order that the desiccation may be thorough. After desiccation the product is reduced to a granular form and put up into suitable cans or packages for the trade.

It is claimed that this concentrated food product will keep in a prime condition for an indefinite length of time, and when used for such purposes as soups, chowders, fritters, sauces, dressings, etc., the original flavor will be retained and greatly augmented by concentration.

One of the most successfully introduced of the proprietary clam extracts is made by the following process:

The uncooked clams are placed in a retort or receptacle, which is preferably air and steam tight, and live steam is admitted into said retort for 20 minutes, or more or less, as may be desired. The steam causes the shell or clam to open, thus liberating the liquid or juice from the solid meat of the clams, and said liquid drops into suitable pans placed for the purpose under the clams, the latter being supported by suitable open racks or gratings. The juice or liquid extract thus obtained is next passed through a suitable filter, and is then boiled to evaporate a part of the water and concentrate the extract, thus making a given quantity of it richer than it would otherwise be. The boiling also cooks the nutritive elements in the liquid sufficiently to prevent ready decomposition when exposed to the air. The liquid is finally put, while hot or cold, into cans or jars and hermetically sealed, the time of processing or cooking the jars or cans, so as to exclude the air and have it keep in any climate, varying as to whether the concentrated juice or extract is filled into the cans or jars hot or cold. (Letters Patent No. 395199, dated December 25, 1888.)

Large quantities of extract are made from soft clams at several points on the Maine and Massachusetts coast after the last-described process, the product being placed on the market in pint, quart, and gallon tin cans and selling at about $2.50 per dozen pint cans, and at proportionate prices for cans of other sizes. The surplus liquor from clams used in the canning factories forms the crude material and this is evaporated and prepared in the manner described above. In discussing the introduction and use of this extract the inventor states:

It has been adopted in very many hospitals, hotels, and large public institutions; it is being prescribed as a valuable stomachic by thousands of physicians, and is already being sold by very many of the leading grocers throughout the United States and foreign countries to families who use it as food in its various forms upon their tables. Another use to which it is getting to be largely put is in making instantaneous hot clam broth or bouillon by dispensers of temperance drinks, being in this way used in conjunction with water or milk, making, with the addition of pepper or salt, a very nutritious and palatable drink. Although this "extract of clams" is of recent introduction, yet the sales have already reached several thousands of cases per year, and the demand is steadily increasing as the people find out the merits of the article. The prices of this new food are reasonable, so that those who are in moderate circumstances can afford to purchase it, the retail price for pint tins not exceeding 30 cents per tin and for the gallon tins not exceeding $1.50 per tin in the United States and the principal European centers of trade.

There are several brands of clam extract or bouillon on the market, made from hard clams or quahogs. These are prepared in various ways, and usually, as in case of extract of soft clams, as a by-product in the canning of the quahogs. At one establishment in New York State the hard clams are steamed for the purpose of opening them, the escaping liquid being saved and placed in tin cans, processed heavily, and sealed, the meats being similarly treated and canned separately. It is claimed by some that this process is objectionable, especially in the manner of opening the clams by steaming. The hot steam coming in contact with the cold shells condenses somewhat and adds to the bulk of the liquor, thereby weakening it; also, when the shells are heated they impart a peculiar flavor to the liquor. As this preparation is not evaporated, diluting it with the condensed steam is especially undesirable. It sells for about $1.75 per dozen 1-pound cans.

At another factory in New York State for preparing clam juice the clams were formerly opened with a knife, all the free liquor being saved and the meats chopped and compressed to obtain additional liquor. This was compressed and the liquid condensed, leaving the clear juice somewhat concentrated. It was soon found more profitable to use the meats in preparing clam chowder, and at present that is one of the principal products of the establishment. The raw clams are opened with a knife, all the liquor being saved. The meats, with sufficient liquor, are then mixed with disks of white potatoes, onions, and other vegetables to suit the taste, just as in preparing chowders at home, placed in tin cans, processed, vented, and hermetically sealed. The surplus liquor is condensed by evaporation and placed in glass jars, which are then sealed. A large spoonful of this juice is sufficient for a cup of bouillon after mixing with water.

Some clam juice is prepared on the North Carolina coast, and occasionally in the Chesapeake region a few cases are prepared experimentally and an attempt is made to market them. This usually results in a loss because of the article being unknown, considerable work being necessary to build up a market for a new brand of clam juice.

In 1897 there was introduced in Scotland a method of preparing an extract of clams or other shellfish mixed with a sufficient quantity of seaweed, such as Irish moss or carrageen, to convert it into a jelly,[*] for use as a "stock" in making soups, sauces, and the like.

The process of preparation is as follows:

Boil a quantity of clams or other shellfish in a close-covered vessel, using the smallest quantity of water necessary for the purpose. After the mollusks are sufficiently cooked, remove the shells, bruise or reduce the flesh to a pulp, and strain off all the extracted liquor from it. To this liquid extract add the water used in cooking and the liquid resulting from opening the mollusks. Then boil this liquid with a quantity of Irish moss or carrageen, or any seaweed having similar properties, which has been well bleached to remove color, apportioning the quantity of Irish moss and timing the boiling operation to obtain a jelly of the desired consistence. Before boiling the moss and liquid, or during that operation, add salt, pepper, and other flavoring condiments desired. Strain the product while hot and store it in stoneware jars or other receptacles, which may be sealed up.

[*] Letters Patent No. 585395, dated June 29, 1897.

MISCELLANEOUS ANTISEPTICS AND ANTISEPTIC PROCESSES.

There are a number of antiseptics which have been brought to the attention of fish-curers for use in preserving their products, in addition to those already noted. The most desirable are those that do not change the texture of the fish, as common salt does, among these being boracic acid, salicylic acid, citric acid, tartaric acid, etc. The first is considered a valuable preservative, as it keeps fish and other food stuffs fresh for a week or more without great injury either to the appearance or quality of the articles preserved, and it is used more extensively than all others, but usually only as an aid to other methods of preservation.

BORACIC ACID.

For many years the value of boracic acid as a preservative agent has been recognized. Its extensive use with articles of food appears to have originated in Norway about 1870. Among the articles preserved by its agency were herring, and the success with them was such that the trade gradually extended beyond the boundaries of the country and in 1885 they were shipped to England in large quantities, successfully competing with fresh herring from Yarmouth and other points of Great Britain.

The general method of application is as follows: The round herring are arranged in layers in a barrel and each tier covered with a thin layer of a mixture made of 5 pounds of boracic acid and 10 pounds of fine salt. When full the barrel is tightened down in the usual way and the contents pickled with a weak solution of boracic acid and fresh water. The fish should then be kept in a cool place at an even temperature. In treating a barrel of herring in this way, 2½ pounds of boracic acid and 5 pounds of salt are required for spreading on the fish during packing, and about 10 ounces of acid for dissolving in the water used for pickling. The cost is about 8 or 10 cents per pound, wholesale.

Some objection has been raised to the use of boracic acid as a preservative because of its alleged injury to health. In opposition to this it is stated that it has been used for years, especially to preserve milk in hot weather, and no evidence has appeared to indicate injurious effects upon the health. The Norwegian herring preserved with boracic acid are said to be of good quality and to be in fair condition when placed on the markets, even after being two weeks out of the water.

In discussing the curing of fish with boracic acid, the British Medical Journal states:

Large quantities of herring preserved with salt and boracic acid being at present imported from Norway and sold in the London and Newcastle markets, attempts have been made to prevent their sale. The National Sea Fisheries Protection Association discussed the question at a recent conference at Fishmongers' Hall, but no decision as to such fish was arrived at. It may, therefore, be worth while to point out that boracic acid being the essential ingredient of our many food preservatives, be it in the form of the acid, of boroglyceride, or of borax, has been used for years, especially to preserve milk in hot weather, and no evidence has ever been brought forward even to suggest injurious effects upon the health; it may, therefore, be taken to be perfectly harmless. The Norwegian herrings preserved with salt and boracic acid are of exceptionally fine quality, are perfectly fresh when brought into the market, and are, of course, subject to the usual process of inspection by the market inspectors, whose power of rejection is almost absolute. If, nevertheless, an outcry is heard against their sale, it is difficult to resist the belief that it is dictated by the jealousy which is notoriously rife in Billingsgate circles. The introduction of cheap food from new sources, welcomed as it always is by the public, is invariably opposed by the trade, who, after all, reap the chief advantage in the long run. One has but to recall the sneers of the meat venders at American and Australian meat to value the agitation against Norway herrings at its proper worth. Hitherto, happily, we have been spared the bitter discussions which have on the Continent led to legislation against certain food preservatives, such as salicylic acid, which we in England admit without hesitation. The question is mainly one of public economy. Shall good food be wasted for want of a preservative, even if certain objections may be urged against their use, or shall we put up with these objections and aim at cheapening food for the masses, provided, always, that nothing which could injuriously affect their health is allowed to be present? A sufficient guaranty is afforded by the vigilance of medical officers, public analysts, and market inspectors against the abuse of antiseptics and food preservatives.

On the other hand, a fish-dealer writes to the *Fish Trades Gazette:*

Hundreds of barrels of herring from Norway out of one cargo were condemned, and there were about 1,500 barrels unsold lying in London at that time. France will not admit Swedish and Norwegian herring, nor any other fish cured by the process named. Many shopkeepers soon find out to their cost that once their customers have tasted herring cured with acid they don't ask for them a second time

A combination of boracic and acetic acids for preserving food products was introduced in the United States in 1877 by C. G. Am Ende, who thus describes his invention:

The invention consists in compounding boracic acid, either in a liquid or pulverous state, with acetic acid, in the proportion of about one drop of acetic acid to every ounce of boracic acid; but the proportion may be varied, according to the nature of the substance to be preserved, and of the atmosphere to which the same is to be exposed. The acetic acid may be used more or less diluted. Other salts may be added to the mixture if desired. The composition is applied to the substances to be preserved in substantially the manner in which preservatives are usually applied. The acetic acid in the composition prevents the formation of fungi, while the boracic acid prevents putrefaction chiefly by hindering the formation of bacteria. (See Letters Patent No. 187079, dated February 6, 1877.)

ROOSEN PROCESS.

This process, invented by August Roosen, of Hamburg, consists in placing the freshly caught fish in an air tight barrel, and then forcing the preservative solution into the tissues of the fish by using a pressure of several atmospheres. The details of the process are as follows:

A strong cask of galvanized iron with an adjustable lid is provided. This resembles somewhat the well-known cans used for conveying milk, but is much larger. As many fresh fish as the cask will conveniently hold are placed therein, and the cask is filled with water and certain proportions of boracic acid and tartaric acid. The purpose of the tartaric acid is to neutralize the taste of the boracic acid, which, however, is quite harmless. The lid, fitted so as to be air-tight, is next adjusted and secured to the cask. A small force pump is connected with a hole in the lid, and additional quantities of a solution of the antiseptic in water are pumped into the cask, expelling all air, which escapes at a second hole in the lid. As soon as the cask is completely full and the air expelled, the liquid begins to flow through the second aperture. An air-tight cap is then screwed tightly over this hole to prevent further escape of the liquid. The pump is again set to work forcing in the mixture until a gauge fixed to the pump indicates a pressure of 90 pounds to the square inch. By means of a stopcock the opening is then closed and the air pump removed. The effect of the high pressure is to force the mixture into the veins and tissues of the fish and thus prevent organic change in any part. The fish are shipped in the cask, and it is stated that they will keep for any reasonable length of time in any climate.

It is claimed that the cost of utensils required for the preservation of fish according to the above process amounts in this country to about $30, the cost of a cask being $20 and of a pump $10. The cask should be made of stout steel, and capable of containing 200 or 300 pounds of fish. The utensils should last five years, and one pump is sufficient for a large number of casks. If the cost of materials be distributed over five years, the average cost will be $6 per year, so that in case the cask is filled only once a year the cost of utensils and materials is a trifle over 2 cents per pound of fish. But as each cask may be filled twenty times a year, and one pump will suffice for a large number of casks, the cost is reduced to one-tenth of 1 cent per pound.

The following extract is from an article in the *Fish Trades Gazette*, of London, July 31, 1886:

The Roosen process is now pretty well known in England, and it is generally accepted as being far the most successful attempt to keep fish not only fresh, but also sweet, wholesome, and attractive for long periods. The process, it may be added, is not confined to fish, but has been applied with equal success to meat, game, fruit, etc. Experiments have been carried out in Scotland, and public demonstrations of the value of the process made in Edinburgh and Glasgow, where its merits have been recognized by the very highest authorities on the subjects of fishing and the fish trade. Messrs. Dufresne & Luders, the agents of Mr. August R. Roosen, of Hamburg, the inventor of the process, lately decided that it would be well to make the process better known in London, and accordingly invited a number of representative guests to witness the opening of several casks of fish preserved by the Roosen process and to taste the same when cooked. There was an excellent response to the invitation, the guests including many famous authorities in science and in medicine, as well as others holding important governmental positions or being connected with commerce, not only in England, but also in the colonies and Indian Empire. Two casks, which had been closed for seventeen days, were opened before the company, and the fish when taken out were found to be perfectly sweet and fresh, bright-looking, and as attractive as the day they were caught. On being eaten they were pronounced excellent, and the advantages of the process were highly commended.

A process somewhat similar to that introduced by Roosen was devised[*] by Magnus Gross, of Washington, D. C., in 1859. Gross's method differed from Roosen's in employing hydrostatic pressure instead of a force pump and in using a strong solution of common salt (100 pounds) mixed with carbonate of soda (4 pounds) and carbonate of potash (2 pounds). This method was intended for the preparation of salted fish, the product being packed in dry salt after the curing process, and it was never used to any commercial extent.

ECKHART PROCESS.

In 1877 John Eckhart, of Munich, Germany, patented[†] a process of preserving fish and meats by introducing a solution of salicylic acid with an apparatus similar to that used in the Roosen process for preserving fresh fish. The solution was made by dissolving half a pound of salicylic acid in 100 pounds of water. A hydraulic pressure of 12 atmospheres was applied for from one to two hours. In 1882 Eckhart introduced another antiseptic compound[‡] for fish, salt, boracic acid, tartaric acid, and salicylic acid being used instead of boracic acid alone. The mixture was composed of 50 per cent salt, 47½ per cent boracic acid, 2 per cent tartaric acid, and ½ per cent salicylic acid. In its application the fish are stripped of skin and bones and mixed with the compound in the proportion of 2 pounds antiseptic to 100 pounds of fish. They are next packed in cases of animal tissue or parchment and put into casks filled with a gelatin solution made in the proportion of 10 pounds of gelatin, 4 pounds of the anti-

---

[*] Letters Patent No. 26427, December 13, 1859.
[†] Letters Patent No. 194550, August 28, 1877.
[‡] Letters Patent No. 251772, January 3, 1882.

septic, and 25 gallons of water. The casks are then headed and connected with a force pump and more of the solution is forced in until the contents are well saturated. The fish may then be shipped in the casks or they may be removed from the cases, sprinkled with dry salt, and marketed dry.

JANNASCH PRESERVATIVE.

A compound patented by Hugo Jannasch, of Germany, and in use at Gloucester and some other ports, is said to be prepared in the following manner:

Equal parts of chloride of potassium, nitrate of soda, and chemically pure boracic acid are dissolved in the proper quantities of water. A solution of chloride of potassium is then heated in a kettle up to the boiling point, and a solution of nitrate of soda added thereto. This solution is kept on a brisk fire until the lye has become perfectly clear. The solution of boracic acid is then added under continual stirring. By the influence of the boracic acid, at a temperature of 212° F., a reaction takes place, which is indicated by the mass assuming a yellowish color, and by the escape of chlorine gas. After the reaction has taken place the solution is slowly evaporated at a low temperature, until a dry salt is obtained, which is composed of a combination of hyponitrate of potash, hypochlorate of soda, borate of soda, borate of potash, and free boracic acid.

In the application of this compound to the preservation of fresh, pickled, smoked, or dry-salted fish the following directions are given:

In preserving fresh fish which are to be shipped or kept on the stand for sale, remove the entrails, sprinkle some of the preservative inside the fish, also in the bottom of the box or barrel in which the fish are to be packed; then place the fish in the box and sprinkle the preservative over each layer. If the entrails and gills are not taken out, insert, according to the size of the fish and the season of the year, more or less of the compound in the mouth of the fish, pushing it down as far as possible; then sprinkle some on the gills, after which treat the fish as above when packed in box or barrel. Use 1 pound of preservative to 100 pounds of fresh fish. Pickled fish, if packed in kegs or barrels, are treated first in the way directed above. The barrels are then headed up and allowed to stand from 4 to 6 hours; then the pickle, which can be made much milder than the present pickle for fish, is added, and the barrels are rolled to facilitate and quicken the dissolving of the preservative. The pickle should be admitted through the bunghole only, to prevent the preservative from being washed off. To prevent the brine from souring and to enable its being used several times over again, it is recommended that to every 6 gallons of brine 1 pound of preservative be added, first dissolving the compound in a gallon of hot water, and after it has cooled off pouring it into the brine. Every time the pickle is used over again add sufficient salt to bring it back to the requisite strength; then use only half the quantity of preservative taken the first time, which would be ½ pound of preservative for every 6 gallons of pickle. By this treatment the pickle will remain sweet and free from slime for a long time, and thus save the labor and expense of making new brine.

MISCELLANEOUS ANTISEPTIC COMPOUNDS.

While boracic acid and other chemicals have not been extensively used in the United States for preserving fresh fish, they have been employed to a considerable extent since 1881 in connection with other processes of preservation. Boracic acid has long been used in a powdered form on dry-salted cod, especially those put up as boneless fish. Its popularity has increased under various names, and it is now employed at several boneless-cod, oyster-shucking, and other establishments. It has been used to some extent in the preservation of caviar, but salicylic acid seems better adapted for this purpose. Most of the preservative antiseptics used are proprietary compounds sold under various trade names, such as "Preservaline," "Rex Magnus," etc. The following antiseptic compounds have been introduced. This summary has no pretensions to completeness, there being scarcely any limit to the number of compounds brought to the attention of fish-curers.

Hydrocarbon gas.—This is substituted for the air which occupies the space in and around the substance to be preserved, subjecting the same to a temperature of about 30° F. The gas is let into the package through a hole in the top and the air escapes through a hole in the bottom, and both holes are then closed. (Letters Patent No. 45765, dated January 3, 1865.)

Sulphides of carbon.—Fish are placed in a receiver and the air exhausted. Gaseous bisulphide, protosulphide, or other sulphide of carbon is then let into the receiver under pressure and permeates the flesh. In combination with the sulphide is used phenic acid, methyl or other product of the destructive distillation of wood. (Letters Patent No. 85184, dated December 22, 1868.)

Gelatin, lime, glycerin, etc.—Put the fish in an air-tight compartment and exhaust the air with a vacuum pump, then by means of a force pump introduce a solution of gelatin and bisulphite of lime. When completely saturated remove the fish and dip them in a concentrated solution of gelatin containing bisulphite of lime, glycerin, sugar, and gum. (Letters Patent No. 90944, dated June 8, 1869.)

Glycerin.—Remove from the fish all the refuse matter, such as skin, bones, etc., and then grind the residue and compress from it the watery portions, blood, and oily matter to whatever extent may be desirable, and then treat it with glycerin, regrinding the material during this process. The fish is then pressed into a compact mass and placed in any suitable wrapper of tin foil or other material, or boxed. (Letters Patent No. 87986, March 16, 1869.)

Glycerin and antiseptic salts.—Oysters, fish, and meats may be preserved by use of a mixture of glycerin with phosphate of soda, or other antiseptic salt in connection with aldehyde, formic ether, or acid in a solution of carbonic acid, water, glycerin, etc., and the preserved substance is then covered with parafin or stearin. (Letters Patent No. 93183, dated August 3, 1869.)

Saltpeter and alum.—The fish, either after or before they have been salted, are placed for 4 hours in a solution of saltpeter and alum, made in proportion of 5 pounds of saltpeter and 4 ounces of alum to 60 gallons of sea water. They are then dried either in the sun or by artificial means. If they are to be smoked, 2 hours in the solution is said to be sufficient. It is claimed that this process removes all tendency to sweat or decay. (Letters Patent No. 95179, dated September 28, 1869.)

Soda and carbolic acid.—After being cleaned the fish are dipped in a solution in proportion of 5 gallons of water, 2 pounds of sulphite or bisulphite of soda, and 2 ounces of carbolic acid in crystals. Oysters, clams, etc., may be dipped in a solution of their own liquor and the chemicals. (Letters Patent No. 86040, January 19, 1869.)

Thymol or thymate salts.—Place the fish, oysters, meats, or other animal substances to be preserved in solutions of thymol, thymic acid, or any of the thymate salts and water, alcohol, or glycerin, etc. (Letters Patent No. 108983, dated November 8, 1870.)

Chloroform and ether, etc.—The meat or fish is placed in air-tight packages, into which is poured a small quantity of chloroform, which becomes vaporized and surrounds the substance with an atmosphere of vapor which acts as a preservative. The cans are then sealed and are ready for shipment. When needed for use the chloroform is removed by means of an air-pump. (Letters Patent No. 128371, dated June 25, 1872.)

Borax, saltpeter, etc.—By the Herzen preserving process, meat is soaked from 24 to 36 hours in a solution of 3 parts borax, 2 boracic acid, 3 saltpeter, and 1 salt, in 100 parts water, then packed in some of the solution. Before use the meat must be soaked 24 hours in fresh water.

Bisulphite of lime.—The Medloch & Bailey method of preserving is said to be one of the most successful of antiseptic processes. The solution used is made of equal parts of water and bisulphite of lime of 105 sp. gr. Fish cured in this solution are claimed not to have an unpleasant flavor.

Acetate of alumina.—Meat and fish are covered with a coating of gum, then immersed in acetate of alumina, then a solution of gelatin, allowing the whole to dry on the surface. The antiseptic acetate of alumina forms an insoluble compound with the gelatin and prevents decomposition by excluding air from the substance.

Benzoin and alum.—In the preservation of meat and fish by the Pagliare process they are immersed in a compound of gum benzoin boiled in a solution of alum, and excess of moisture is driven off by a current of hot air, leaving the antiseptic on the surface of the fish or meat.

Salicylic acid and alcohol.—To 50 grams of salicylic acid is added 300 grams of rectified alcohol. White blotting paper is well saturated in this mixture and left to dry. By this mixture the paper becomes full of little red pricks and has a sweetish taste. The fish are wrapped in this paper and packed rather loosely in dry hay. By this method it is claimed that fish or game can be transported at any time during the summer without danger of spoiling.

Bicarbonate of soda and saccharine matter.—Take 40 parts of bicarbonate of soda and 60 parts of saccharine matter, such as sugar, and mix them in enough water to form a thick paste or sirup, which is applied with a brush to the surface of the fish to be preserved. The fish so coated are suspended in a shady place for an hour or so and then exposed to an air current until the surface is thoroughly dried. By soaking the fish in water for 3 hours or more the coating is dissolved, when the fish may be prepared for the table. (Letters Patent No. 474581, dated 1892.)

Fluoride of sodium and chloride of sodium.—A mixture made of 80 parts fluoride of sodium with 20 parts of common salt gives the best results, but the proportions may be varied according to conditions. This may be used either in the form of a powder or dissolved in water. When the fish are to be preserved a considerable length of time they should be soaked in the antiseptic solution; but when they are to be preserved for a short time only they may be sprinkled with the powder. It is claimed that this antiseptic does not exert an injurious influence on the digestive fluids, but on the contrary is rather beneficial.

Miscellaneous.—The following is said to be the composition of a number of proprietary antiseptics used in Europe and to some extent in this country:

Composition.	Per cent.	Composition.	Per cent.
Sozolithe:		The "Minerva" Chinese preservative:	
Sulphite of ammonia	37.3	Chloride of sodium	25
Sulphurous acid	39.7	Boric acid	17.7
Soda	21	Sulphate of soda	38.8
Water	2	Sulphite	9.2
Concentrated berlinite:		Water	9.3
Crystallized borax	82.7	Australian salt:	
Boracic acid	9.8	Crystallized borax	94
Chloride of sodium	7.5	Chloride of sodium	5.5
Paechel berlinite:		Some hydrocarburet	.5
Chloride of sodium	45.9	Rugers barmenide:	
Nitrate of potash	32.3	Boric acid	50
Boric acid	19.3	Chloride of sodium	50
Water	2.5		

MOSS WATER.

Among the "Papers in Colonies and Trade" for 1820 a somewhat novel method of preserving herring by means of moss water was described by J. Fred. Denovan, one of the pioneers in developing the pickled-herring trade of Scotland. His description is as follows:

Having often observed the strong antiseptic powers of moss water on vegetable and animal substances, I conceived that it might be used with effect in the cure of herrings, particularly of those intended for a warmer climate, and I resolved to try the experiment on a small scale: I first cured a few kegs of the later herrings (in October, 1818) in the usual way; but instead of throwing away the gut, gills, and bloody part, as is customary, I put them into a small cask with a proportionate quantity of Lisbon salt, and pressing down the whole by means of an iron plate a dissolution of the salt took place in a few hours, and a strong red pickle was produced, on the top of which the fixed oil was floating. After carefully skimming off the oil, I added one-third of strong brown moss water taken from a natural pond formed in the moor near Eyemouth; and having taken the herrings out of the original pickle, I packed them anew and filled up the kegs with this pickle. On opening them some months afterwards I not only found they were in excellent preservation, but that the scales (which always proves the quality of the pickle) were as bright as when the fish were taken out of the water.—("Papers in Colonies and Trade" for 1820, p. 195.)

PRESERVATION BY COMPRESSED AIR.

Various experiments have been made in preserving meat and fish by compressed air. One of the most important processes is that known as Brandt's method, devised by Martin Brandt, of Denmark. A brief review of this process and its importance to the fish trade appears in the *Deutsche Fischerei-Zeitung* July 8, 1884, from which we quote the following:

Martin Brandt's new method is said to have this advantage, that it does not change the shape, looks, and flavor of the fish, and prevents the development of fungus. It is done by compressed air. It may be continued for an unlimited period and be employed in the holds of vessels, railroad cars, warehouses, etc. For lining the rooms where the fish are kept metal or cement is used. The preserving medium weighs very little, as 1,000 cubic feet of compressed air weigh but 10 pounds. In Mr. Brandt's warehouse a pipe runs along the wall from the floor to the ceiling, and back again, twisting several times, and finally ending on the floor. The machine or development apparatus consists of an iron cylinder connected with a so-called vacuum air filter. The cylinder is filled with air compressed by about 200 atmospheric pressure. By means of the vacuum apparatus the machine is connected with the pipe in the warehouse, and the compressed air flows, after a valve has been opened, with great velocity through the filter and the pipes. New air is also introduced in the vacuum apparatus through cotton filters, thus purifying it of all matter apt to decay, and, united with the stream of compressed air, it continues to pass through the pipes. As the air expands it loses some of its warmth and is gradually cooling off. When let out of the pipes the air, which has now become quite cool, rises evenly throughout the room and drives the warm air, filled with germs or fungi, through an opening in the ceiling. As the inventor claims, fish and meat can be kept fresh for an unlimited period in rooms whose air has been purified in the manner described above. (Translated in U. S. Fish Commission Bulletin, 1884.)

INDEX.

I

○

www.ingramcontent.com/pod-product-compliance
Lightning Source LLC
Chambersburg PA
CBHW021529210326
41599CB00012B/1432